MICRO & NANO-ENGINEERING OF FUEL CELLS

# Sustainable Energy Developments

*Series Editor*

Jochen Bundschuh
*University of Southern Queensland (USQ), Toowoomba, Australia*
*Royal Institute of Technology (KTH), Stockholm, Sweden*

ISSN: 2164-0645

Volume 11

# Micro & Nano-Engineering of Fuel Cells

*Editors*

Dennis Y.C. Leung

*Department of Mechanical Engineering, The University of Hong Kong, Pokfulam Road, Hong Kong*

Jin Xuan

*Institute of Mechanical, Process and Energy Engineering, School of Engineering and Physical Sciences, Heriot-Watt University, Edinburgh, UK*

**CRC Press**
Taylor & Francis Group
Boca Raton  London  New York

CRC Press is an imprint of the
Taylor & Francis Group, an **informa** business

A BALKEMA BOOK

CRC Press
Taylor & Francis Group
6000 Broken Sound Parkway NW, Suite 300
Boca Raton, FL 33487-2742

First issued in paperback 2017

*CRC Press/Balkema is an imprint of the Taylor & Francis Group, an informa business*

© 2015 by Taylor & Francis Group, LLC

Typeset by MPS Limited, Chennai, India

No claim to original U.S. Government works

ISBN-13: 978-0-415-64439-6 (hbk)
ISBN-13: 978-1-138-74864-4 (pbk)

*Library of Congress Cataloging-in-Publication Data*

Micro & nano-engineering of fuel cells / editors, Dennis Y.C. Leung,
Department of Mechanical Engineering, The University of Power Engineering,
Pokfulam, Hong Kong, Jin Xuan, School of Mechanical and Power Engineering,
East China University of Science and Technology, Shanghai, China.
    pages cm. – (Sustainable energy developments)
  Includes bibliographical references and index.
  ISBN 978-0-415-64439-6 (hardback) – ISBN 978-1-315-81507-7 (ebook)
1. Fuel cells.   I. Leung, Dennis Y.C.   II. Xuan, Jin.   III. Title: Micro and nano-engineering
of fuel cells.

  TK2931.M523   2015
  621.31′2429–dc23

                                                                    2015005650

Published by:   CRC Press/Balkema
                P.O. Box 11320, 2301 EH Leiden, The Netherlands
                e-mail: Pub.NL@taylorandfrancis.com
                www.crcpress.com – www.taylorandfrancis.com

**Visit the Taylor & Francis Web site at
http://www.taylorandfrancis.com**

**and the CRC Press Web site at
http://www.crcpress.com**

# About the book series

Renewable energy sources and sustainable policies, including the promotion of energy efficiency and energy conservation, offer substantial long-term benefits to industrialized, developing and transitional countries. They provide access to clean and domestically available energy and lead to a decreased dependence on fossil fuel imports, and a reduction in greenhouse gas emissions.

Replacing fossil fuels with renewable resources affords a solution to the increased scarcity and price of fossil fuels. Additionally, it helps to reduce anthropogenic emission of greenhouse gases and their impacts on climate change. In the energy sector, fossil fuels can be replaced by renewable energy sources. In the chemistry sector, petroleum chemistry can be replaced by sustainable or green chemistry. In agriculture, sustainable methods can be used that enable soils to act as carbon dioxide sinks. In the construction sector, sustainable building practice and green construction can be used, replacing, for example, steel-enforced concrete by textile-reinforced concrete. Research and development and capital investments in all these sectors will not only contribute to climate protection but will also stimulate economic growth and create millions of new jobs.

This book series will serve as a multi-disciplinary resource. It links the use of renewable energy and renewable raw materials, such as sustainably grown plants, with the needs of human society. The series addresses the rapidly growing worldwide interest in sustainable solutions. These solutions foster development and economic growth while providing a secure supply of energy. They make society less dependent on petroleum by substituting alternative compounds for fossil-fuel-based goods. All these contribute to minimize our impacts on climate change. The series covers all fields of renewable energy sources and materials. It addresses possible applications not only from a technical point of view, but also from economic, financial, social and political viewpoints. Legislative and regulatory aspects, key issues for implementing sustainable measures, are of particular interest.

This book series aims to become a state-of-the-art resource for a broad group of readers including a diversity of stakeholders and professionals. Readers will include members of governmental and non-governmental organizations, international funding agencies, universities, public energy institutions, the renewable industry sector, the green chemistry sector, organic farmers and farming industry, public health and other relevant institutions, and the broader public. It is designed to increase awareness and understanding of renewable energy sources and the use of sustainable materials. It also aims to accelerate their development and deployment worldwide, bringing their use into the mainstream over the next few decades while systematically replacing fossil and nuclear fuels.

The objective of this book series is to focus on practical solutions in the implementation of sustainable energy and climate protection projects. Not moving forward with these efforts could have serious social and economic impacts. This book series will help to consolidate international findings on sustainable solutions. It includes books authored and edited by world-renowned scientists and engineers and by leading authorities in economics and politics. It will provide a valuable reference work to help surmount our existing global challenges.

Jochen Bundschuh
(Series Editor)

# Editorial board

# Table of contents

# List of contributors

| | |
|---|---|
| Ulises Cano-Castillo | Non Conventional Energies Department, Instituto de Investigaciones Eléctricas, Cuernavaca, Mexico |
| Junseok Chae | Electrical, Computer and Energy Engineering, Arizona State University, Arizona, USA |
| Kwong-Yu Chan | Department of Chemistry, The University of Hong Kong, Hong Kong, China |
| Takemi Chikahisa | Graduate School of Engineering, Hokkaido University, Hokkaido, Japan |
| Gonzalo García | Department of Physical Chemistry, Institute of Materials and Nanotechnology, University of La Laguna, Tenerife, Spain |
| Maria Victoria Martínez Huerta | Institute of Catalysis and Petrochemistry, Spanish National Research Council (CSIC), Madrid, Spain |
| Ming Hou | Fuel Cell System and Engineering Laboratory, Dalian Institute of Chemical Physics, Dalian, China |
| Balaiah Kuppan | National Centre for Catalysis Research and Department of Chemistry, Indian Institute of Technology – Madras, Chennai, India |
| Dennis Y.C. Leung | Department of Mechanical Engineering, The University of Hong Kong, Hong Kong, China |
| Michael K.H. Leung | Ability R&D Energy Research Centre, School of Energy and Environment, City University of Hong Kong, Hong Kong, China |
| Keaton Lesnik | Department of Biological and Ecological Engineering, Oregon State University, Oregon, USA |
| Li Li | Ability R&D Energy Research Centre, School of Energy and Environment, City University of Hong Kong, Hong Kong, China |
| Hong Liu | Department of Biological and Ecological Engineering, Oregon State University, Oregon, USA |
| Meng Ni | Department of Building and Real Estate, The Hong Kong Polytechnic University, Hong Kong, China |
| Srinivas Palanki | Department of Chemical & Biomolecular Engineering, University of South Alabama, Alabama, USA |
| Romeli Barbosa-Pool | Engineering Department, University of Quintana Roo, Chetumal, Mexico |
| Hao Ren | Electrical, Computer and Energy Engineering, Arizona State University, Arizona, USA |
| Parasuraman Selvam | National Centre for Catalysis Research and Department of Chemistry, Indian Institute of Technology – Madras, Chennai, India |
| Yiyi She | Ability R&D Energy Research Centre, School of Energy and Environment, City University of Hong Kong, Hong Kong, China |
| Nicholas D. Sylvester | Department of Chemical & Biomolecular Engineering, University of South Alabama, Alabama, USA |
| Yutaka Tabe | Graduate School of Engineering, Hokkaido University, Hokkaido, Japan |
| Bin Wang | Ability R&D Energy Research Centre, School of Energy and Environment, City University of Hong Kong, Hong Kong, China |

Huizhi Wang — Institute of Mechanical, Process and Energy Engineering, School of Engineering and Physical Sciences, Heriot-Watt University, Edinburgh, UK

Yao Wang — CAS Key Laboratory of Materials for Energy Conversion, Department of Materials Science and Engineering, University of Science and Technology of China, Hefei, China

Changrong Xia — CAS Key Laboratory of Materials for Energy Conversion, Department of Materials Science and Engineering, University of Science and Technology of China, Hefei, China

Yu Xiao — Department of Energy Sciences, Lund University, Sweden, & Fuel Cell System and Engineering Laboratory, Dalian Institute of Chemical Physics, Dalian, China

Jin Xuan — Institute of Mechanical, Process and Energy Engineering, School of Engineering and Physical Sciences, Heriot-Watt University, Edinburgh, UK

Jinliang Yuan — Department of Energy Sciences, Lund University, Lund, Sweden

Hao Zhang — State-Key Laboratory of Chemical Engineering, School of Mechanical and Power Engineering, East China University of Science and Technology, Shanghai, China

# Preface

Energy shortage and environmental issues are two major problems confronting our society today. As the dominant sources for energy consumption so far, fossil fuels contribute more than 80% of the total world energy needs. However, fossil fuel supplies are limited and cannot keep pace with demand. Sustainable and clean energy sources, along with high efficiency energy conversion technologies, are urgently needed to handle this dilemma.

Conventional electricity is generated from heat engines through burning fossil fuels like coal, oil and natural gas. Since the process involves the energy conversion from chemical to thermal, from thermal to mechanical, and then from mechanical to electrical energy, the efficiency of this process meets an instinctive limit (Carnot efficiency), according to the second law of thermodynamics. Compared with a heat engine, a fuel cell is a kind of electrochemical reactor, which directly converts the chemical energy of the fuel into electricity with high efficiency.

The first fuel cell prototype was developed by Sir William Grove in 1839, one year after the principle was proposed. In 1959, the first practical 5kW fuel cell based on an alkaline electrolyte was demonstrated by British engineer Dr. Francis Bacon at the University of Cambridge. At around the same time, efforts were made to make commercial use of the fuel cell. In the late 1950s and early 1960s, NASA began to develop a fuel cell system to meet the electricity demand for manned space applications. A 1.5 kW alkaline fuel cell (AFC) was developed by International Fuel Cells in the Apollo space program for power generation and water production. In the 1970s, 12 kW AFC systems were used to supply onboard reliable power for all space missions. The oil crisis and emergence of increasing environmental awareness in 1970s compelled the oil importing countries to become concerned about the energy efficiency problem. Then came the boom in fuel cell technologies. To date, as dominant fuel cells, proton exchange membrane fuel cells (PEMFC) and molten carbonate fuel cells (MCFC), are applied in the portable, stationary and transport sectors, providing a total capacity of about 180 MW around the world.

A fuel cell can satisfy society's demand in almost every aspect. Portable fuel cells promise to be the next generation power source for consumer electronics by enabling longer operation time. Compared with lithium-ion batteries, a fuel cell, which is fed with liquid fuel, enables larger power capacity and miniaturization at the same time. Integrating the advantages of the heat engine and the battery, the uninterruptible and high efficiency power supply makes the fuel cell a perfect driving force for vehicles. It has always been the wish of big car manufacturers in the world to push fuel cell electric vehicles forward to commercialization. On the other hand, stationary applications of the fuel cell are expected to replace the traditional electricity from power plants by independent systems supplying stable and clean electricity for homes, communal areas and even cities. An MCFC system, with a capacity of 40 MW, has already been installed in South Korea by POSCO Energy since 2011 for this purpose.

Hydrogen, a carbon free fuel, is the most promising energy source for a fuel cell since it produces neither pollutants nor greenhouse gases. However, high cost and inconvenient storage become the biggest restrictions faced by a hydrogen application. Onboard hydrogen generation and other hydrocarbon fuels, which are of low cost or high volume energy density, like natural gas and methanol, are regarded as alternative choices. Sustainable biomass energy, which is renewable, carbon neutral and widely available, is also a good energy source for fuel cells. Fuel cells can be categorized into the following five types according to different electrolytes: proton exchange membrane fuel cell (PEMFC), phosphoric acid fuel cell (PAFC), alkaline fuel cell (AFC), molten carbonate fuel cell (MCFC) and solid oxide fuel cell (SOFC).

**Proton exchange membrane fuel cell** – Oxidation and reduction reactions are separated by electrolyte membrane, but proton transport is allowed. Commonly, a typical PEMFC works under 100°C; water and heat management systems are important to maintain the performance of the membrane. Hydrogen is the preferred fuel for PEMFC. Organic fuel, like methanol and formic acid, can also be used however at the risk of catalyst poisoning by its intermediate by-products such as carbon monoxide. By using a mineral-based membrane instead of a polymer membrane, PEMFC can be operated up to 200°C. Several benefits can be achieved with PEMFC operated at high temperatures: (1) the electrochemical kinetics are enhanced; (2) the carbon monoxide tolerance is increased; (3) the generated steam can be utilized as a heat source.

**Phosphoric acid fuel cell** – PAFC is the first fuel cell to have reached commercialization. Phosphoric acid is used as the electrolyte. More than 75 MW PAFC systems are in operation in 85 cities in 19 countries so far. Similar to the high temperature PEMFC, the operating temperature of a PAFC is higher than 200°C. This helps increase the tolerance of the Pt catalyst against carbon monoxide. Hydrocarbon reforming gas is a promising fuel source for PAFC. The overall efficiency of PAFC can reach 80% if the cogenerated heat is harnessed. Due to these characteristics of the PAFC, its main application focuses on stationary power source.

**Alkaline fuel cell** – The first practical fuel cell made in history was the alkaline fuel cell contributed by Dr. Francis T. Bacon in 1959. The AFC also served as the energy supply in NASA's space programs. Due to the alkaline electrolyte, AFC is sensitive to carbon dioxide. Therefore, an additional carbon dioxide separation system is needed to filter carbon dioxide from the air; the need for pure hydrogen is another disadvantage of AFC. Efficiency as high as 60% can be obtained in AFC. Although research enthusiasm for the alkaline fuel cell has cooled down since the 1970s, fuel cell based on alkaline electrolyte has started attracting attention again in recent years due to its non-noble catalyst compatibility.

**Molten carbonate fuel cell** – The typical running temperature of a MCFC is around 650°C. Molten carbonate salt is used as the electrolyte. Due to the high working temperature, a MCFC can be fuelled with various kinds of fuel such as derived gas and natural gas without an external reformer. The corrosion effect of sulfur should be taken into consideration when choosing the type of fuel. High efficiency and non-noble catalyst are the major advantages of MCFC. To date, MCFC has retained more than 40% of the fuel cell market share.

**Solid oxide fuel cell** – Similar to a MCFC, a SOFC is also a kind of high temperature fuel cell working at around 1000°C. Solid ceramic material like yttria-stabilized zirconia is employed as the electrolyte. The internal reform reaction of fuel makes SOFC flexible to various hydrocarbon fuels without using a fuel reform system. However, the sulfur tolerance of SOFC is poor. Because of the slow start up and high operation temperature, SOFC is suitable to large-scale stationary power applications with heat cogeneration.

With lots of efforts being made, great progress has been achieved on fuel cell technology. Vehicles and electronics devices powered by PEMFC have been demonstrated to be usable. Power plants based on MCFC and SOFC are in operation in Asia and the U.S.A. New types of fuel cells with high efficiency are emerging, like the microfluidic fuel cell, the aluminum-air fuel cell and the microbial fuel cell. The fuel cell technology is booming and has come to the critical stage of commercialization. Nevertheless, many existing obstacles are still impeding the advancement of fuel cells. The cost of a fuel cell remains at a high level. Reliability and durability of fuel cells needs to be further developed. Issues of hydrogen generation and storage also need to be resolved.

In this book, Micro- and Nano-engineering, with emphasis on the miniaturization of fuel cells and fundamental mechanism understanding in a micro- and nano-scale, are proposed to deal with the challenges of fuel cell development.

Research on miniaturization of fuel cells is motivated by a great interest in portable applications. In order to power electronic devices like laptops and cell phones, fuel cells with a compact structure and high energy-to-weight ratio are required. Studies on miniaturized fuel cell systems have recently been boosted with the development of technologies for micro-size fabrication.

Miniaturizing a fuel cell is not as simple as reducing the geometrical dimension. With the size decreasing, the surface-to-volume ratio increases significantly. This provides a large benefit for a heterogeneous reaction like an electrochemical reaction, which directly relies on the active area. A decrement in dimension will result in a reduction in mass and heat transport resistance due to the shortening of transport pathways. Besides, it is much easier for a micro fuel cell to manipulate the reaction process rapidly according to the changing power demand.

Fast and low-cost micro-fabrication processes are of great importance for the commercialization of miniaturized fuel cells. Most fuel cells apply a membrane electrode assembled (MEA) structure, which compacts electrode, catalyst and membrane as an ensemble. Various kinds of materials like silicon, glass, stainless steel, polymethylmethacrylate (PMMA) can be used for micro fuel cells manufacturing. Etching and printing are usually employed in the manufacture of large-scale microchannels. Nevertheless, complex multi-step manufacture keeps costs high and makes it difficult to industrialize. Direct shaping methods, like laser cutting and casting of polymer, are promising approaches permitting large-scale production. However, due to the temperature limitation, the application of organic material is restricted.

Catalyst coating is another important process in fabricating micro fuel cells. Uniform and reliable catalyst coating is extremely crucial for mini type fuel cells. The preparation method and compatibility between catalyst and substrate affect the activity and durability of the catalyst layer.

Finally, good sealing and a rigid enclosure are vital for the safe and durable operation of micro fuel cells.

Hydrogen is a perfect fuel regardless of the dimensions of a fuel cell. However, the generation and storage problem of hydrogen makes it impossible to apply directly in a micro fuel cell operation. Alternatively, onboard continuous hydrogen production through reforming reaction is considered as a practical choice. Liquid hydrocarbon fuel, like methanol, is a good hydrogen source for the fuel processor. However, the integration of the processor still needs to be developed and optimized.

A micro aluminum-air fuel cell is a novel kind of fuel cell that harnesses the energy of metal. Compared with conventional hydrocarbon fuels or hydrogen, no harmful and greenhouse gas is emitted during the cell operation. Aluminum is the most abundant crustal metal and an aluminum-based cell possesses high open-circuit potential of 2.73 V with an energy density reaching 2.8 kWh/kg. With such a high energy density and simple structure, the micro aluminum-air cell is competitive compared with the traditional fuel cell.

The performance of a fuel cell is closely related to the transport and reaction phenomenon at the electrode/electrolyte interface. For example, porosity and tortuosity affect the effective diffusivity significantly, as well as the triple phase boundary (TPB) area in a SOFC. This will impact the polarization loss, and changes in the microstructure of the electrode will severely affect the performance of fuel cell. The apparent performance of a fuel cell is a statistical result of every single active site at the catalyst layer. Nevertheless, in the absence of an inner view of the transfer process in the porous electrode, most of the studies, either numerical or experimental, only focus on the overall characteristics of fuel cells such as the J-V curve and the electrical impedance spectroscopy (EIS). A comprehensive understanding of the behavior and mechanisms of a fuel cell is still needed.

The most extensive approach to the investigation of fuel cells numerically so far, applies continuum equations and global relations to predict the characteristics of fuel cells on a system level. For porous material, the homogeneous properties of porous material, like porosity and tortuosity, are used to calculate an equivalent result. This strategy reduces the complexity of modelling within the microstructure but sacrifices the precision of modeling. Moreover, detailed information of the transfer process at the electrode/electrolyte interface is missing.

Several length and time scales exist in a fuel cell system simultaneously, in the range from $10^{-9}$ to $10^2$ m in length and from $10^{-15}$ to $10^1$ s in time. Conventional modeling methods can only cover the macro-scale range, above $10^{-6}$ m. At the porous region and catalyst/electrode interface, like the diffusion layer and Pt/C interface, the characteristic dimension is down to a mesoscale (below $10^{-6}$ m) and even a microscale ($\sim 10^{-9}$ m); continuum equations and homogenous charge transfer relations are not precise enough to describe the transport phenomenon. Many methods have been developed to model a fuel cell in microscale and mesoscale like molecular dynamics and lattice Boltzmann methods.

However, although the global properties of a fuel cell can be obtained by summarizing micro and nano-scale modelling, it is costly to use microscale methods to cover all the scales. To model the fuel cell accurately and economically, a suitable modelling strategy should be chosen according to different length scales and a multi-scale method is needed to bridge the gap between segregated scales.

Some experimental techniques are also developed in order to obtain a detailed description of the morphology of the microstructure within the porous electrode and catalyst layer. Reconstruction technologies, which are used to reproduce the inner structure of porous material, have been invented and widely applied to investigate the real microstructure of the electrode and catalyst layer. The 3D reconstruction data does not only provide the interesting material parameters, like porosity, tortuosity, specific area etc., but can also be used as a geometry input for further numerical research. The most widely used reconstruction methods are focused on ion beam/scanning electron microscopy (FIB/SEM) and X-ray computed tomography. No matter which method is applied, several steps are commonly employed. Firstly, sequential slices of a sample are required. Then, post-processing steps are needed to address the 2D images and distinguish the different phases, followed by an alignment of consecutive slices. Finally, a quantification of different parameters gives the essential parameters of interest.

Based on an understanding of the mechanism of an electrode, an optimization of the electrode structure and catalyst loading are of great significance for fuel cell improvement and cost reduction. A novel electrode composed of core-shell particles with a less costly catalyst presents a much better performance than conventional ones since a core-shell structure provides a large specific area and better catalyst dispersion.

In summary, a fuel cell is a kind of clean and efficient energy conversion device expected to be the next generation power source. Through more than seventeen decades of R&D, various types of fuel cells have been developed to meet different energy demands and application requirements. Intensive efforts have been devoted to the commercialization of the fuel cell. However, several issues ranging from mechanism studying to system integration are still need to be researched prior to massive application. This book focuses on the Micro- and Nano-engineering aspects to remove the challenges faced by fuel cell technically and theoretically. Miniaturization is one of the main bottlenecks for fuel cell advancement into our daily lives. Low cost manufacturing techniques are important for a compacted fuel cell. On the other hand, theories focusing on micro and nano-scale mechanisms should be established, which will bring great benefit for the optimization of fuel cells of all sizes.

# About the editors

**Prof. Dennis Y.C. Leung**, born in Hong Kong in 1959, received his BEng in 1982 and PhD in fluid dynamics in 1988, both from the Department of Mechanical Engineering at the University of Hong Kong. He had worked in the Hongkong Electric Co. Ltd., for five years leading the air pollution section of the company. He joined the University of Hong Kong in 1993 as a lecturer in the Department of Mechanical Engineering. He developed the environmental engineering and energy engineering program of the department after joining the university and was then very active in conducting research in various energy related researches, in particular fuel cells. Professor Leung is now a full professor of the department specializing in air pollution, renewable and clean energy. He has published more than 400 articles in this research area including 200 peer reviewed SCI journal papers with high impact factors. He was invited to write more than 10 review articles for various journals and received high citations. His current h-index is 42 and total citations are more than 8000. He is one of the top 1% highly cited scientists in the world in energy area since 2010 (Essential Science Indicators). Prof. Leung has delivered more than 30 keynote and invited speeches in many conferences as well as public lectures. He serves as a chief editor of Frontiers in Environmental Sciences and editorial board member of many other journals including Applied Energy, Journal of Power & Energy, Sustainable Energy etc.

Prof. Leung is a chartered engineer, a Fellow of the Institution of Mechanical Engineers, Energy Institute and the Hong Kong Institute of Acoustics. He also serves/served in the board of directors of the Friends of the Earth (HK), Hong Kong Quality Assurance Agency, Asia Pacific Hydrogen Association and Hong Kong Institution of Science. With his rich experience in energy and environmental research, he had been appointed as chairman of many government and non-government committees including the Energy Institute (HK Branch), ISO 14001 Technical Committee of Hong Kong Quality Assurance Agency, the Task Force on Greenhouse Gas Validation and Verification, just name a few. He is/was an Appeal Board Panel Member of the Gas Safety Ordinance and Noise Control Ordinance of the HKSAR Government, and sits in various committees of the Hong Kong Environmental Protection Department for cleaning up the air in Hong Kong. Prof. Leung received numerous awards including the Outstanding Earth Champion Hong Kong award in 2008 in recognizing his contributions in protecting the environment.

**Dr. Jin Xuan** was born in Zhejiang, China in 1984. He is currently an Assistant Professor at School of Engineering and Physical Sciences, Heriot-Watt University, United Kingdom. Dr. Xuan obtained his BSc degree in Building Environment and Service Engineering from Zhejiang Sci-Tech University in 2007, and PhD degree in Mechanical Engineering from The University of Hong Kong in 2011. After that, he joined East China University of Science and Technology as a Faculty Member in School of Mechanical and Power Engineering. In 2014, he moved his academic career to UK, taking his current position in Heriot-Watt University. Dr. Xuan is also a visiting researcher in The University of Hong Kong and City University of Hong Kong during 2011–2014. Dr. Xuan's research interests include fuel cells, renewable energy and process intensification. He has completed several big fuel cell related projects including solid oxide fuel cell, microfluidic fuel cell and photocatalytic fuel cell. He has published 3 book chapter, 30+ journal papers and 20+ conference papers. He was the invited speaker of several conferences. He has also received a number of awards including American Chemical Society ENVR Certificate of Merit Award, Shanghai Pujiang Talent Award and Honorable Mention of Hong Kong Young Scientist Award.

# CHAPTER 1

## Pore-scale water transport investigation for polymer electrolyte membrane (PEM) fuel cells

Takemi Chikahisa & Yutaka Tabe

### 1.1 INTRODUCTION

Polymer electrolyte membrane (PEM) fuel cells could potentially become promising power sources for vehicles and stationery combined heat and power (CHP) generators in the near future, as they have higher efficiency and cleaner emission characteristics than conventional engines. Although hydrogen fuel for the cells is mainly produced from natural gas with a conversion efficiency of around 75% at present, well-to-wheel efficiency is still superior to conventional engines. When hydrogen is produced from renewable energy sources, such as solar cells and windmills, PEM fuel cells will be the most powerful devices to convert hydrogen into power with the highest efficiency.

Due to extensive research and development of fuel cell vehicles, PEM fuel cells are almost ready for commercial applications with a driving performance superior to conventional vehicles. In order for fuel cell vehicles to become highly competitive with conventional vehicles, the major issues of cost and durability need to be resolved. To reduce costs, the platinum amounts employed must be decreased. It will also be necessary to reduce component costs. Improving durability of membranes and catalysts are also important to increase the lifetime of PEM fuel cells. Water management is also an important area to research as it strongly relates to the efficiency, power, and durability of the cells. Drying of the membrane reduces cell lifetime due to membrane thinning and pinhole formation, whereas flooding accelerates degradation of the catalyst and porous transport layers due to hot spots caused by non-uniform current density distributions.

The chapter details the water transport phenomena in PEM fuel cells from the macroscale in the gas channels to the microscale in the catalyst layers. As there are a substantial number of papers on water transport research and review papers dealing with this, this chapter gives a summary of our major work and findings. Details of experimental conditions and other information are in the original papers in the references (Chikahisa, 2013; Chikahisa et al., 2012; Naing et al., 2011; Tabe et al., 2009a; 2009b; 2011; 2012; 2013).

### 1.2 BASICS OF CELL PERFORMANCE AND WATER MANAGEMENT

Polymer electrolyte membrane fuel cells have the basic structure illustrated in Figure 1.1, which shows a unit cell. By overlaying repeated layers of cells in series, a battery stack is formed, producing sufficient voltage. In the middle of a unit cell there is a polymer electrolyte membrane, which permits protons to pass through the membrane. At both sides of the membrane, the unit cell has catalyst layers as electrodes, forming a membrane electrode assembly (MEA), which is sometimes termed a catalyst coated membrane (CCM). The thickness of the catalyst layer (CL) is ca. 20 $\mu$m, and the polymer electrolyte membrane has a thickness of ca. 50 $\mu$m. The MEA is sandwiched by gas diffusion layers (GDL), about 200 $\mu$m thick, and these layers are further sandwiched by bipolar plates, also termed separators. Figure 1.2 illustrates the structure near the GDL with images of the liquid water distribution. Hydrogen is supplied to the anode separator channel, and oxygen or air is supplied to the cathode channel. The hydrogen reacts in the anode catalyst layer to decompose into protons and electrons. The proton transfers into the membrane to

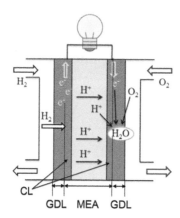

Figure 1.1.   Basic structure of fuel-cells.

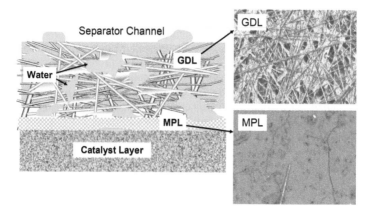

Figure 1.2.   Schematic cross-section of the cathode layers and photos of the GDL and MPL layers.

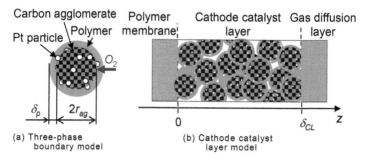

Figure 1.3.   Schematic outline of the PEM model.

reach the cathode catalyst layer, whereas electrons pass through the external circuit to reach the cathode. The water produced in the cathode catalyst layers is transported through the micro-porous layer (MPL) and GDL to reach the gas channels.

Figure 1.3 illustrates the cathode catalyst layer. In the cathode catalyst layers, protons transfer in the polymer (ionomer), a material similar to the membrane, which covers the surface of carbon agglomerates. Electrons transfer in the carbon agglomerates in a chain structure. Oxygen transfers

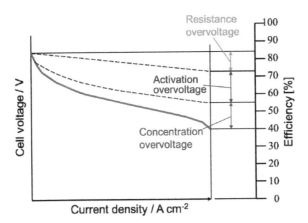

Figure 1.4.   Cell voltage and overpotentials.

in the pores in the catalyst layer and dissolves into the ionomer. The three species (protons, electrons, and oxygen) combine at the surface of platinum particles, which are distributed on the carbon-agglomerate particles, to produce water.

The transport of protons in the membrane accompanies the water molecules by osmotic force, resulting in a water flux from the anode side to the cathode side in the membrane. This produces a gradient in the water content in the membrane with more water on the cathode side. The water gradient in the membrane also produces a back diffusion flux of water to the anode side. When the humidity of the anode gas is low and insufficient to supply water for the osmotic transfer, the membrane dries out, and the proton conductivity in the membrane becomes low: the so-called "dry-out" phenomenon. The water transferring through the membrane to the cathode and the water produced in the cathode catalyst layer must be removed smoothly from the reaction zone to the separator channels through the GDL. When the water interferes with the oxygen diffusion into the reaction zone, the cell voltage drops: the so-called "flooding". The major effort of cell water management is to control the water behavior to prevent dry-out and flooding in the fuel cell.

To improve water management, the GDL commonly has a micro-porous layer (MPL) of ca. 70 µm on one side in contact with the catalyst layer as shown in Figure 1.2. As suggested by the illustrations, the GDL consists of a coarse random mesh of carbon fibers with pore diameters of around 10–30 µm and a porosity of ca. 75%, whereas the MPL is a fine layer with pores of around 0.1–0.5 µm. The MPL is responsible for protecting the catalyst layers from the uneven and irregular GDL fibers, providing improved contact resistance at the interface, and better water management characteristics for dealing with flooding and dry-out events. However, the mechanisms required for optimum water management characteristics by the MPL are not well understood at present. As shown in Figure 1.2, water is produced in the catalyst layer in the cathode and transfers to the separator channels through the various layers. Here there is no general agreement on the phase of the water transferring through the MPL, i.e. whether it is in liquid or vapor form. It is also unclear how the liquid water distributes in the cell layers and what kind of layer structure would achieve optimal water management.

Fuel cells generally generate a voltage curve for the current density as shown in Figure 1.4. The open circuit voltage (OCV) corresponds to the Gibbs free energy of the fuel and oxidant, and it is about 1.1 volt for a hydrogen-air fuel cell. The voltage drops due to overpotentials when the current increases. As the cell voltage corresponds to the thermal efficiency of the cell, it is important to maintain a high voltage and to extend the curve to higher current densities to increase power.

The polarization curve can be decomposed into overpotential (overvoltage) elements as shown in Figure 1.4. The activation overpotential is due to the chemical reactions in the catalyst layers and

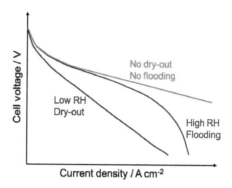

Figure 1.5.   Dry-out and flooding.

the cathode overpotential is the dominant part, compared to the anode overpotential. Reduction of the activation overpotential is dependent on the reactivity and structure of the catalyst layers. The ohmic loss overpotential and the resistance overvoltage in the figure are due to the electric resistance in the cell components. The resistance in the membrane is the major part of this overpotential, together with the contact resistance. When the water content in the membrane decreases, the membrane resistance increases, resulting in the dry-out phenomenon. This tends to happen in the area close to the air inlet at operating conditions of medium current densities with low humid and high temperature gases. At high current density conditions, a large amount of water is produced, which interferes the oxygen diffusion into the reaction zone. This is the so-called flooding and it causes a significant voltage drop, which corresponds to the concentration overpotential (Fig. 1.4).

Figure 1.5 is an example of cell voltage changes with the humidity of the anode gas. When the humidity is low, the cell voltage drops significantly, due to dry-out. When the humidity is high, the cell voltage decreases and starts fluctuating; this is a typical feature of flooding. The dry-out and flooding are strongly dependent on the cell temperature, gas flow rate, and size of the fuel cell.

Extensive research has been conducted to establish optimal water management technology, for example, controlling the wetability of the contact layers between catalyst and GDL, controlling the pore size distribution in the MPL and GDL, designing optimal gas channel configurations, and others. The GDL of PEM fuel cells supplies the reactant gases to the catalyst layer and removes the produced water from the catalyst layer; here the GDL plays a crucial role in the water management and is important for mitigating the water flooding phenomenon.

A variety of experimental studies have been carried out to investigate the effects of the GDL thickness, porosity, permeability, and wettability on the performance of PEM fuel cells (Liu *et al.*, 2005; Noponen *et al.*, 2004; Wang, 2004). Investigations of liquid water behavior inside the GDL of PEM fuel cells have also been reported, as well as a number of *in situ* visualization methods, including neutron radiography/tomography (Kramer *et al.*, 2005; Manke *et al.*, 2007; Satija *et al.*, 2004), X-ray micro tomography (Sasabe *et al.*, 2010; Sinha *et al.*, 2006), and direct imaging/visualization (Tüber *et al.*, 2003; Yang *et al.*, 2004; Zhang *et al.*, 2006).

## 1.3   WATER TRANSPORT IN THE CELL CHANNELS

### 1.3.1   *Channel types*

The channels in the separators function to supply reactant gases and remove the produced water from the cell. The separators must have the characteristics of high electric and thermal conductivity

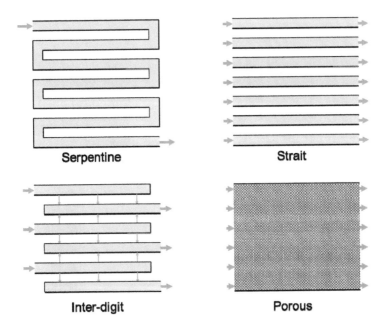

Figure 1.6.   Types of separator channels.

with corrosion resistance. They are generally made of carbon, but because of cost reduction aspects, pressed metals coated by anti-corrosion substances have also been used in recent years.

There is a variety of types of channels: straight, serpentine, inter-digit, and porous, as shown in Figure 1.6. Although it suffers from relatively high pressure losses, the serpentine type is most commonly used as it has good water removal characteristics with high gas-velocity. The straight type is simple and is subject to lower pressure losses, but it may give rise to uneven flow rates and non-uniform water accumulation among channels. However, the maldistribution characteristics may be alleviated by the unique gas flow in the GDL as will be discussed later. Inter-digit type flow channels have also been investigated for a variety of configurations, but they are not very common at present. Porous type flow channels display good characteristics for the prevention of flooding and can be operated with lower stoichiometric ratios than serpentine channels. However, they suffer from disadvantages in their cooling performance due to the lower thermal conductivity of porous separators.

Depending on the size of the cell and the materials used in the separators, the optimal type of channel design, channel width and depth, and the wettability of the surfaces are different. Additionally, the current density and gas flow rates also affect the design. To establish general guidelines for optimal channel designs for these and related parameters, more research is still necessary. This section will present some results of the water behavior in cells and the characteristics of porous separators.

### 1.3.2   *Observation of water production, temperatures, and current density distributions*

Figure 1.7 shows the experimental setup for measuring the local current density, pressure, and liquid water behavior discussed here. One separator with straight channels was used on the anode side, and separators with straight or serpentine channels were used on the cathode side. The width of the channels and lands (sometimes called "ribs") of both cathode separators was 2.0 mm. The cell has a glass window in the cathode end, through which the GDL surfaces can be observed. The cathode separators were made of copper overlaid with gold and the collected current was transmitted through the lands to the end metal plates on both sides. The thickness of the separator,

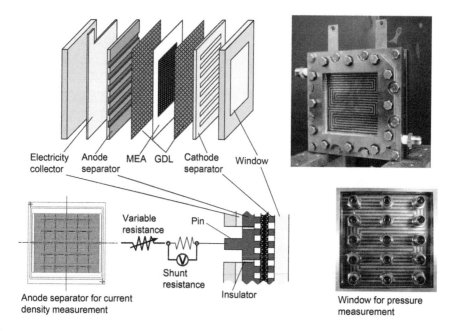

Figure 1.7.   Experimental apparatus for visual observations, current, and pressure measurements.

which corresponds to the channel height, was 0.3 or 0.5 mm. Further, 0.3 mm thick carbon papers were used for the GDLs. The current density distribution was measured with 25 pins placed in the anode separator. The pin diameter was 5.5 mm, and the pins were electrically insulated as illustrated in Figure 1.7 (bottom left). Each pin was connected to a shunt resistance of 0.1 $\Omega$ to measure the current. To compensate for contact resistance variations among the pins and to ensure uniform resistance over the whole area, calibration was made with a variable resistance connected to each shunt resistance. Additionally, thin carbon cloth was placed over the GDL paper on the anode side to minimize variations in the contact resistance among pins. The pressure drop distributions were measured by replacing the glass window by an acrylic window with 15 holes of 0.3 mm diameter along a single serpentine channel, as shown in Figure 1.7 (bottom right).

In general serpentine channels result in better cell performance than straight channels at low stoichiometric ratio, as the gas flow speed is higher and it blows off accumulated liquid water better. It may be thought that the gas flows along the channel, but actually, some amounts of gas bypass the channel in the serpentine layout. Figure 1.8 shows the flow rate distributions in two separators with different channel heights. The simulation model is based on the flow resistance network, where constant flow resistance is imposed on the channel flow, corner flow, and the flows in the GDL respectively (Tabe *et al.*, 2009a). It is composed of 24 areas of resistance to the gas flow along the channel, including the flow in the GDL, and 16 areas of resistance to the gas flow through the GDL under the land. The resistance in a channel was calculated by the equation for a laminar tube flow with a rectangular cross-section while the resistance in a GDL and in bends in the channel was determined experimentally. The model was validated by the experiment with the measured pressure drop distribution along the channel.

The variations in flow rate through the channel are expressed by the diameters of the circles. The shortcut flow rates under the lands are expressed as percentages of the inlet flow rate. Figure 1.8 clearly shows that the flow rate through the channel decreases due to shortcut flow at the center of the cell. The rate of the shortcut flow has a maximum value at the corners and decreases along the channel in the flow direction. This is due to the larger pressure difference between adjacent channels at the corners.

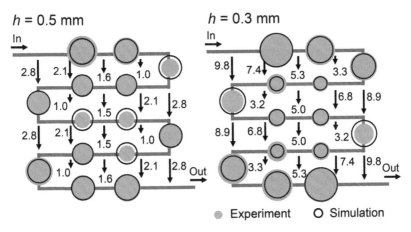

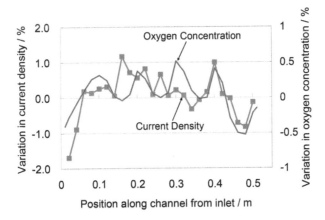

Figure 1.8.   Variations in the flow rate through channels, expressed by circle diameters, and shortcut flow rates through the GDL under lands, expressed as percent of the total flow rate. The channel heights, *h*, are 0.5 mm (left) and 0.3 mm (right).

Figure 1.9.   Variations in the measured current density and the estimated oxygen concentration at a 0.6 MPa clamping pressure.

Figure 1.9 plots the variations in the measured current density and the estimated oxygen concentrations from the average rates of decrease in oxygen for the current density. The oxygen concentration was calculated from the flow rate distribution by the circuit model, assuming the rate of oxygen consumption to be equivalent to the average power generation. The variations in current density are similar to those of oxygen concentration. The oxygen concentration decreases in the middle parts of the straight channel segments. It may be postulated that these slight variations in oxygen concentration affect the local and overall current density distributions. The analysis was conducted under the condition where the effect of condensed water is negligible because it is difficult to estimate the gas-flow resistance under flooded conditions. Generally, the flooding increases the gas flow resistance through the channels and the effect of shortcut flows on the cell performance can therefore be anticipated to increase under flooded conditions.

In straight channel separators a noteworthy phenomenon has sometimes been observed, results have shown relatively good voltage performance obtained even when the channels were filled almost entirely by liquid water. As in Figure 1.10, the cell voltage and pressure drop change after the start of the power generation. During the operation, the cell voltage at the oxygen

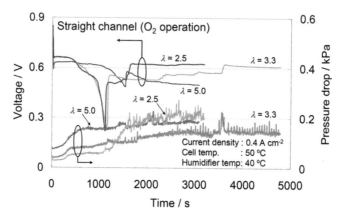

Figure 1.10.   Cell voltage and pressure variations in the cathode flow in the flooded condition for three different stoichiometric ratios of oxygen, $\lambda = 2.5$, 3.3, and 5.0, in the straight channel cell.

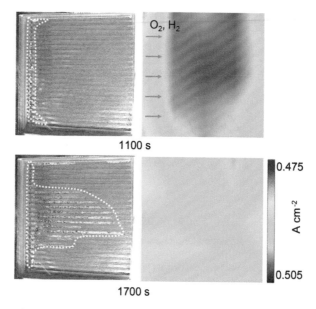

Figure 1.11.   Direct view (left) and current density distribution (right) in the cathode channels at two different times in the Figure 1.10 experiment for the 2.5 stoichiometric ratio. Dotted outlines enclose areas with gas phase.

stoichiometric ratio, $\lambda$, of 2.5 gradually decreases and reaches a minimum value of 1100 s. Figure 1.11 is a direct view of the cathode channel and the current density distribution at two different times after the start of the operation. At 1100 s in Figure 1.11 (top), the channel paths are filled with liquid water from the downstream (right-hand) side and there is only a small region with gas phase (white area enclosed with a dashed line) at the left end. The other parts of the cell are filled with liquid water, which appears as dark in the picture. After 1100 s, the voltage increases abruptly and this may be explained by the gas phase pushing out the liquid water and an expansion of the gas-filled region from the left. With this change, the area with high current density increases around the gas phase region and the cell voltage increases abruptly as shown in Figure 1.10. At 1700 s, the current distribution becomes uniform even though the region of the channel filled with condensed water remains, as shown in Figure 1.11; a similar phenomenon

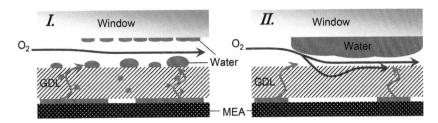

Figure 1.12.   Experimentally suggested liquid water distribution and gas flow paths: (I) the early stage of the voltage drop, and (II) the stabilized high-voltage stage with large pressure variations.

has also been observed at the other stoichiometric ratios. The pressure drop increases after the cell voltage recovers as shown in Figure 1.10. The cell voltage stabilizes at a relatively high level, when the pressure drop becomes high and the pressure variation also becomes larger at this stage. The final voltage, after the recovery, is higher for the smaller gas flow rates, i.e. smaller $\lambda$.

Based on the water behavior in the cathode channel, the changes in the cell voltage and the pressure drop, the phenomena inside the GDL were hypothesized to be as illustrated in Figure 1.12. In the early stage of the operation (I), the produced water flows out to the channel through the relatively wide paths in the GDL and accumulates at the surface of the GDL. Because of the hydrophilic properties of the glass window and metal separator, liquid water also spreads on the window and the separator surface. The volume of condensed water at the surface, and maybe inside the GDL, gradually increases and the cell voltage drops correspondingly. At this stage, there is still adequate space in the channel to allow gas flow and maintain an almost even pressure drop (phase I, left in Fig. 1.12). Then a slight increase in the pressure drop pushes the liquid water in the channel to form water films along the window, establishing gas paths under the liquid water on the window (phase II). This situation is further reinforced with a number of gas paths to the reaction area of the MEA, resulting in a stabilized satisfactory performance with a large pressure drop. Overall, this results in a sudden increase in the cell voltage as shown above.

This explanation suggests the possibility of improving cell performance with smaller stoichiometric ratios under conditions where the path is filled with liquid water; liquid water on the window attracts water from the GDL and may play an important role in improving and stabilizing the cell performance. This would suggest that the formation of a liquid layer near the channel surface is advantageous under flooded conditions, and such considerations may be applied to realize stable air operation with low stoichiometric ratios. However, it should be noted that this unique phenomenon was observed only with the oxygen and not with the air operation; it was confirmed that the difference was due to the gas flow rates with the two gases and that a similar operation was possible with air, when the gas was switched from oxygen, after the establishment of a steady state following a temporary drop in voltage (Tabe *et al.*, 2009a).

### 1.3.3   *Characteristics of porous separators*

Further right in Figure 1.13 is a photograph of the porous separator used in the experiments in the following section. The porous media were made of nickel, 1.4 mm thick with porosity 97%. In the experiments, the cell was set vertically and the experiments were conducted with pure hydrogen as the anode gas and air or pure oxygen as the cathode gas. All the gases were supplied with 100% relative humidity.

Figure 1.14 shows the polarization curves of the porous and serpentine channels with the cell arrangement of Figure 1.13. The flow rate of the cathode air was varied at two different stoichiometric ratios, $\lambda = 1.4$ and 2.0. The polarization curve for the serpentine type could not be determined at the lower stoichiometric ratio, 1.4, however with the porous separator it can realize operations up to $0.7\,\mathrm{A\,cm^{-2}}$. At the higher stoichiometric ratio of 2.0, the cell voltage of the serpentine separator is higher but drops in the high current density region, whereas the

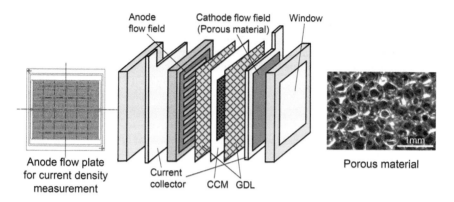

Figure 1.13.   Experimental arrangement for visual observations and current density measurements.

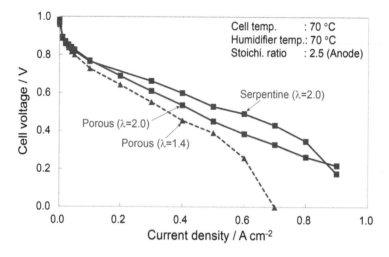

Figure 1.14.   Polarization curves for cells with the porous flow field and serpentine channels.

cell voltage of the porous separator is less affected by the concentration overpotential. Here, the lower voltage in the medium current density region is mainly caused by the larger cell resistance of the porous separator. This is considered to be due to the high contact resistance between the porous separator and the cathode electric collector. These results indicate that a cell with porous separator has advantages in high current density and low stoichiometric operations.

Figure 1.15 shows direct views of the liquid water in the cross-section of the porous separator (Fig. 1.13) at four different times after the start of water droplets emerging from the GDL. In the first instance (a), one side of the porous separator was open for direct observation of the end surface. The overall current density was $0.4\,A\,cm^{-2}$ and the stoichiometric ratio of the cathode air was 2.7. The humidity of both gases was 100% with the cell temperature at 45°C. The liquid water emerges from the GDL surface through the relatively large pores (b). After growing (c) and touching the fibers of the porous media, water drops are attracted to the porous medium and spread along the pores to form a liquid water film (d). This is because the porous material has hydrophilic characteristics. It was estimated that the liquid water along the porous wall drained by the air flow and that a dynamic equilibrium is maintained, which causes the tolerance to flooding in the porous separator. These observations support the fact that the porous separator is superior in drainage performance from the surface of the GDL. They also show that the hydrophilic properties of porous separators provide an important condition for this tolerance to flooding.

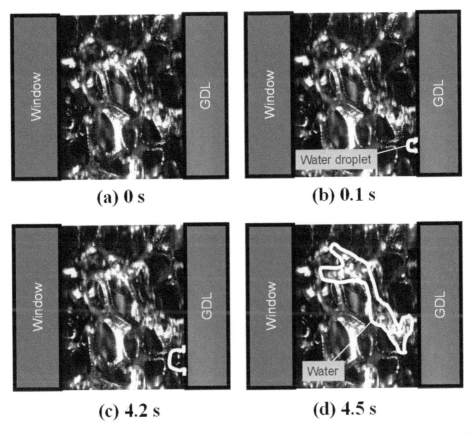

(a) 0 s        (b) 0.1 s

(c) 4.2 s        (d) 4.5 s

Figure 1.15.   Direct view of liquid water transport in a cross-section of the hydrophilic porous material, with times after start of droplets emerging from the GDL.

To confirm the transport mechanism and its effect on the cell performance, experiments were conducted using a hydrophobic porous separator. The results showed that the voltage with the hydrophobic porous separator fluctuated, whereas a stable operation was achieved with the hydrophilic porous separator. This is related to the behavior of the droplet emerging from the GDL surface as suggested in Figure 1.15. It may be concluded that a hydrophilic porous separator spreads the liquid water and drains it efficiently from the surface of the GDL, resulting in good tolerance to flooding.

## 1.4   WATER TRANSPORT IN GAS DIFFUSION LAYERS

The water transport in the gas diffusion layer (GDL) is very important, as the region is at the intersection of the water and oxygen flows. Additionally, water tends to accumulate under lands and deteriorates the oxygen supply to these areas. Thus, it would be desirable to have a GDL structure, which smoothly guides liquid water to the separator channels from the MPL surface and from the regions under the lands.

The authors investigated the effect of anisotropic fiber directions of the GDL and conducted numerical simulations of the water flow in porous media with different wettability gradients in the media. The major results of the research in these two investigations are presented in this section.

Table 1.1.    Structural properties of anisotropic GDL.

| | |
|---|---|
| Thickness | $278\,\mu m$ @0 MPa, $210\,\mu m$ @2 MPa |
| Porosity | 74% |
| Permeability (as fiber direction) | $0.61 \times 10^{-12}\,m^3\,Pa^{-1}\,s^{-1}$ @1.8 MPa |
| Permeability (transverse direction) | $0.39 \times 10^{-12}\,m^3\,Pa^{-1}\,s^{-1}$ @1.8 MPa |
| Fiber diameter | $\Phi\,12.5\,\mu m$ |
| Maximum pore size | $\Phi\,35\,\mu m$ |

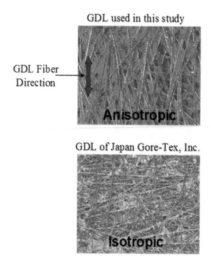

Figure 1.16.    Anisotropic (above) and isotropic (below) GDLs.

### 1.4.1   *Water transport with different anisotropic fiber directions of the GDL*

Commonly used GDLs have random and isotropic carbon fiber directions. However, the GDL used in the investigation described next has partially oriented fiber directions. The structural properties of the anisotropic GDLs here are detailed in Table 1.1. The upper panel of Figure 1.16 shows the surface of the anisotropic GDL and the red arrow indicates the general fiber direction. Comparing the picture of the anisotropic GDL with the picture of isotropic GDL in the panel below, the differences in the fiber orientations in the two GDLs are apparent from the figure. The ratio of the gas permeability in two different directions is 1.6, showing the degree of anisotropy of the GDL. In the following, one configuration is termed "perpendicular", where the averaged fiber direction is perpendicular to the gas channel direction; the other configuration is termed "parallel", where the averaged fiber direction is parallel to the gas channels as shown in Figure 1.17. There are micro-porous layers of the GDLs in contact with the MEA.

The characteristics of the cell performance with the two carbon fiber directions were investigated at various current densities using a $25\,cm^2$ active area cell with the cathode separators of both serpentine and straight channels. Figure 1.18 shows the cell voltages for the two fiber directions at different current densities using the two separator types, with serpentine and straight channels. Here, air was used as the cathode gas and pure hydrogen as the anode gas. The stoichiometric ratios were 3.0 for the cathode and 2.0 for the anode at $1.0\,A\,cm^{-2}$. The humidity of both gases was set to saturation at 60°C with the cell temperature at 58°C. The overall cell resistance in both cells was set to be equal before the cell operation. The results show that the performances of the two cells are very similar at the lower current densities. However, the cell voltages of the perpendicular fiber direction are higher than those with the parallel fibers. The differences in

GDL Fiber
direction

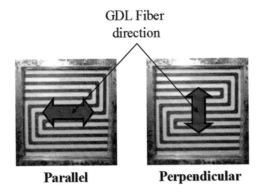

**Parallel**          **Perpendicular**

Figure 1.17.   Two types of configuration of the anisotropic GDLs used in this investigation, perpendicular and parallel fiber directions.

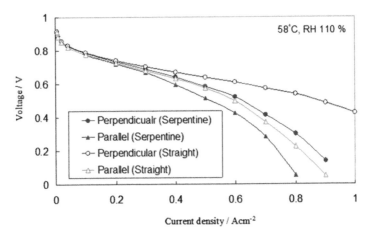

Figure 1.18.   Polarization curves for the two orientations of anisotropic GDL, perpendicular and parallel, with two separator types, serpentine and straight.

the cell performances become more apparent when the current densities increase. These results lead us to conclude that the perpendicular fiber direction GDL is superior to the cell with parallel fibers. The differences become even clearer when using the straight channel type cell. It was also confirmed by the results previously mentioned that the perpendicular fiber-direction GDL showed less flooding characteristics than the parallel fiber-direction GDL.

The characteristics of the water removal from the GDL to the flow channel were investigated using the 25 cm² active area cell. A comparison was made using photos of the growth of the liquid water film in the perpendicular cell and of a droplet in the parallel cell. The photos showed that water film flows from the side of the rib (land) to the channel with the perpendicular fiber direction. This shows that the accumulated liquid water inside the GDL under the ribs flows to the channels using the fiber direction perpendicular to the channel. However, with the parallel fiber direction there was a liquid droplet at the center of the channel and there was no apparent liquid flow from under the ribs. The behavior of the water indicates that the water may accumulate under the ribs when the fiber direction is parallel to the channels and that this accumulated water may affect the cell performance.

An extensive *ex situ* study of the liquid water distribution in the anisotropic GDL was conducted with a cell freezing method. After the steady state operation was achieved, the cell operation was stopped and the cell assembly was cooled to below freezing temperature within 4 minutes.

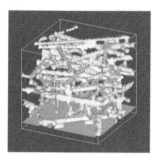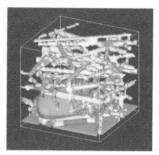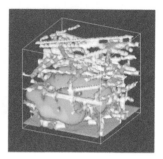

Figure 1.19.   Example of a three dimensional LBM simulation showing liquid water growth in GDL fibers. Liquid water is supplied from a pore in the bottom surface.

The accumulated water in the frozen state was then observed for the cross-sectional GDL. It was confirmed from the experiments that the ice distribution under the channels and ribs were very similar in the perpendicular case, while there was more ice under the ribs than that under the channels in the parallel case (Naing *et al.*, 2011). This suggests that the accumulated water can flow out more easily to the channels along the fiber direction in the perpendicular case.

In the PEM fuel cell operation, the liquid water generated in the cell accumulates inside the GDL and is expelled through the openings between the fibers in the GDL to the channel. In the case with fibers perpendicular to the flow direction, the accumulated water under the ribs can migrate toward the channels along the GDL fiber direction. In the case of the parallel fiber direction, the water under the ribs cannot easily flow out to the channels because of the fiber direction. The growth in the volume of accumulated water obstructs the supply of reactant gases and, as a result, the supply of reactant gases to the MEA under the ribs becomes insufficient. This explanation is in agreement with the experimental results in which liquid droplets grew from the GDL surface around the center of the channel with the parallel fiber cell.

### 1.4.2   *Water transport simulation in GDLs with different wettability gradients*

Liquid water behavior in complex geometries can be simulated with the lattice Boltzmann method (LBM). The LBM for two-phase flows with large density differences has been applied to the simulation of liquid water and air flows in a PEM fuel cell and the effect of wettability and cross-sectional shape on the liquid water behavior in the gas flow channels was reported by us (Tabe *et al.*, 2009b). Figure 1.19 is an example of the liquid water simulation of the situation inside GDLs. It shows the behavior of the liquid water accumulating in the GDL fibers, when liquid water is supplied from a pore at the bottom of the simulation surface. The hydrophobic and hydrophilic properties of the fibers can be simulated with this method. As a 3D simulation for a larger space is under development at present, only findings obtained for the more simple 2D models are introduced in this section.

To ensure a favorable distribution of liquid water and a reliable air flow, a cell with a porous separator without channels was simulated. Figure 1.20 shows a schematic outline of the computation domain, the obstacles simulating the porous separator and fiber sections of the GDL. The whole domain is divided into $100 \times 50$ cells in the x and y directions. The vertical length of the simulated domain is $H = 1.0\,\mathrm{mm}$. The top is a wall. Uniform air flow is provided at the left inlet side, at $x = 0\,\mathrm{mm}$. There is unobstructed outflow at the right outlet side, $x = 2.0\,\mathrm{mm}$. A uniform flow of liquid water is also provided at a bottom pore of the MPL. The velocities of inlet air and liquid water are $0.05\,\mathrm{m\,s^{-1}}$ and $0.02\,\mathrm{m\,s^{-1}}$, respectively. To discuss the water and air flow distributions in the porous medium, the domain is divided into three regions as shown in Figure 1.20; (I) lower and (II) intermediate sections are in the GDL, and (III) the upper section corresponds to the porous separator. Two cases with different wettabilities of the porous medium

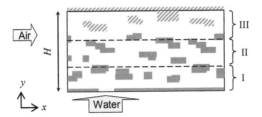

Figure 1.20. Schematic outline of the computation domain for the liquid water and air flow behavior in a porous medium. The (I) lower and (II) intermediate sections are in the GDL. Section (III), the upper section, corresponds to the porous separator. Sections (I) and (II) in the GDL are set to be hydrophobic. Section (III) of the porous separator is computed for two cases: hydrophobic and hydrophilic.

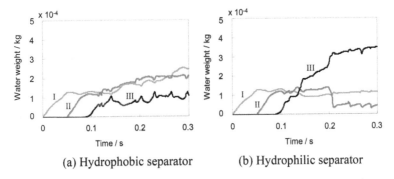

(a) Hydrophobic separator      (b) Hydrophilic separator

Figure 1.21. Changes in water weight in the three sections in the (a) hydrophobic porous separator and (b) hydrophilic separator.

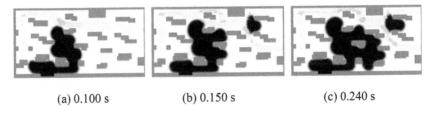

(a) 0.100 s      (b) 0.150 s      (c) 0.240 s

Figure 1.22. Behavior of liquid water and air flow in the hydrophobic porous separator.

were simulated; in the first case, the solid surfaces were all hydrophobic (contact angle 150°). In the second case, the upper shaded portions were changed to hydrophilic (contact angle 40°).

Figure 1.21 shows the changes in water weight in the three regions: (a) is the result with the hydrophobic porous separator, and (b) is with the hydrophilic porous separator. With the hydrophobic separator, the liquid water weight in region I increases first, before 0.100 s, followed by increases in region II as shown in Figure 1.21a. In the case of the hydrophilic separator in (b), the liquid water weight grows in region III later on in the process.

Figure 1.22 shows the appearances of the areas of liquid water (black) and air flows in the hydrophobic porous separator. The liquid water proceeding upward is broken up and drained by the strong air flow in region III, establishing a dynamic equilibrium. Thus, the increase in the water weight in region III is suppressed and the oscillations were caused by the breakup of the liquid water body.

Figure 1.23 shows the changes in water distribution in the hydrophobic GDL with the hydrophilic porous separator. The liquid water volume grows in region III, attracted to the

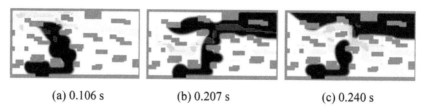

(a) 0.106 s      (b) 0.207 s      (c) 0.240 s

Figure 1.23.   Behavior of liquid water and air flow in the hydrophilic porous separator.

hydrophilic porous separator. This induces the increase in water weight in region III after 0.106 s and the average air flow in region III is decreased, forming a liquid water region along the top wall as shown in the figure. In region II, the liquid water weight decreases and the average air flow rate increases sharply. The liquid water is broken up in region II and a dynamic equilibrium is maintained after 0.240 s. This air path in region II can be anticipated to contribute to a better cell performance because the reaction gas is supplied to the reaction area under the bottom boundary with only little obstruction.

Overall, the simulation results show the possibility that controlling the wettability of the porous separator and GDL in through plane direction may be effective in realizing an optimum two-phase distribution of liquid water.

## 1.5   WATER TRANSPORT THROUGH MICRO-POROUS LAYERS (MPL)

As illustrated in Figure 1.2, one side of the cathode GDL facing to the catalyst layer is generally provided with a micro-porous layer (MPL) of polytetrafluoroethylene (PTFE) with hydrophobic characteristics. It has much smaller pores and a much smoother surface than the GDL. The MPL plays an important role in the water management of polymer electrolyte fuel cells; however, details of the mechanism that works to suppress water flooding are not fully understood. Questions still remain about whether the water transfer in the MPL occurs in the liquid or vapor phase.

A report on the function of the MPL by Hizir *et al.* (2010) has indicated that the interface between the CL and MPL may act as a site for water accumulation and prevent the reactant gases from reaching the reaction zone. Turhan *et al.* (2010) investigated the effect of the interface on the water transport and flooding numerically. The amount of liquid water in the space between the CL and MPL was investigated and the result showed that the MPL reduces the saturation level of water on the CL surface (Hartnig *et al.*, 2008; Nam *et al.*, 2009; Swamy *et al.*, 2010). These studies mainly focus on the structure of the interface; water transport phenomena are not discussed in detail. Owejan *et al.* (2010) compared cell performance with and without an MPL and analyzed the vapor flux transported through the MPL based on saturation vapor pressure-gradients between the two sides of the MPL. The results showed that large vapor fluxes might pass through the MPL. However, there is no research clearly establishing which phase of water is transferred through the MPL or what the effect the MPL has on the transfer.

The following section focuses on the function of the MPL and introduces work by the authors showing the mechanism affecting the water transport phenomena in the MPL.

### 1.5.1   *Effect of the MPL on the cell performance*

Figure 1.24 shows polarization curves for GDLs with and without MPL. For the GDL with an MPL, the cell performance is superior to that without the MPL at both relative humidity conditions investigated here. The difference is more significant under higher current density conditions and also under the higher relative humidity condition. This indicates that the difference can be attributed to the concentration overpotential change due to the MPL. The GDL with the MPL tends to suppress flooding, and the oxygen supply is maintained even under the higher water

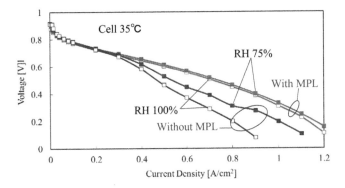

Figure 1.24. Polarization curves with and without the MPL.

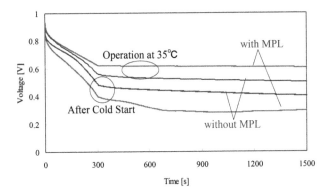

Figure 1.25. Voltage changes before and after a cold start, when the cell is operated at 35°C.

production conditions. Without an MPL, the produced water causes a deterioration of the oxygen diffusion, resulting in a significant voltage drop at the higher current densities under the higher humidity condition.

Although cells with an MPL generally show better cell performance than cells without MPL, deteriorated performance may be observed after cold starting. Figure 1.25 shows voltage differences for cells with and without MPL before and after cold starting. The cell performance was first measured at 35°C as a reference and then after complete purging the cell was cooled to −10°C or −20°C for the cold start operation. After the shut down in the cold start operation, the cell was heated slowly to 35°C over 10 minutes and the cell performance was measured again at the same conditions as for the reference test before the cold start. These results are shown as "after cold start" in the figure. Figure 1.25 also shows that the cell voltage is higher with the MPL than without an MPL before the cold start, which is consistent with the discussion above. After the cold start, the cell voltage becomes lower than before the cold start, and the voltage deterioration is larger with the MPL than without. The deteriorated voltage recovered to the original 35°C value when the cell temperature was raised above 45°C. The observed voltage drop was smaller after the cold start at −20°C.

These phenomena may provide a clue to elucidate the function of the MPL, as the difference in the performance appears to be caused by liquid water accumulating at the interface between the CL and MPL (or GDL) surfaces. This will be discussed in the next section.

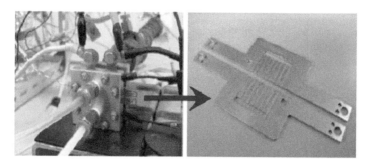

Figure 1.26.    Experimental cell for observations of cross sections of the GDL and MEA. It has bipolar plates consisting of two separate sections for each side of the cathode and anode, and the two sections are connected by a silicon sheet.

### 1.5.2  *Observation of the water distribution in the cell*

To investigate the mechanism of the performance differences with and without an MPL, the water distribution in the cell was observed using the freezing method. In this method, the operation of the cell is stopped and the cell is cooled to $-30°C$ in a thermostatic chamber in about 30 minutes to capture details of the position of the liquid water in ice form. The cell was cooled to $0°C$ in less than 4 minutes and it was confirmed that the water location did not change during the cooling process (Naing *et al.*, 2011). The cell was disassembled into its component parts in the thermostatic chamber, and then kept at $-30°C$ until the MEA was removed from the cell. The surface of the MEA was then observed under a microscope installed in the thermostatic chamber at the same temperature. Details of the apparatus and procedures are explained in Section 1.7. For the microscopic observations, the cell shown in Figure 1.26 was custom-made with bipolar plates each consisting of two separate sections for the cathode and anode sides; the two separate sections are connected by a silicon sheet. This cell allows observation of the water distribution in the form of ice inside the GDL. For the CRYO-SEM observations, the MEA and an adjoining GDL were removed from the cell and were then rapidly immersed in liquefied nitrogen in the thermostatic chamber. The combined set was cut to pieces of approximately $5 \times 5$ mm size and these were placed on sample holders in the liquefied nitrogen. Then, the sample was moved to the preparation chamber kept at $-160°C$ in the CRYO-SEM and the air in the chamber was evacuated. The sample was cut with a cold knife in the chamber and the surfaces of the cut section were coated by platinum vapor. Finally, samples were observed by a CRYO-SEM at $-160°C$ with an acceleration voltage of 5 kV.

Figure 1.27 shows photographs at the surface of the CL and image-processed pictures with and without the MPL. The cell had been operated at a current density of $0.7\,A\,cm^{-2}$ with relative humidity of 100% for 1 hour. Table 1.2 gives the average percentages of ice-covered areas in the image-processed pictures. The table shows that the ice-covered area is clearly larger without the MPL than with the MPL. The area increases with the current density and relative humidity. Without the MPL, the ice-covered areas under the channels and under the lands are similar. However, with the MPL the ice-covered area is significantly smaller under the lands than under the channels. This indicates that the MPL limits the water amount at the interface between the CL and MPL and that the water amount is more significantly limited under the lands than under the channels due to the higher contact pressure under the lands. It may be concluded that without the MPL, more liquid water accumulates at the interface between the CL and GDL; this difference in the water amounts at the interface would appear to be a cause for the differences in cell performance, i.e. the cell voltage becoming higher with an MPL than without an MPL.

Figure 1.28 shows CRYO-SEM photographs of sections of the MEA, MPL, and GDL layers. There are large vacant spaces between the layers due to the cutting process involved in the sectioning. The cell here was operated at $0.7\,A\,cm^{-2}$ and humidity 100% for 60 minutes at 35°C.

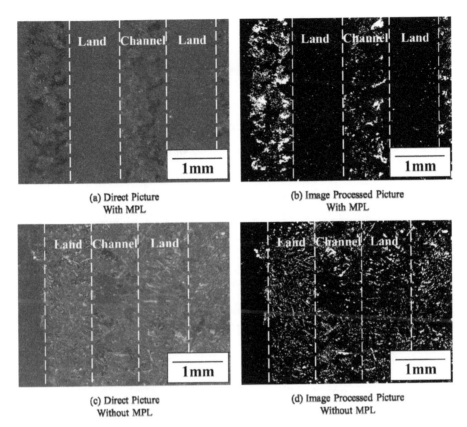

Figure 1.27.   Water distribution on the CL captured in ice form with the freezing method at 35°C operation; the operating condition is $0.7\,A\,cm^{-2}$ with relative humidity of 100% for cathode gas.

Table 1.2.   Area percentages of ice on the CL in the image-processed pictures.

|  |  | Channel [%] | Land [%] | Total [%] |
|---|---|---|---|---|
| $0.5\,A\,cm^{-2}$ |  |  |  |  |
| RH75% | with MPL | 4.9 | 0.8 | 2.9 |
|  | without MPL | 5.4 | 5.8 | 5.6 |
| RH100% | with MPL | 5.0 | 1.3 | 3.1 |
|  | without MPL | 4.4 | 8.3 | 6.4 |
| $0.7\,A\,cm^{-2}$ |  |  |  |  |
| RH75% | with MPL | 4.9 | 2.9 | 3.9 |
|  | without MPL | 8.2 | 8.1 | 8.2 |
| RH100% | with MPL | 7.8 | 1.6 | 4.7 |
|  | without MPL | 7.5 | 8.9 | 8.2 |

The water distribution was then captured with the freezing method detailed above. The pictures indicate that ice formed at the interface between the CL and MPL and no ice can be observed in the MPL pores. This suggests that there is no liquid water in the pores in the MPL although there is liquid water at the interface. The capillary pressure of water in the pores can be estimated to be in the order of 300 kPa and it would appear to be difficult for water to flow into the pores in liquid form.

Figure 1.28.    CRYO-SEM photograph at cross sections of the MEA and GDL layers for the operation at 35°C.

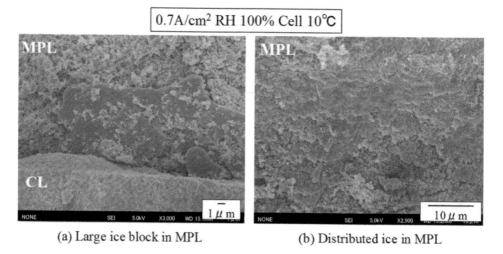

(a) Large ice block in MPL                                    (b) Distributed ice in MPL

Figure 1.29.    CRYO-SEM photographs of the MPL after operation at 10°C.

Figure 1.29 shows photographs for a cell after operation at 10°C. In this case, a large volume of ice can be seen in cracks in the MPL and in places, ice fills the small pores in the MPL. These pictures clearly suggest the possibility of the presence of liquid water in the small pores in the MPL depending on the temperature conditions of the operation. No ice was observed in the small pores in the photograph of the 35°C operation; this is clearly different from the 10°C operation, suggesting that water may be transported in vapor form in the 35°C operation.

### 1.5.3   *Analysis of water transport through MPL*

Owejan *et al.* (2010) analyzed the water vapor flux though the MPL as the water vapor flux caused by the saturation pressure gradient in a porous layer as a function of temperature.

The water vapor flux by the gradient of the saturation pressure at the two sides of an MPL is expressed as:

$$f_w \frac{i}{2F} = \frac{D^{\text{eff}}}{RT} \left( \frac{dP^{\text{sat}}}{dT} \right) \left( \frac{dT}{dx} \right) \tag{1.1}$$

where the left side of the equation indicates the fraction of produced water transported through the MPL. The parameters and their values are listed in Table 1.3.

The heat flux through the MPL is:

$$f_q \times i(U_H - V) = k \left( \frac{dT}{dx} \right) \tag{1.2}$$

where $f_q$ is the fraction of the generated heat at the CL, $i(U_H - V)$, transported through the MPL.

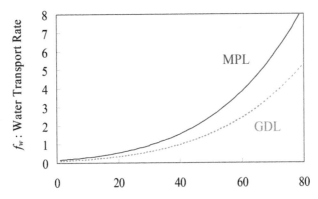

Figure 1.30.   Vapor flux rate for the transport through the MPL and GDL layers plotted against the water production.

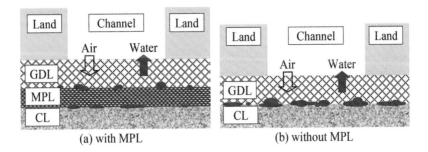

(a) with MPL                                    (b) without MPL

Figure 1.31.   Schematic representation of the water distribution for cells with and without an MPL.

By combining Equations (1.1) and (1.2) to eliminate the $dT/dx$ term, one obtains:

$$f_w = \frac{2FD^{\text{eff}}}{RT} \left(\frac{dP^{\text{sat}}}{dT}\right) \frac{f_q(U_H - V)}{k} \tag{1.3}$$

This shows the water vapor flux relative to the water production rate. When the value is greater than unity, all of the produced water can be transported through the MPL in vapor form.

Table 1.3 shows the properties of the MPL and GDL used in the experiments in Figure 1.24. With these values, the water transfer rate, $f_w$, can be calculated for the MPL and GDL, for the measured cell voltage used in the analysis. Figure 1.30 is the result of the analysis, where the current density is $0.7\,\text{A\,cm}^{-2}$ for the corresponding voltage value. The figure shows that a sufficient vapor flux can be obtained at MPL temperatures above 30°C. It also indicates that the MPL has a larger vapor transport flux than the GDL. This is because the thermal conductivity of the MPL is smaller than that of the GDL. Here the porosity of the MPL and GDL was set to 0.5 and 0.8 respectively for the calculation of $D^{\text{eff}}$. As the pore diameters of the GDL are quite large, the formula shown above may not be valid for a GDL, because water can transport easily in the liquid form through the pores. However, the analysis suggests the possibility of water being transported in the vapor phase through the MPL.

### 1.5.4   *Mechanism for improving cell performance with an MPL*

Figure 1.31 is the experimentally suggested situation of the liquid water distribution in cells with and without an MPL. The water produced in the CL may accumulate in liquid form at the interface

Table 1.3.  Parameters and their values as used in the analysis.

| Parameters | | Values |
|---|---|---|
| $F$ | Faraday's constant [C mol$^{-1}$] | 96500 |
| $R$ | Universal gas constant [J mol$^{-1}$ K$^{-1}$] | 8.314 |
| $f_q$ | Fraction of released heat though the layer | 0.5 |
| $U_H$ | Enthalpy potential for fuel cell reaction [V] | 1.48 |
| $k$ | Through-plane thermal conductivity [W m$^{-1}$ K$^{-1}$] | 0.097 (MPL) |
| | | 3.1 (GDL) |
| $f_w$ | Water transfer rate | |
| $i$ | Current density [A m$^{-2}$] | |
| $D^{\text{eff}}$ | Effective water vapor diffusion coefficient [m$^2$s$^{-1}$] | |
| | $D^{\text{eff}} = 0.22 \cdot (T/273)^{1.75} \varepsilon^{1.5} \cdot 10^{-4}$ [m$^2$ s$^{-1}$] | |
| $\varepsilon$ | Porosity | |
| $T$ | Temperature [K] | |
| $p^{\text{sat}}$ | Water vapor saturation pressure [Pa] | |
| $V$ | Cell voltage [V]: measured value | |

between the CL and MPL particularly under high current density conditions. Because of the low thermal conductivity of the MPL, a steep temperature gradient is created across the MPL and a large flux of water vapor flow occurs through the MPL. The vapor passing through the MPL may condense in the GDL. As the temperature at the CL is increased with the low thermal conductivity of the MPL, the water in the CL is more likely to be in vapor form with the MPL than in the case without an MPL. Considering this mechanism, the MPL appears to play a role in separating liquid water accumulation and keeping water apart from the CL surface. Additionally, the limited space in the interface, due to the closer contact of the CL with an MPL, limits the volume of accumulated liquid water at the interface. This would explain why a smaller amount of water is observed under the lands than under the channels. The finer, closer contact also contributes to the reduction in contact resistance.

Without an MPL, liquid water can form close to the CL surface, as the temperature gradient is flatter than with an MPL. Here, the liquid water is distributed in the larger space existing both under the channels and under the lands, and this causes flooding more easily than in the case with an MPL.

When a cell is operated at sub-zero temperatures, the produced water accumulates in the interface between MPL and CL and as the liquid water transport is limited by the MPL, the water accumulates in the smaller space at the interface and increases the volume of the space after freezing. When the cell is then heated to room temperature, a large amount of liquid water remains at the interface and this causes a deterioration in the oxygen diffusion at the cathode. Until the liquid water has disappeared and the enlarged space has contracted to the original state (that before the cold start), the deteriorated performance continues. These processes help explain the performance change after the cold start shown in Figure 1.25. In the case without an MPL, ice is formed in the wide space in the GDL at cold starting. The liquid water after melting does not cause as much deterioration in oxygen diffusion as in the case with the MPL.

## 1.6  TRANSPORT PHENOMENA AND REACTIONS IN THE CATALYST LAYERS

### 1.6.1  *Introduction*

Although water transport is not specifically considered in this section, structures appropriate for effective platinum (Pt) utilization are discussed. Because electrons, protons, and oxygen are all necessary for the cathode reaction, ensuring an optimum structure of the electrode catalyst layer and efficient transport of reactants would be effective for reducing the amount of Pt catalyst used in cells.

Determination of the optimum compositions and structures of the catalyst layer (CL) has been researched using time-consuming and expensive experiments, as well as computational modeling approaches which assist in the elucidation of details of the processes in the CL (Chisaka *et al.*, 2006; Jaouen *et al.*, 2002; Mukherjee and Wang, 2006; Secanell *et al.*, 2007; Song *et al.*, 2005). Catalyst layer models can be classified into three categories: pseudo-homogeneous film models (Secanell *et al.*, 2007; Song *et al.*, 2005), agglomerate models (Chisaka *et al.*, 2006; Jaouen *et al.*, 2002; Tabe *et al.*, 2011), and pore-scale models (Mukherjee and Wang, 2006). To elucidate further details of the optimum structure to achieve optimum Pt utilization, wide ranging research with the various models is necessary.

Using the models and the calculated results, a simplified equation, termed an evaluation equation in this section, was derived to show the relationships among major parameters in the cathode catalyst structure (Tabe *et al.*, 2011). From the analysis of the evaluation equation, optimal structural parameters were identified. The results indicate that the dominant parameters of the CL structure are the polymer electrolyte thickness covering carbon agglomerates and the CL thickness.

### 1.6.2  *Analysis model and formulation*

The simplified agglomerate model in Figure 1.3 was applied as the model for the cathode three-phase boundary, which is the reaction site of the Pt surface where electrons, protons, and reaction gases can be supplied and removed. It was assumed that a Pt-supported carbon agglomerate is covered with a uniform thick polymer electrolyte and that the cathode reaction occurs only due to dissolved oxygen in the polymer at the carbon agglomerate surface. Furthermore, it was assumed that the mass transport losses inside the carbon agglomerate are negligible. Figure 1.3b shows a schematic diagram of the cathode CL model, here identical carbon agglomerates with polymer electrolyte films (as in Fig. 1.3a) are dispersed uniformly in the CL. On the $z$-axis along the thicknesses direction, the origin (i.e. $z = 0$) and $z = \delta_{CL}$ are, respectively, the interfaces between the polymer electrolyte membrane and the CL and between the CL and the gas diffusion layer (GDL). In this one-dimensional model of the cathode catalyst layer, the surface areas of the agglomerates and Pt catalysts were calculated using the number of the agglomerates per unit volume in the catalyst layer, $n_{ag}$. The phenomena in the cathode CL were analyzed under the assumptions that: (a) the CL and the GDL are isothermal, isobaric, and in a steady state condition, (b) the electric potential of the carbon agglomerate is constant, and (c) the generated water is in the gas phase.

The proton current in the CL is generated by the oxygen consumption, and the change in the proton current density, $i_{H+}$, is satisfied with the balance expressed by Equation (1.4). The change in the cathode overpotential, $\eta$, corresponds to the voltage decrease due to the proton transport; it is expressed by Equation (1.5) using the proton current density, $i_{H+}$, and the effective proton conductivity, $\kappa^{eff,CL}$:

$$\frac{di_{H+}(z)}{dz} = 4F \cdot n_{ag} S_{ag} J_{O_2}^{ag} \tag{1.4}$$

$$\frac{d\eta(z)}{dz} = \frac{i_{H+}}{\kappa^{eff,CL}} \tag{1.5}$$

Here, $F$ is Faraday's constant, and $S_{ag}$ is the surface area of one agglomerate. The oxygen consumption, $J_{O_2}^{ag}$, corresponds to the oxygen spherical diffusion flux at the carbon agglomerate surface, and $J_{O_2}^{ag}$ was calculated using the relationship between the oxygen diffusion flux, according to Fick's law in the polymer, and the oxygen reduction constant of the reduction reaction, $k_{ORR}$, as expressed by Equation (1.6). The reaction constant, $k_{ORR}$, was given by a simplified

Table 1.4.   Simulation parameters and properties for the standard condition here.

| Parameter | Value |
|---|---|
| Temperature, $T$ [K] | 353.15 |
| Tafel slope, $b$ [V] | 0.11589 |
| Reference exchange current density, $i_0$ [A cm$^{-2}$] | $1.6 \times 10^{-7}$ |
| Reference oxygen concentration, $C_{0,O_2}$ [mol m$^{-3}$] | 0.41 |
| Oxygen diffusion coefficient in polymer, $D_{O_2}^p$ [m$^2$ s$^{-1}$] | $8.7 \times 10^{-10}$ |
| Oxygen concentration ratio at polymer-gas interface, $Y$ | 0.093 |
| Platinum amount, $m_{Pt}$ [mg cm$^{-2}$] | 0.4 |
| Agglomerate radius, $r_{ag}$ [nm] | 500 |
| Platinum particle radius, $r_{Pt}$ [nm] | 1.5 |
| Catalyst layer thickness, $\delta_{CL}$ [μm] | 10 |
| GDL thickness, $\delta_{GDL}$ [mm] | 0.3 |
| Polymer thickness, $\delta_p$ [nm] | 70 |
| Catalyst layer porosity, $\varepsilon_{CL}$ | 0.3 |
| GDL porosity, $\varepsilon_{GDL}$ | 0.7 |

Butler-Volmer expression, Equation (1.7):

$$J_{O_2}^{ag} = -C_{O_2}^g \gamma \left[ \frac{r_{ag}\delta_p}{(r_{ag} + \delta_p)D_{O_2}^p} + \frac{1}{k_{ORR}} \right]^{-1} \tag{1.6}$$

$$k_{ORR} = \frac{Ai_0}{4FC_{0,O_2}} \exp\left( -\frac{2.303\eta}{b} \right) \tag{1.7}$$

Here, $C_{O_2}^g$ is the oxygen concentration in the gas phase at the polymer surface, $\gamma$ is the ratio of the oxygen concentration in the polymer to that in the gas phase, $r_{ag}$ is the radius of the carbon agglomerate, $\delta_p$ is the thickness of the polymer film, $D_{O_2}^p$ is the oxygen diffusion coefficient in the polymer, $A$ is the area ratio of the reaction site of the Pt surface to the surface of the carbon agglomerate, $i_0$ is the exchange current density at the standard oxygen concentration, $C_{0,O_2}$ is the standard oxygen concentration in the polymer, and $b$ is the Tafel slope. The distributions of the proton current density, $i_{H+}(z)$, and cathode overpotential, $\eta(z)$, were calculated by setting $\eta(\delta_{CL})$ as a boundary condition at the interface between the catalyst layer and the GDL, together with $i_{H+}(\delta_{CL}) = 0$.

In the catalyst layer, the effective proton conductivity of the polymer electrolyte and the effective diffusion coefficient in the pores of the catalyst layer depend on the catalyst layer structure. These properties, $\kappa^{eff,CL}$ and $D_i^{eff,CL}$, were given using the Bruggeman correction factor as follows:

$$\kappa^{eff,CL} = \kappa \left[ (1 - \varepsilon_{CL}) \left\{ 1 - \left( \frac{r_{ag}}{r_{ag} + \delta_p} \right)^3 \right\} \right]^{1.5} \tag{1.8}$$

$$D_i^{eff,CL} = D_i^g \varepsilon_{CL}^{1.5} \tag{1.9}$$

Here, $\kappa$ is the proton conductivity of the polymer electrolyte, $D_i^g$ is the diffusion coefficient of the species $i$ in the CL pores, and $\varepsilon_{CL}$ is the porosity of the CL.

The cathode gas in the inlet of the channel was set to that of air at 1 atm, 80°C, and 80% relative humidity, and the stoichiometric ratio was 2.5. The following values were used as the standard parameters: the CL thickness, $\delta_{CL} = 10 \, \mu m$; the diameter of Pt catalyst, 3 nm; Pt loading, 0.4 mg cm$^{-2}$; the porosity of the CL, $\varepsilon_{CL} = 0.3$; and the electrolyte film thickness, $\delta_p = 70$ nm. The parameters used in this study for the standard conditions are listed in Table 1.4.

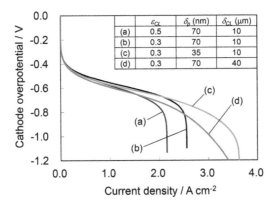

| | $\varepsilon_{CL}$ | $\delta_p$ (nm) | $\delta_{CL}$ (µm) |
|---|---|---|---|
| (a) | 0.5 | 70 | 10 |
| (b) | 0.3 | 70 | 10 |
| (c) | 0.3 | 35 | 10 |
| (d) | 0.3 | 70 | 40 |

Figure 1.32.   Polarization curves for four structures of the catalyst layer (CL).

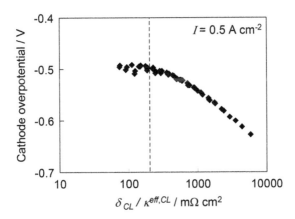

Figure 1.33.   Effect of proton conductivity on cathode overpotential: the abscissa is the parameter repre-senting the electric resistance within the catalyst layer, $\delta_{CL}/\kappa^{eff,CL}$, and the ordinate is the cathode overpotential, $\eta(0)$, at $I = 0.5\,\mathrm{A\,cm^{-2}}$. The porosity of the catalyst layer was varied from 0.3 to 0.5, the polymer thickness from 18 nm to 140 nm, and the catalyst layer thickness from 5 µm to 40 µm. The broken line represents the 200 mΩ cm² threshold in Equation (1.10).

### 1.6.3   *Results of analysis and major parameters in CL affecting performance*

The effects of some of the parameters of the CL structure were analyzed and as an example, the polarization curves with four different CL structures are shown in Figure 1.32. The curve (b) shows the polarization curve under standard conditions. Reducing the thickness of the polymer electrolyte also decreases the oxygen diffusion resistance in the polymer electrolyte; the cell performance improves as shown by curve (c). Inversely, in the (a) condition, the limiting current density becomes lower, when the porosity of the CL increases. This is because the higher CL porosity increases the oxygen flux per agglomerate but that also increases the proton diffusion resistance in the polymer electrolyte. In the (d) condition, the limiting current density increases, but the cell performance decreases even in the low current density region. This is due to the uneven distribution of the cathode overpotential in the CL, which is caused by the thicker CL.

To examine the performance deterioration due to the inadequate proton conductivity in the low current density region, the results when changing the porosity of the catalyst layer from 0.3 to 0.5, the polymer thickness from 18 nm to 140 nm and the catalyst layer thickness from 5 µm to 40 µm are shown in Figure 1.33. The abscissa is the parameter representing the electric resistance within the catalyst layer, $\delta_{CL}/\kappa^{eff,CL}$, and the ordinate is the cathode overpotential, $\eta(0)$, under a current density of $0.5\,A\,cm^{-2}$. The results where the limiting current density is below $1.0\,A\,cm^{-2}$ are disregarded because of the too large effect of the concentration overpotential at these overpotential values. Figure 1.33 shows that the data generally falls on a smooth curve and that the parameter of the abscissa provides a measure of the deterioration due to the low proton conductivity in the low current density region. The cathode overpotential at $0.5\,A\,cm^{-2}$ is maintained approximately constant where $\delta_{CL}/\kappa^{eff,CL}$ is below $200\,m\Omega\,cm^2$ and it decreases above $200\,m\Omega\,cm^2$ (In Fig. 1.33, the broken line is at this threshold, $200\,m\Omega\,cm^2$.) The results in Figure 1.33 suggest that the catalyst layer structure must satisfy the following relationship to prevent performance deterioration at low current densities:

$$\delta_{CL}/\kappa^{eff,CL} \leq 200\,m\Omega \cdot cm^2 \tag{1.10}$$

A further equation to evaluate the limiting current density determined by the mass transfer alone was developed using the following assumptions based on the results in the previous section: (a) the effect of advection in the catalyst layer and in the GDL is negligible, (b) the oxygen concentration at the surface of the carbon agglomerates (reaction sites) at the interface between the catalyst layer and the PEM is zero at the limiting current density, and (c) the oxygen concentration distribution in the catalyst layer pores is approximated by a quadratic function of $z$ because the distribution of the reaction is uniform in the catalyst layer. Using these assumptions, the limiting diffusion current density, $I_{LD}$, can be expressed by the following equation:

$$I_{LD} = 4FC_{O_2}^{Ch} \left[ \frac{1}{n_{ag}\delta_{CL}} \cdot \frac{1}{4\pi r_{ag}(r_{ag}+\delta_p)\gamma} \cdot \frac{\delta_p}{D_{O_2}^p} + \frac{\delta_{CL}}{3D_{O_2}^{eff,CL}} + \frac{\delta_{GDL}}{D_{O_2}^{eff,GDL}} \right]^{-1} \tag{1.11}$$

Here, $C_{O_2}^{Ch}$ is the oxygen concentration in the gas channel, and it is given as an average concentration of the inlet and outlet channel. The $\delta_{GDL}$ term is the GDL thickness and $D_{O_2}^{eff,GDL}$ is the effective oxygen diffusion coefficient in the GDL. In the bracket on the right hand side of Equation (1.11), the first, second, and third terms represent the oxygen diffusion resistance in the polymer electrolyte, in the catalyst layer pores, and in the GDL respectively. Further, Equation (1.11) can be transformed into Equation (1.12), where $n_{ag}$ and $D_{O_2}^{eff,CL}$ are expressed by the relationship with the catalyst layer structure:

$$I_{LD} = 4FC_{O_2}^{Ch} \left[ \frac{(r_{ag}+\delta_p)^2}{3(1-\varepsilon_{CL})\delta_{CL}r_{ag}\gamma} \cdot \frac{\delta_p}{D_{O_2}^p} + \frac{1}{3\varepsilon_{CL}^{1.5}} \cdot \frac{\delta_{CL}}{D_{O_2}^g} + \frac{\delta_{GDL}}{D_{O_2}^{eff,GDL}} \right]^{-1} \tag{1.12}$$

Equation (1.12) explains, quantitatively, that the limiting current density increases with decreasing polymer thickness, $\delta_p$, and increasing catalyst layer thickness, $\delta_{CL}$, as suggested in Figure 1.32, because the oxygen diffusion resistance in the polymer electrolyte (the first term in the right side bracket of Equation (1.12)) decreases. With decreasing catalyst layer porosity, the oxygen diffusion resistance in the catalyst layer pores (the second term in the bracket) increases, but the diffusion resistance in the polymer decreases. The effect of the decrease in the resistance in the polymer is dominant under the conditions used in this study, as shown in Figure 1.32; this effect can be evaluated by Equation (1.12).

Figure 1.34 is a plot of the limiting current densities calculated by Equation (1.12) (abscissa) and the agglomerate model by Equation (1.11) (ordinate) for the various structures as in the analysis investigating the effect of the proton conductivity in Figure 1.33. The straight line in Figure 1.34 shows the case where the value calculated by Equation (1.12) is the same as that calculated by the agglomerate model simulation. The filled diamonds are for the conditions satisfying Equation (1.10), and the open diamonds for those not satisfying Equation (1.10). The filled diamonds distribute very nearly on the straight line; this shows that Equation (1.12) can estimate the limiting current density well under the conditions satisfying Equation (1.10).

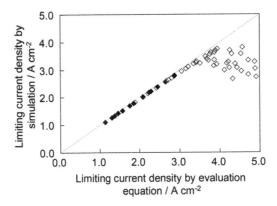

Figure 1.34.   Plot of the limiting current density by simulation vs. that by Equation (1.12), which is termed the evaluation equation in the text. The porosity of the catalyst layer varies from 0.3 to 0.5, the polymer thickness from 18 nm to 140 nm, and the catalyst layer thickness from 5 μm to 40 μm. The straight line shows the case where the value calculated by Equation (1.12) is the same as that by the agglomerate model simulation by Equation (1.11). The filled diamonds are for the conditions satisfying Equation (1.10), and the open diamonds for those not satisfying Equation (1.10).

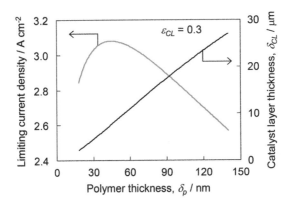

Figure 1.35.   Catalyst layer thickness for the equality condition in Equation (1.10) (right ordinate) and the limiting current density calculated by Equation (1.12) (left ordinate) plotted against polymer thickness. The porosity of the catalyst layer is set to 0.3.

Under the conditions used in this study, a lower porosity in the catalyst layer is advantageous, because the porosity has a stronger influence on the oxygen diffusion resistance in the polymer than on that in the catalyst layer pores, as expressed in Equation (1.12). The catalyst layer porosity was set to 0.3 as the minimum value for the optimum structure in the following analysis. Figure 1.35 shows the limiting current density and the catalyst layer thickness *vs.* the thickness of the polymer, with the catalyst layer thickness selected to be the maximum value satisfying Equation (1.10) (to ensure adequate proton conductivity and to achieve a maximum limiting current density). A thinner polymer electrolyte is effective to increase the limiting current density, but at the same time, a thinner catalyst layer thickness is needed to ensure sufficient proton conductivity, which then decreases the limiting current density. The model suggests an optimum parameter setting, where the advantages of the thin polymer thickness and the disadvantages of the thin catalyst layer thickness become dominant, at polymer thicknesses above and below about 40 nm, respectively.

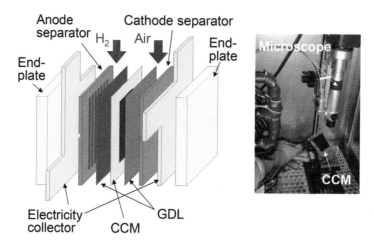

Figure 1.36.   Experimental apparatus and photo of the inside of the thermostatic chamber.

For the conditions used in the study in Table 1.4, there is a maximum limiting current density, around $3.0\,\mathrm{A\,cm^{-2}}$, for a catalyst layer structure of $\varepsilon_{CL} = 0.3$, $\delta_p = 45\,\mathrm{nm}$, $\delta_{CL} = 7.8\,\mathrm{\mu m}$.

The above shows that it is effective to reduce the thickness of the polymer electrolyte and to decrease the CL thickness to maintain a low oxygen diffusion resistance in the polymer electrolyte and a high proton conductivity in the CL. Thus, it is possible to investigate dominant parameters and evaluate suitable structures of the CL for various conditions using the evaluation Equations (1.10) and (1.12). Additionally, previous studies by the authors have shown that the ratio of oxygen concentration in the polymer electrolyte to that in the gas phase at the polymer-gas interface, $\gamma$, is a highly critical parameter (Tabe *et al.*, 2011).

## 1.7   WATER TRANSPORT IN COLD STARTS

Characteristically, PEM fuel cells are less affected by deterioration at sub-zero temperature operations than ordinary batteries, however, the low-temperature startability issues with PEM fuel cells remain an obstacle for the use of PEM fuel cells in cold areas. Freezing of the water produced by the cathode reaction may induce shutdowns during cold starts in sub-zero temperatures. This section introduces results of an investigation of the cold start characteristics of PEM fuel cells.

### 1.7.1   *Cold start characteristics and the effect of the start-up temperature*

The experimental apparatus and a photo of the working arrangements inside the thermostatic chamber are shown in Figure 1.36. A single cell with an active area of 25 cm$^2$ was used for the experiments. The procedure of the cold start experiments consists of four steps: preconditioning, gas purge, cooling down, and then the cold start. In the gas purging process, the initial water conditions in the cell were carefully controlled, because residual water in the cell strongly affects the cold start characteristics (Tabe *et al.*, 2012; Tajiri *et al.*, 2007). A dry N$_2$ gas purge was conducted to eliminate water in the cell and it was continued until the measured cell resistance increased to a specific value. A wet N$_2$ gas purge was then conducted until an equilibrium of the MEA (membrane electrode assembly) was achieved, where the resistance was steady. These procedures lasted about 5 hours. After this, the cell and chamber temperature were cooled to the target temperature of $-10°C$ or $-20°C$, and the cell operation was started at a constant current density until the cell shut down due to freezing.

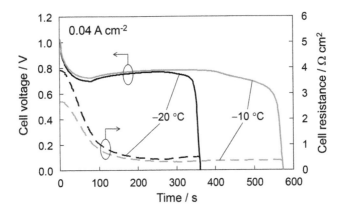

Figure 1.37.   Cell voltages and resistances for the 0.04 A cm$^{-2}$ cold start operation at $-20°$C and $-10°$C.

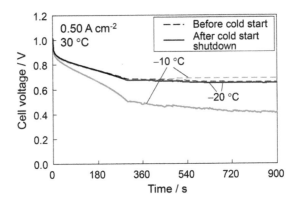

Figure 1.38.   Plot of the cell performance at 30°C before and after the occurrence of a shutdown due to cold start at $-20°$C or $-10°$C.

The cold start characteristics from freezing temperatures, for 0.04 A cm$^{-2}$ at $-10°$C and $-20°$C, were investigated. Figure 1.37 shows the behavior of the cell voltage and resistance in the cold start operation. The cell resistance decreases initially because the generated water humidifies the membrane due to back diffusion. At $-20°$C, the back diffusion terminates at around 270 seconds; here the cell voltage starts decreasing while the cell resistance increases slightly. At 350 seconds the operation shuts down. At $-10°$C, the cell voltage gradually decreases after around 400 seconds and the operation shuts down around 580 seconds. These voltage drops are caused by the freezing of the produced water in the cell. The results show that the freezing characteristics in the cell are strongly influenced by the start-up temperature.

After the shutdown from the cold start, as detailed in the previous section, there is a deterioration of the cell performance, which may affect subsequent operation at relatively low temperatures above zero, for example, at 30°C. This temporary deterioration was recovered by operating at a high temperature, i.e. 70°C. Figure 1.38 shows the results before and after the cold start shutdown. After shutting down from the cold start, the cell was heated to 30°C and operation resumed with the cell temperature kept fairly constant. Here the amounts of residual water on the cathode side after the $-20°$C and $-10°$C shutdown were very similar. The applied current density was increased to 0.50 A cm$^{-2}$ during the first 300 s of the operation. The broken lines in Figure 1.38 are the cell voltages before the cold starting and the solid lines are those after. The plots clearly show that the cell voltage after shutting down from the $-10°$C cold start does not recover to the value before the cold start, while there is little difference between the voltages before and after the

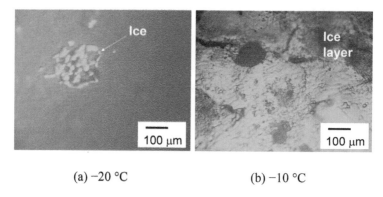

(a) −20 °C                                (b) −10 °C

Figure 1.39.   Micrographs of the cathode catalyst layer surface after the shutdown at −20°C and −10°C.

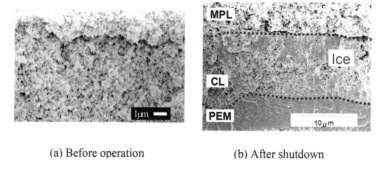

(a) Before operation                 (b) After shutdown

Figure 1.40.   Pictures of cross-sections of the cathode catalyst layers, (a) before and (b) after cold start operation with air at $0.04\,A\,cm^{-2}$ and −20°C.

cold start shutdown at −20°C. This deterioration after the −10°C shutdown could be eliminated by a dry nitrogen purge for more than 5 minutes or by heating the cell to 70°C. These results are due to differences in the ice formation locations with the −20°C and −10°C cold starts, as will be shown in the next section.

### 1.7.2   *Observation of ice distribution and evaluation of the freezing mechanism*

After the shutdown in the cold start, the cell was disassembled and the MEA surface of the cathode side and cross-sections of the MEA were observed as detailed in the previous Section, 1.5.2. Figure 1.39 shows pictures of the cathode catalyst surface at the two temperatures. At −10°C, there are numerous ice layers across the MEA surface, while few ice crystals were observed at −20°C. These results indicate that at −10°C the produced water transfers through the catalyst layer (CL) and freezes at the CL/MPL interface. At −20°C, most of the produced water appears to freeze in the CL. From the transport balance of water it was verified that the accumulated water amount was the same for a current density of $0.08\,A\,cm^{-2}$ at −10°C and one of $0.02\,A\,cm^{-2}$ at −20°C. Therefore, it may be concluded that the differences in the ice formation characteristics are caused, not by the water amount produced, but by the differences in the start-up temperatures. To determine more details of the ice formation in the catalyst layer, cross-sections of the cathode catalyst layer were observed by CRYO-SEM.

For the CRYO-SEM tests, the cell before operation and after the cell shutdown at −20°C was used. Cross-sections of the cathode catalyst layer were observed. Figure 1.40 shows micrographs of the dry pre-operation MEA catalyst layer and the MEA catalyst layer after the cold start operation. Compared with the dry catalyst layer, ice formation is clearly observed at the upper

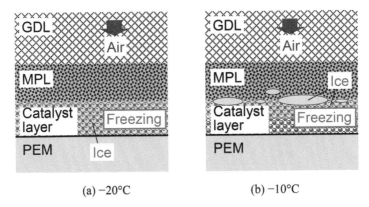

(a) −20°C                     (b) −10°C

Figure 1.41.   Experimentally suggested representations of the cross-sectional distribution of ice at the shutdown after cold start at −20°C and −10°C.

half of the cathode catalyst layer (Fig. 1.40b). The ice may be considered to obstruct the supply of oxygen to the catalyst surface and cause the shutdown. More ice was observed in the catalyst layer after the cold start at lower current densities, when the accumulated water volume was larger due to the longer operation before the shutdown. This is because the operation at the lower current density tolerates a larger amount of ice formation due to the smaller free area necessary for a reaction to proceed.

Figure 1.41 is a schematic representation of the cross-sectional ice distribution and the freezing mechanism suggested by the experimental results. The produced water during the cold start at −20°C freezes only inside the cathode catalyst layer, Figure 1.41a, and this causes the shutdown of the cell operation. The produced water at the −10°C cold start reaches the interface between the catalyst layer and the MPL, where it remains as a super-cooled liquid water film for some time before it freezes (Fig. 1.41b). This is the reason why the released heat of solidification due to the freezing can be observed by infrared thermography (Ishikawa *et al.*, 2007). The freezing at the interface triggers the operation shutdown and in total the freezing mechanism and cold start characteristics differ, depending on whether super-cooled water is present in a cell. The water produced by the cathode reaction passes through the cathode catalyst layer and freezes from the super-cooled state at the interface between the catalyst layer and the MPL at temperatures nearer zero, for example, at −10°C, while at lower temperatures, for example, at −20°C, the produced water freezes immediately near the reaction sites inside the cathode catalyst layer. This behavior can be explained as the space, which allows the super-cooled water to accumulate at the interface between the catalyst layer and the MPL, becomes larger at the −10°C cold start. Sometimes it was observed during the experiments that the freezing of a larger amount of liquid water significantly increases the contact resistance between the catalyst layer and the MPL. This also explains the differences in the cell performance shown in Figure 1.38 after recovery from the shutdown. The water melted from the ice at the interface between the MEA and MPL in the −10°C cold start limits the gas diffusion in the operation at 30°C, and the ice formed in the catalyst layer at the cold start at −20°C does not affect the gas diffusion after the temperature recovery at 30°C.

### 1.7.3   *Strategies to improve cold start performance*

Experiments with operation at three relative humidities of the nitrogen in the wet purge process, i.e. 22, 35 and 76%, were conducted to investigate the relationships between the cell resistances just before the cold start operation and the operation periods (expressed by the amount of water produced) under these three wet conditions of the polymer membrane. The experimental results of the 0.04 A cm$^{-2}$ cold start at −20°C and −10°C are plotted in Figure 1.42. The durations of the operation and current density are used here to estimate the amount of water produced by

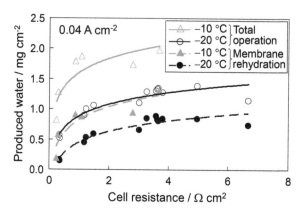

Figure 1.42.   Amount of water produced in the total operation and in the membrane rehydration at $-20°C$ and $-10°C$; the abscissa is the cell resistance before the cold start.

the cathode reaction. The amounts of water produced during the whole period of operation and during the membrane rehydration (this could also be termed "water uptake by the membrane" or "water back diffusion into the membrane") period were plotted against the cell resistance before the cold start. The period of the membrane rehydration was defined as the period when the cell resistance was decreasing. The abscissa is the cell resistance just before the cold start; the leftmost two data points for each condition (two of each of open and solid triangles and circles) are by the RH 76% purge, the following two by the RH 35%, and the remaining (rightmost data points) by the RH 22% purges. The estimated amounts of the produced water during the membrane rehydration (shown as solid plots [dots and triangles] and the broken lines in Fig. 1.42) become larger as the cell resistances before the cold start increase, both at $-20°C$ and $-10°C$, and their changes appear to be correlated. This would be because the dryer membrane with the higher initial cell resistance has a larger capacity to absorb water produced by the operation. Comparing the water amounts produced during the membrane rehydration at $-20°C$ and $-10°C$, the amounts at $-10°C$ are larger with the same cell resistance before the cold start. One factor contributing to this difference would be the temperature dependence of the membrane resistance as mentioned above; the initial cell resistance at $-10°C$ is about $1\ \Omega\,cm^2$ lower than that at $-20°C$ under the same wetness condition. Even allowing for this temperature dependence, the amounts of water at $-10°C$ are still larger, suggesting that the freezing at $-20°C$ occurs before the membrane is fully saturated. The estimated amount of produced water during the complete operation at $-20°C$ (open circles in Fig. 1.42) also increases to about $1.4\ mg \cdot cm^{-2}$ with an increase in the initial cell resistance, while that at $-10°C$ (open triangle in Fig. 1.42) reaches $2.0\ mg\,cm^{-2}$ but with some variation.

To examine the characteristics from the completion of the membrane rehydration to the shut-down, Figure 1.43 shows the estimated amounts of produced water following the membrane rehydration in the experiments. The amounts after the membrane rehydration at $-20°C$ are very similar for the various cell resistances before the cold start operation, but there is great variability among the amounts with the $-10°C$ operation. These results indicate that after the end of the membrane rehydration, the produced water that is not absorbed on the membrane immediately freezes at $-20°C$ while this water may remain as super-cooled water at $-10°C$, and this is then reflected in the larger amounts of produced water and the larger variations at $-10°C$ in Figure 1.43.

The above results indicate that the produced water initially diffuses into the membrane. After the membrane is saturated, water starts to accumulate in the cell and freeze. At temperatures below $-20°C$ the accumulating water starts freezing immediately in the catalyst layer. It is therefore not possible to control the amount of water freezing in the locations where the ice does not affect the oxygen diffusion. This indicates that one solution to improve the cold start is to dry the

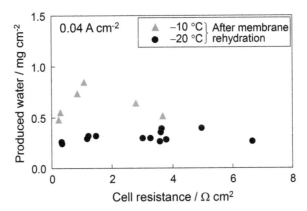

Figure 1.43.  Amount of water produced during the period from the end of the membrane rehydration until the time of the stop of operation at $-20°C$ and $-10°C$.

membrane before operation to ensure sufficient capacity for accommodating the back-diffusing water. This may be accomplished by conducting an adequate dry purge at the end of operation and preparing for the next cold start. In the cold start, it is important to raise the cell temperature above sub-zero temperatures during the period when water is back-diffusing into the membrane. For cold starts at temperatures above $-10°C$, there may be ways to ensure that the water may accumulate apart from the catalyst layers before it freezes.

1.8  SUMMARY

This chapter details the general idea of water transport phenomena in PEM fuel cells ranging from the nano-scale in the catalyst layer to the millimeter scale in the separator channels. Different from a review paper, the chapter is written largely as a summary of work performed by the authors. Some of the results still need to be proven to be generally applicable, but the readers will become aware of the phenomena and be able to consider the mechanisms involved from the point of view of the authors.

The major findings presented in the chapter may be summarized as follows:

1. Although serpentine channels are most commonly used, straight channels may give better performance with lower stoichiometric ratio, if an optimum water distribution structure is established as discussed in Section 1.3. Porous separators also have the potential to enable the operation of cells without flooding at high current densities with smaller stoichiometric ratios. Here, attention must be paid to the cooling due to the low thermal conductivity of the porous separators.
2. The design details of GDLs are very important to ensure a smooth and efficient removal of liquid water from the catalyst layer to the channels and from the region under the lands, where liquid water tends to accumulate. A gradient distribution of the wettability of the GDL fibers in the through plane direction may be effective for the water removal. Anisotropic orientations of the GDL fibers also offer good possibilities for improvement of water removal characteristics.
3. The MPL functions to limit the water amounts at the interface between the catalyst layer and the MPL, resulting in better flooding characteristics. There is the possibility of vapor phase water transport through the small pores of the MPL at temperatures above 30°C; this is due to the steep gradient in the saturation pressure caused by the temperature differences between the two sides of the MPL.
4. One of the major parameters in the cathode catalyst layer affecting the cell performance is the oxygen solubility in the enclosing ionomer. It is important to employ an ionomer structure,

which permits large oxygen concentrations close to the platinum particles as well as allowing a high proton conductivity.

5. In cold starts, the produced water freezes inside the catalyst layer at ambient temperatures below −20°C, whereas at −10°C it freezes at the interface between the catalyst layer and the MPL. To ensure a sufficient period of time for starting below −20°C, there are not many measures that can be taken other than ensuring that the membrane is in a dry state before a cold start.

## ACKNOWLEDGEMENTS

The authors thank to Prof. Yukio Hishinuma, who first introduced fuel cell research in their laboratory, and many graduate students in Hokkaido University, who were involved in the research presented here for their significant contributions to the research. The authors also express their thanks to the Grants-in-Aid for Scientific Research Program of the Japan Society for the Promotion of Science for the financial support of the research.

## REFERENCES

Chikahisa, T.: Microscopic observations of freezing phenomena in PEM fuel-cells at cold starts. *Heat Transfer Eng.* 34:2–3 ( 2013), pp. 258–265.

Chikahisa, T., Tabe, Y. & Kadowaki, K.: Study on water transport phenomena through micro-porous layers in PEFC. *Proceedings of the ASME 10th International Conference on Nanochannels, Microchannels, and Minichannels*, ICNMM2012-73063, 2012, pp. 1–7.

Chisaka, M., and Daiguji, H.: Design of ordered-catalyst layers for polymer electrolyte membrane fuel cell cathodes. *Electrochem. Comm.*, 8 (2006), pp. 1304–1308.

Hartnig, C., Manke, I., Kuhn, R., Kardjilov, N., Banhart, J. & Lehnert, W.: Cross-sectional insight in the water evolution and transport in polymer electrolyte fuel-cells. *Appl. Phys. Lett.* 92 (2008), 134106.

Hizir, F.E., Ural, S.O., Kumbur, E.C. & Mench, M.M.: Characterization of interfacial morphology in polymer electrolyte fuel-cells: Micro-porous layer and catalyst layer surfaces. *J. Power Sources* 195 (2010), pp. 3463–3471.

Ishikawa, Y., Morita, T., Nakata, K., Yoshida, K. & Shiozawa, M.: Behavior of water below the freezing point in PEFCs. *J. Power Sources* 163 (2007), pp. 708–712.

Jaouen, F., Lindbergh, G. & Sundholm, G.: Investigation of mass-transport limitations in the solid polymer fuel-cell cathode. *J. Electrochem. Soc.* 149:4 (2002), pp. A437–A447.

Kramer, D., Zhang, J., Shimoi, R., Lehmann, E., Wokaun, A., Shinohara, K. & Scherer, G.G.: In situ diagnostic of two-phase flow phenomena in polymer electrolyte fuel-cells by neutron imaging – Part A. Experimental, data treatment, and quantification. *Electrochim. Acta* 50 (2005), pp. 2603–2614.

Liu, Z., Mao, Z., Wu, B., Wang, L. & Schmidt, V.M.: Current density distribution in PEFC. *J. Power Sources* 141 (2005), pp. 205–210.

Manke, I., Hartnig, C. Grünerbel, M., Kaczerowski, J., Lehnert, W., Kardjilov, N., Hilger, A., Banhart, J., Treimer, W. & Strobl, M.: Quasi–in situ neutron tomography on polymer electrolyte membrane fuel-cell stacks. *Appl. Phys. Lett.* 90 (2007), 184101.

Mukherjee, P.P. & Wang, C.-Y.: Stochastic microstructure reconstruction and direct numerical simulation of the PEFC catalyst layer. *J. Electrochem. Soc.* 153:5 (2006), pp. A840–A849.

Naing, K.S.S., Tabe, Y. & Chikahisa, T.: Performance and liquidwater distribution in PEFCs with different anisotropic fiber directions of the GDL. *J. Power Sources* 196 (2011), pp. 2584–2594.

Nam, J.H., Lee, K.-J, Hwang, G.-S, Kim, C.-J. & Kaviany. M.: Microporous layer for water morphology control in PEMFC. *Int. J. Heat Mass Transf.* 52 (2009), pp. 2779–2791.

Noponen, M., Ihonen, J., Lundblad, A. & Lindbergh, G.: Current distribution measurement in a PEFC with net flow geometry. *J. Appl. Electrochem.* 34 (2004), pp. 255–262.

Owejan, J.P., Owejan, J.E., Gu, W., Trabold, T.A., Tighe, T.W. & Mathias, M.F.: Water transport mechanisms in PEMFC gas diffusion layers. *J. Electrochem. Soc.* 157:10 (2010), pp. B1456–B1464.

Sasabe, T., Tsushima, S. & Hirai, S.: In-situ visualization of liquid water in an operating PEMFC by soft X-ray radiography. *Int. J. Hydrogen Energy* 35 (2010), pp. 11,119–11,128.

Satija, R., Jacobson, D.L., Arif, M. & Werner, S.A.: In situ neutron imaging technique for evaluation of water management systems in operating PEM fuel-cells. *J. Power Sources* 129 (2004), pp. 238–245.

Secanell, M., Carnes, B., Suleman, A. & Djilali, N.: Numerical optimization of proton exchange membrane fuel-cell cathodes. *Electrochim. Acta* 52 (2007), pp. 2668–2682.

Sinha, P.K., Halleck, P. & Wang, C.-Y.: Quantification of liquid water saturation in a PEM fuel-cell diffusion medium using X-ray microtomography. *Electrochem. Solid-State Lett.* 9:7 (2006), pp. A344–A348.

Song, D., Wang, Q., Liu, Z., Eikerling, M., Xie, Z., Navessin, T. & Holdcroft, S.: A method for optimizing distributions of Nafion and Pt in cathode catalyst layers of PEM fuel-cells. *Electrochim. Acta* 50 (2005), pp. 3347–3358.

Swamy, T., Kumbur, E.C. & Mench, M.M.: Characterization of interfacial structure in PEFCs: Water storage and contact resistance model. *J. Electrochem. Soc.* 157:1 (2010), pp. B77–B85.

Tabe, Y., Kikuta, K., Chikahisa, T. & Kozakai, M.: Basic evaluation of separator type specific phenomena of polymer electrolyte membrane fuel-cell by the measurement of water condensation characteristics and current density distribution. *J. Power Sources* 193 (2009a), pp. 416–424.

Tabe, Y., Lee, Y., Chikahisa, T. & Kozakai, M.: Numerical simulation of liquid water and gas flow in a channel and a simplified gas diffusion layer model of polymer electrolyte membrane fuel-cells using the lattice Boltzmann method. *J. Power Sources* 193 (2009b), pp. 24–31.

Tabe, Y., Nishino, M., Takamatsu, H. & Chikahisa, T.: Effects of cathode catalyst layer structure and properties dominating polymer electrolyte fuel-cell performance. *J. Electrochemical Soc.* 158:10 (2011), pp. B1246–B1254.

Tabe, Y., Saito, M., Fukui, K. & Chikahisa, T.: Cold start characteristics and freezing mechanism dependence on start-up temperature in a polymer electrolyte membrane fuel-cell. *J. Power Sources* 208 (2012), pp. 366–373.

Tabe, Y., Nasu, T., Morioka, S. & Chikahisa, T.: Performance characteristics and internal phenomena of polymer electrolyte membrane fuel-cell with porous flow field. *J. Power Sources* 238 (2013), pp. 21–28.

Tajiri, K., Tabuchi, Y. & Wang, C.-Y.: Isothermal cold start of polymer electrolyte fuel-cells. *J. Electrochem. Soc.* 154:2 (2007), pp. B147–B152.

Tüber, K., Pócza, D. & Hebling, C.: Visualization of water buildup in the cathode of transparent PEM fuel-cell. *J. Power Sources* 124 (2003), pp. 403–414.

Turhan, A., Kim, S., Hatzell, M. & Mench, M.M.: Impact of channel wall hydrophobicity on through-plane water distribution and flooding behavior in a polymer electrolyte fuel-cell. *Electrochim. Acta* 55 (2010), pp. 2734–2745.

Wang, C.-Y.: Fundamental models for fuel-cell engineering. *Chem. Rev.* 104:10 2004), pp. 4727–4766.

Yang, X.G., Zhang, F.Y., Lubawy, A.L. & Wang, C.-Y.: Visualization of liquid water transport in a PEFC. Electrochem. *Solid-State Lett.* 7:11 (2004), pp. A408–A411.

Zhang, F.Y., Yang, X.G. & Wang, C.-Y.: Liquid water removal from a polymer electrolyte fuel-cell. *J. Electrochem. Soc.* 153:2 (2006), pp. A225–A232.

# CHAPTER 2

## Reconstruction of PEM fuel cell electrodes with micro- and nano-structures

Ulises Cano-Castillo & Romeli Barbosa-Pool

## 2.1 INTRODUCTION

Despite the fact that the fuel cells' operation principle, including reactions, is quite simple, their design and engineering are rather complex. From a simplistic point of view, there are great new materials available and there exists a great deal of useful engineering knowledge to build successful devices. However, in practice, these approaches have not fully succeeded in providing satisfying answers to the economic challenges and technical requirements faced by these technologies.

On the other hand, nanotechnology now offers a huge range of new nano-materials with new properties and the possibility of creating new structures and synergies when they are combined. That is why scientists and engineers are finding ways to design new structures with nanotechnologies, apparently improving performance with the promise of more energy efficient devices and an extended life. These professionals are mainly aiming to reduce technology costs, for example by developing new but cheap and abundant catalyzers and more corrosion-resistant supporting components.

Despite such new scientific developments, engineers are faced with additional challenges as they need to design practical fuel cell generators based on new materials, new structures and even materials with properties explained by "less conventional" theories. Fuel cell design on the other hand should typically include preprocessing and manufacturing stages for such materials, where properties may alter and where synergies appear, giving place to distinctive new functionality features. In order to reach broad commercial use, fuel cell technologies require not only cheap and effective individual materials, but also such components need an effective interplay and synergy functionality when used together. This refers not only to compatibility from the physical and chemical point of view, but also refers to added qualities for optimum functionality to allow fuel cell processes to take place in an improved fashion. This is true at all levels, however the main processes in a fuel cell start to happen in the electrodes at the nanoscale level and in a very localized region, i.e. the porous electrodes.

### 2.1.1 *The technology: complex operational features required*

As mentioned above, the operation of fuel cells strongly depends on processes that take place in very small regions with specific features, in the so-called catalyst layer (CL). This CL, a few tens of a micron thick, requires several features to facilitate and promote the main processes for the operation of fuel cells: mass transport, electrons' (negative electrical current) transfer and transport, ion transport, and even heat transport, not to mention other complex physical phenomena such as adsorption of reactants on catalyst surfaces, secondary reactions, etc. Mass transport is required as CLs are the place that both reactants and products of the electrochemical reactions in the fuel cell should reach or leave respectively. Anodic CLs need an adequate structure to allow the access of hydrogen gas (in the case of a PEMFC) and sometimes the exit of liquid water. The water comes from the cathode side due to its diffusion through the membrane or from

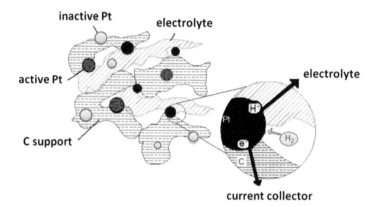

Figure 2.1.   CLs showing complex structural and compositional requisites as different phases have specific functions that could interfere with the roles of other phases.

the excess liquid water in a humidified hydrogen stream. At different levels the CL needs, besides adequate operating parameters, a porous structure to allow for these processes to occur.

Once molecular hydrogen reaches the electrode, which is formed by a catalyzer, such as platinum, and another conductive phase, like carbon graphite, it transfers two electrons per molecule becoming then two protons, which should travel to the other side of the cell through a solid electrolyte and helped, in an apparent solvated way, by water molecules. The proton transport is accomplished at the electrodes by an ionic path formed through the addition, in the anodic and cathodic CLs, of an ion exchange membrane in a soluble version that extends up to the membrane itself, which separates both electrodes, serving as the ion conductive bridge between both the anode and cathode, thereby completing the charge transport circuit. To sum up, besides a proper structure for access of reactants and exit for products, a region commonly referred to as a three-phase zone should be provided at the CL for the main processes to occur (Fig. 2.1): charge transfer (the electrochemical reaction promoted by the catalyst), current collection (through carbon conductive support), and proton transport (through the ion conductive paths described). A membrane, with an anode electrode on one side and a cathode electrode on the other, form the so-called membrane electrodes assembly (MEA).

### 2.1.1.1   *Nano-technology to the rescue?*

In the last decades, nano-materials have revolutionized our society as they offer new or enhanced properties. Some of the properties that have been exploited in fuel cells normally refer to the increase in effective surface area, conductivity and electro-activity as catalyzers, and of course their size, which in some applications intensifies a specific property, as well as its cost benefits. Such benefits, coming from the use of nano-technology, have been used in many other industries including food, medicine, biology, electronics, textile, and cosmetics, as well as many others and, most importantly for this work, in the energy sector.

In low temperature fuel cells, such as PEM fuel cells, the CLs are basically composed of a microscopic conductive material that supports a nano-structured electro-catalyzer, both arranged in a three dimensional structure. After cell fabrication, the electrode's structure appears as agglomerates leaving empty volume in the form of pores of different sizes that actually help to allow reactants reach reactive sites and products leave to the outside of the catalyst region. As mentioned above, a network of ion conductive paths and electron conductive paths needs to be constructed to facilitate charge transfer. What is interesting is the fact that the effective properties (some sort of global properties) of CLs depend strongly on its composition, on the individual component properties, as well as the final structure that such components form after their fabrication. For example, if catalyst nano-particles end up inside agglomerates, their role as reaction promoters will not be efficient. If the amount of the soluble ionic component is too large, it will cover the platinum (the most commonly used catalyzer in low temperature fuel cells) and the carbon particles, producing a decrease in the available active sites and in the electron conductive sites, making the overall process inefficient. If this situation was not complex enough, alternatives studied by researchers include the use of nano-structures for catalyst support which brings new and different interactions, new and more complex structures and different effects on CLs' properties and eventually different complexity of fuel cells.

It is pertinent at this point to say that experimental observations at the micrometer scale suggest that CL is formed by agglomerates in a porous matrix (Makoto, 1996). The concept of agglomerates was developed some years ago for electrochemical systems and it has been extensively applied to PEMFC (Broka and Ekdunge, 1997; Kamarajugadda and Mazumder, 2008; Wang *et al.*, 2004; Yan and Wu, 2008). Nevertheless, obtaining experimental information at the nanoscale is very difficult using current experimental characterization techniques, which do not allow for the capture of images at that scale without modifying the original structure (Scheiba *et al.*, 2008; Xie *et al.*, 2005). This limitation has led to the formulation of different theories; the most popular of which considers agglomerates as small spheres covered with a thin layer of ionomer and an inside of Pt/C particles (Secanell *et al.*, 2007; Seung and Pitsch, 2009; Shah *et al.*, 2006; Sun *et al.*, 2005).

### 2.1.1.2 *Challenges: technical and economic goals still remain*

To be economically successful, fuel cells require good performance and low costs according to their applications. For example, transportation applications require a durability of 5000 hours while stationary ones need generators that can last up to 40,000 hours. In terms of cost, it is recognized that components such as membranes and catalyzers, the latter based on the precious metal platinum, need more competitive prices either by improving manufacturing processes, substituting materials, or in the case of the catalyzer, lowering the platinum content. There exist other requirements that can contribute to improve both performance and costs. If durability improves, an economic impact will show. One example is the substitution with a more corrosion-resistant catalyst support, as carbon black-based (CB) supporting materials are known to corrode under electrode conditions encountered in fuel cells.

Other components, such as bipolar plates or balance of plant auxiliaries, should contribute to more competitive costs. For example, metal bipolar plates are proving to be a cost competitive alternative to carbon-based plates. Nevertheless, in some applications, fuel cells are being competitively advantageous and engineers and scientists are playing an important role as they are also proposing new approaches to develop, study and provide advice about solutions to overcome the remaining challenges. That is, they still need to study component structures and their relation with individual or overall effective properties by developing and using computational tools and simulating expected processes and performance. In doing so, computational resources and time, being both part of the economic equation, should be kept at a minimum.

## 2.2 CATALYST LAYERS' STRUCTURE: A REASON TO RECONSTRUCT

We have previously mentioned that a series of factors alter the effectiveness of CLs as electrochemical reactors, i.e. there are many complexities involved in the so-called heart of fuel cell technology. For this reason, it has been important for scientists and engineers not only to simulate the different processes that take place in those regions, but also to assume, reconstruct, and even design structures of CLs. Different approaches and tools have been developed to reconstruct, simulate, and predict CL properties, but what is recognized is the fact that once the components have been put together, with a certain composition of components with particular properties, by using a specific deposition technique, the final CL may have different properties associated with its heterogeneous nature (see Fig. 2.2), where its components are often randomly distributed. For that reason, a determination of the effective properties of should take this into consideration.

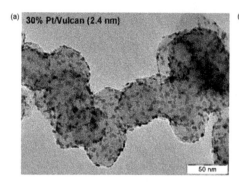

Figure 2.2.   TEM micrographs of apparently similar structures of a nano-Pt catalyst supported on (a) Vulcan and (b) Ketjen black (see Arico *et al.*, 2008). Different supports promoted different nano-particle size leading to a large difference in the electrochemical specific area (ECSA).

### 2.2.1   *Heterogeneous materials*

Heterogeneity arises from the fact that a CL is constituted by different phases, a phase being an identifiable domain with its own particular properties that differentiates it from the rest of the other phases in the CL (i.e. voids, gases, liquids, solids, or representative structures).

Proportionality coefficients for mass, energy, and charge transport in a Random Heterogeneous Material (RHM) are strongly affected by: (i) the properties of the phases from which it is composed, (ii) the volume fraction composition of such phases, and (iii) their contributing structure. For that reason, an effective transport coefficient (ETC) needs to be defined for a heterogeneous material as a proportionality coefficient, which characterizes the domain of the material. For a randomly formed heterogeneous material with M phases, a general ETC can then be represented by an effective coefficient named $K_e$ using the following relation:

$$K_e = f(K_1, K_2, \ldots, K_M; \; \emptyset_1, \emptyset_2, \ldots, \emptyset_M; \Omega) \tag{2.1}$$

where the subscript of variables indicates a respective phase, $K$ is the proportionality constant for that phase, $\emptyset$ is the phase volume fraction, and $\Omega$ is the micro-structural information of the domain (Torquato, 2002).

The right value of ETC is indispensable for an adequate numerical simulation of systems composed of one or more components with characteristics of heterogeneous media. An incorrect ETC value will alter the results and consequently it will provide a wrong interpretation of the transport phenomena.

### 2.2.2   *First steps for the reconstruction of catalyst layers*

Once the heterogeneous nature of CLs is recognized, other properties need to be considered before simulating the transport processes. As mentioned above, the final structure produced by the manufacturing technique used in the preparation of a CL may serve as the basis for the computation reconstruction. Many fuel cell researchers have explained CL performance based on the structural features that their product shows, for example, porosity and agglomerates sizes etc., are a result of the manufacturing route as well as the particular conditions during the production stage. For example, screen printing as a manufacturing technique, would normally use a more dense catalyst ink, while the spraying technique can be used with an ink that is less dense. Depending on the particular ink application conditions (e.g. temperature, humidity, solvent evaporation rate, etc.), the structure could be a more compact CL in the first technique. Thin layer electrode manufacturing techniques (Wilson, 1993) are among the most popular approaches used. Many variants to these techniques are reported in the literature, for example, Barbosa *et al.* (2007),

Escobar *et al.* (2010), Ledesma-García *et al.* (2009), Litster and McLean (2004), Passalacqua *et al.* (2001); however most of them involve mixing Pt/C and the ionomer in a liquid solution that is physically homogenized. The resultant mixture (catalytic ink) is then deposited on a substrate. Furthermore, some processes may use thermomechanical stages (applying a specific load during certain times and at a specific temperature) during the fabrication of MEAs, which also will alter the final structure.

### 2.2.2.1 *Structural features matter*

Although the nano-metric scale information of CLs is limited, structural and compositional information in advance does exist and it can be obtained before, during, and after manufacturing the electrodes or MEA. Design parameters that are used and controlled before MEA's manufacturing typically include: (i) platinum load ($\gamma_{Pt}$) in mg Pt cm$^{-2}$; (ii) ionomer load ($\beta_N$) which is a relation of Nafion® weight to total electrode weight, and (iii) platinum-to-carbon weight ratio ($\theta_{Pt}$), which is a characteristic of the catalytic material synthesis. From such information and from other physical properties of its components (densities), volume fractions of phases in a CL can be calculated.

On the other hand, we know that the solid phase volume ($V_S$) of any mixture is equal to the total sum of individual phase mass ($m$) divided by the density ($\rho$) of each $i$ element in the mixture:

$$V_S = \sum \frac{m_i}{\rho_i} \qquad (2.2)$$

By using the design parameters mentioned above ($\gamma_{Pt}, \beta_N, \theta_{Pt}$) in Equation (2.2), we can obtain the volume occupied by each primary element in the solid phase of the CL. Equations (2.3)–(2.5) can then be used to calculate platinum ($V_{Pt}$), carbon ($V_C$) and nafion ($V_N$) volume, respectively:

$$V_{Pt} = \frac{V_S}{\left[ \frac{\beta_N(1+\theta_{Pt})}{\theta_{Pt}\rho_N(1-\beta_N)} + \frac{1}{\rho_{Pt}} + \frac{1}{\theta_{Pt}\rho_C} \right]\rho_{Pt}} \qquad (2.3)$$

$$V_C = \left( \frac{1}{\theta_{Pt}\rho_C} \right)(V_{Pt} * \rho_{Pt}) \qquad (2.4)$$

$$V_N = \left( \frac{\beta_N}{(1 - \beta_N)\rho_N} \right)\left( \frac{[V_{Pt} * \rho_{Pt}](1 + \theta_{Pt})}{\theta_{Pt}} \right) \qquad (2.5)$$

where $\rho_{Pt}, \rho_C, \rho_N$ are platinum, carbon, and nafion densities, respectively.

On the other hand, Equation (2.6) relates the electrode total porosity ($\Phi_T$) with the solid phase volume ($V_S$) and the electrodes total volume ($V_T$):

$$\Phi_T = 1 - \frac{V_S}{V_T} = 1 - \frac{\sum \frac{m_i}{\rho_i}}{V_T} \qquad (2.6)$$

Other important structural information that can be obtained before manufacturing the CL are the Pt and the C particle sizes. Pt is typically supported on C during catalyst material synthesis, i.e. before manufacturing the CL, which also conditions the final electrode structure. The CL manufacturing method determines to a certain degree its final structure, but such effects (synthesis and manufacturing) are not specifically considered in this work. After manufacturing the CL and using modern techniques such as porosimetry and high resolution microscopy, one can get approximate size and structure of agglomerates, total porosity and pore size distribution. In this work, we assumed that the physical properties of the primary components are not modified during manufacture. This assumption is valid when the composite material is fabricated using physical techniques in which chemical reactions do not occur which is normally the case in state-of-the-art MEAs manufacturing techniques. It is important to say that regardless of the catalytic ink composition and type of substrate, it is assumed that the primary elements are effectively distributed randomly (given a complete homogenization stage).

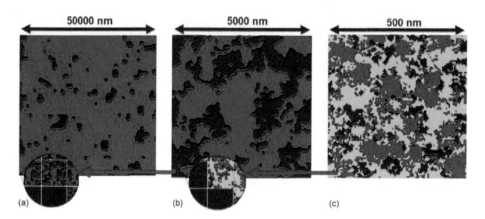

Figure 2.3.    Different scales of domains used to study a CL: for description see main text.

During the manufacturing process of CLs, agglomerates composed of primary components plus a number of pores in a number of different sizes are formed. These structural features of the CL allow us to define internal substructures at different scale levels, and with that we can establish a scaling strategy for numerical simulation.

### 2.2.2.2    *Scaling – a matter of perspectives*

Before establishing the scaling strategy, let us explain these apparent structural differences at different scale levels by picturing ourselves at infinity, getting closer and closer into the CL structure. When the vision scale is ~50,000 nm, one will observe a homogenous dispersion of a mesoporous structure; at a scale of ~5000 nm and focusing on areas where meso-pores are absent, the structure is defined by agglomerates and micropores; finally at a vision scale of ~500 nm and focusing at an agglomerate, the observer will distinguish a structure formed by a random distribution of Pt/C particles, ionomer and probably nano-metric pores. This last observation actually is a hypothesis due to the fact that available information from experimental techniques is limited at this scale. Nevertheless, it is a more realistic hypothesis compared to the most commonly used theory where agglomerates are considered spheres with Pt/C particles inside and covered with a thin layer of ionomer on the outside.

In Figure 2.3 the previously described scales are graphically represented for a PEMFC CL. Figure 2.3a refers to the so-called "meso-porous" scale with a ~50,000 nm domain, where the blue color represents the pseudo-solid phase while black represents the empty phase; Figure 2.3b represents the "agglomerate and micro-pores" with a domain scale of ~5000 nm, here the blue color represents the agglomerates while black shows the empty areas; finally Figure 2.3c refers to the "inside of an agglomerate" scale with a ~500 nm domain where the primary elements, Pt (~5 nm, in red), carbon agglomerate (~50 nm, in blue) and ionomer (in light blue) can be distinguished.

In this work, the sample that simulates the inside of an agglomerate is the only one that contains the primary elements, while samples of ascending scale contain only the sample of the immediately smaller scale and its corresponding porosity. We define the relative porosity as the volume fraction occupied by the empty spaces at that scale, ignoring that the pseudo-solid phase (or smaller scale) also contains some porosity. The volume fractions of the solid phase of each sample at different scales can be related mathematically by:

$$S_R = \prod_1^n S_i \qquad (2.7)$$

where $S_R$ is the volume fraction of the solid phase in the entire CL, $S_i$ is the individual volume fraction of the solid phase in each of the samples and the Greek letter Pi in capital represents the product of $S_i$ terms. By substitution of Equation (2.6) in Equation (2.7), we can obtain a mathematical relation that calculates the "total" porosity in the complete CL with the relative porosities of each of the samples:

$$\Phi_T = 1 - \prod_1^n (1 - \Phi_i) \tag{2.8}$$

where $\Phi_T$ represents the total porosity of the complete CL and $\Phi_i$ the relative porosity of each individual samples.

A technique that takes into consideration the different scaling effects and available information can facilitate the reconstruction of structures as this allows a significant reduction in computational resources needed for simulation. Such scales are defined based on different observable structures in scales of different orders of magnitude. In addition, they allow the use of information available before and after the manufacture of CLs. It is for these reasons that reconstruction techniques and their features (Barbosa *et al.*, 2011a; 2011b) are used in this chapter.

### 2.2.3 *Stochastic reconstruction – scaling method*

Although a method called "annealing method" (Capel *et al.*, 2009; Jiao *et al.*, 2008; Patelli and Schüeller, 2009; Torquato, 2002; Zhao *et al.*, 2007) can also be applied, for simplicity, our reconstruction method considers, for now, only one filter: the two-point correlation function (volume fraction). Later in this chapter the annealing method is also used in conjunction with statistical information from real structures.

In this section, we will use a powerful technique called "stochastic reconstruction" to determine the effective properties of random heterogeneous materials, of which the first application is attributed to Quiblier (1984). This technique is based on the computational generation of a mesh that characterizes the real micro-structure of the heterogeneous material, mathematically described by statistical functions and referred to as "correlation functions". Torquato (2002) proposes a methodology to characterize micro-structures, as well as a fundamental theory to estimate effective properties. Recent works offer modifications of Torquato's methodology in order to optimize the quality in the stochastic reconstruction or to reduce convergence time during computing (Capel *et al.*, 2009; Jiao *et al.*, 2008; Patelli and Schüeller, 2009; Zhao *et al.*, 2007).

When the structure of a sample is composed of more than two phases, subroutines repeat each of them in such a manner that the structure is generated in stages and where each stage corresponds to the complete generation of a phase.

The input variables in the micro-structural reconstruction algorithm are: (i) platinum load ($\gamma_{Pt}$); (ii) Nafion® load ($\beta_N$); (iii) weight percent of carbon-supported platinum ($\theta_{Pt}$); (iv) total porosity and relative porosity of $n - 1$ samples; (v) domain and control volume dimensions at each scale; and (vi) average size of representative elements at each scale.

The domain is fractioned in finite control volumes (FCV), where each volume is identified by an index number randomly distributed and computer-generated by a random number generator (Matsumoto and Nishimura, 1998).

The reconstruction method of each phase comprises three formation stages: (1) the center of the representative elements at each scale (i.e. primary elements, agglomerates, meso-pores, isles, etc.) are stochastically distributed; (2) around such centers and in one single step, a previously configured specific tridimensional geometry is generated (i.e. amorphous sphere structures, ellipses, tubes, etc.); (3) in a random manner the surroundings are filled until the required volume fraction is fulfilled. The centers of such structures are denominated seeds and the number of seeds distributed in the computing domain is calculated using Equation (2.9):

$$NS = \frac{VC_{Ti}}{VC_{Ui}} \tag{2.9}$$

where $NS$ in the integer number of seed control volumes, $VC_{Ti}$ is the number of total control volumes necessary to fulfill the volume fraction of the representative element in the whole computing domain, and $VC_{Ui}$ is the number of control volumes occupied by one single representative element. The index function ($I_s$) that distributes the seeds has the following general form:

$$I_s(x) = \begin{cases} 1, & \text{if } 1 \leq x \leq NS \\ 0, & \text{otherwise} \end{cases} \tag{2.10}$$

The structure of representative elements is previously defined and designed pixel by pixel by a subroutine according to the known morphological features of that element and then systematically generated around all the seeds. The growth of the surroundings of representative elements is controlled on the basis of volume fractions as described by:

$$E = \frac{|V_R - V_A|}{V_R} \tag{2.11}$$

where $E$ is the volume error that exists during the assigning, $V_A$ is the assigned volume, and $V_R$ the reference volume for the specific phase. $V_R$ is obtained for each reconstructing phase. At the "inside an agglomerate" scale, $V_R$ is obtained for platinum, carbon, and ionomer by using Equations (2.3), (2.4) and (2.5), respectively. As mentioned earlier, at the "meso-pores" and "agglomerate and micro-pores" scales, $V_R$ is obtained by Equation (2.6). The range of available adjacent FCVs (RID) for the expansion is limited by the following function:

$$RID = VCT * \exp\left(\frac{E - 1}{T}\right) \tag{2.12}$$

where VCT is the total number of FCVs in the sample domain and $T$ is a parameter that regulates the RID's value. This is an empirical relation used to guarantee that $V_R$ converges to $V_A$. The index function ($I_e$) for this assignment is:

$$I_e(NI) = \begin{cases} 1, & \text{if } NI \leq RID \\ 0, & \text{otherwise} \end{cases} \tag{2.13}$$

where $NI$ is the random FCV's index number. After the expansion, the structure is characterized. As one can see, the reconstruction technique employed here allows the use of information available both before and after the manufacture of CLs.

### 2.2.3.1  Statistical signatures

The reconstructed structures are characterized by a two-point correlation function and their pore size distribution. The two-point correlation function ($S_2$) of an homogeneous medium can be obtained by randomly "tossing" a line segment of $r$ length with a specific orientation and counting the number of times that the beginning ($x$) and the end ($x + r$) of the line fall in phase $j$, as described by:

$$S_2(x,r) = \langle I_j(x)I_j(x+r) \rangle \tag{2.14}$$

where the angular parenthesis refers to the statistical average when the whole domain is evaluated, and $I_j$ is equal to 1 when the point belongs to phase $j$ as described by the following identity function:

$$I_j(x) = \begin{cases} 1, & \text{if } x \text{ is in the phase } j \\ 0, & \text{otherwise} \end{cases} \tag{2.15}$$

Using orthogonal coordinates, the two-point correlation function is employed to characterize the generated structures and it becomes:

$$S_2(r) = \frac{1}{3N^2}\left[ \sum_{j,k=1}^{N} S_{2,i}(r) + \sum_{k,i=1}^{N} S_{2,j}(r) + \sum_{i,j=1}^{N} S_{2,k}(r) \right] \tag{2.16}$$

where $N$ is the computing domain cubic length and $S_{2,i}(r)$ is the two-point correlation function along $i$ direction:

$$S_{2,i}(r) = \frac{1}{N-r}\sum_{i=1}^{N-r} I(i,j,k) * I(i+r,j,k) \tag{2.17}$$

Similarly $S_{2,j}$ and $S_{2,k}$ can also be defined.

The pore size is determined by the distribution of spheres of different radii in the porous structure, where the pore radius is equal to the radius of the sphere that in its inside is formed by the "empty" phase (Schulz et al., 2007). To avoid that large spaces are fractioned into smaller spaces, the spheres' radii begin to be modified from a maximum limit to unity. The characterized spaces are identified by the following index function:

$$I_{tp}(x) = \begin{cases} 0, & \text{if } B \subseteq X \\ 1, & \text{otherwise} \end{cases} \tag{2.18}$$

where $X$ represents the porous spaces and $B$ the sphere. The maximum sphere's radius is determined by a similar function evaluated in the plane $i-j$ along $k$. The pore size distribution in each sample at different scale is defined by:

$$F_{r_a,M_n} = \frac{VP_{r_a,M_n}}{VP_{M_n}} \tag{2.19}$$

where $F_{r_a,M_n}$ is the volume fraction that characterizes the occupied volume by pores of diameter $r_a$ ($VP_{r_a,M_n}$) over the total volume occupied by the empty space ($VP_{M_n}$) in each of the $M_n$ samples. For scaling, this equation can be extended to find the pore size distribution along the CL according to:

$$FT_{r_a}(F_{r_a,M_n}) = \frac{F_{r_a,M_n}}{\Phi_T}(\Phi_i)\prod(1-\Phi_{i-1}) \tag{2.20}$$

where $FT_{r_a}$ is the volume fraction that characterizes the occupied volume by pores of diameter $r_a$ over the total occupied volume by the empty phase in the whole CL, $\Phi_T$ is the CLs total porosity, and $\Phi_i$ the sample's relative porosity. For example, when $n=1$, $\Phi_{i-1}=0$.

### 2.2.4 Let's reconstruct

Once a structure has been reconstructed, we will determine the electronic and ionic conduction efficiency of a CL with different structural features: (1) ionomer electrode load ($\beta_N$) in the range of 20% to 80%wt; (2) CL porosity in the range of 20% to 50%, and (3) pore size distribution by modification of the relative porosity. However, first the reconstruction should be done. All generated structures will have the following characteristics: 20% wt. Pt/C; 0.5 mg Pt cm$^2$; average Pt particle diameter, C diameter ($ER_C$), agglomerate diameter ($ER_M$), and meso-pore diameter ($ER_P$) of 3 nm, 50 nm, 500 nm and 600 nm, respectively. Density of materials forming the solid structure: Pt, C, and ionomer are 21,450, 1800 and 2000 mg cm$^{-3}$, respectively. Each one of the resulting structures is generated ten times by a different random series and then their values are averaged.

Every CL is then studied at three different scales, namely:

- Inside an agglomerate ($M_3$): the phase's reconstruction sequence in the computing domain is: (i) carbon agglomerate generation, (ii) platinum on carbon placement-generation, (iii) the rest is ionomer.
- Agglomerates and micro-spores ($M_2$): the reconstruction sequence is: (i) agglomerate generation, (ii) the rest is empty room.
- Meso-pores ($M_1$), reconstruction sequence is: (i) meso-pores generation, (ii) the rest is a pseudo-solid structure.

Table 2.1.  Dimensions of the domain and control volume for the studied scales.

| Sample | $N_x = N_y = N_z$ [nm] | $D_x = D_y = D_z$ [nm] |
|--------|------------------------|------------------------|
| $M_3$  | 300                    | 3                      |
| $M_2$  | 5000                   | 50                     |
| $M_1$  | 45000                  | 300                    |

Table 2.2.  Porosity of simulated structures.

| $\Phi_T$ [%] | $\Phi_\Phi$ [%] | $\Phi_{M_1}$ [%] | $\Phi_{M_2}$ [%] |
|--------------|-----------------|------------------|------------------|
| 0.50         | D1              | 7.50             | 45.95            |
| 0.50         | D2              | 10.00            | 44.44            |
| 0.50         | D3              | 12.50            | 42.86            |
| 0.50         | D4              | 15.00            | 41.18            |
| 0.40         | D1              | 6.00             | 36.17            |
| 0.40         | D2              | 8.00             | 34.78            |
| 0.40         | D3              | 10.00            | 33.33            |
| 0.40         | D4              | 12.00            | 31.82            |
| 0.30         | D1              | 4.50             | 26.70            |
| 0.30         | D2              | 6.00             | 25.53            |
| 0.30         | D3              | 7.50             | 24.32            |
| 0.30         | D4              | 9.00             | 23.08            |
| 0.20         | D1              | 3.00             | 17.53            |
| 0.20         | D2              | 4.00             | 16.67            |
| 0.20         | D3              | 5.00             | 15.79            |
| 0.20         | D4              | 6.00             | 14.89            |

Table 2.1 summarizes the dimension of the computing domain ($N_x$, $N_y$, $N_z$) and the FCVs ($d_x, d_y, d_z$) for each of the scales ($M$). These values are based on average diameters of representative elements of the previously mentioned scaling method.

In this method, the pore size distribution is controlled indirectly by defining the relative porosity of the so-called "meso-porous" scale ($\Phi_{M_1}$) as a percentage of total porosity:

$$\Phi_\Phi = \frac{\Phi_{M_1}}{\Phi_T} \qquad (2.21)$$

where $\Phi_\Phi$ is the total porosity percentage assigned to $\Phi_{M_1}$ and $\Phi_T$ is the total electrode porosity. From a physical point of view, $\Phi_\Phi$ is directly proportional to the pore size distribution because it determines the volume occupied by meso-pores. It should be mentioned that the relative porosity of the $M_2$ scale is conditioned by Equation (2.8). Table 2.2 specifies relative porosities used in this work, where $D1$, $D2$, $D3$, and $D4$ refer to $\Phi_\Phi$ equal to 0.15, 0.20, 0.25, and 0.30, respectively.

The present methodology proposes the study of the following CLs: 4 porosities, 4 pore size distributions and 13 different ionomer loads (from 20 to 80 in steps of 5%). As each structure will be generated by ten different random series, we will have a total of 2080 analyzed structures. Using a single personal computer with 3 GB of RAM memory and a 2.4 GHz processor (rather limited resources), the computing time is of approximately 100 hours, with the use of scaling and not including the time for establishing the problem and for the analysis of results. The scaling method not only reduces computer processing time but also allows a detailed study of the CL at the nano-scale level. The domain lattice of the cubic simple simulating the CL ($M_1$) is 45,000 nm, while the lattice of the sample's control volume detailing the inside of an agglomerate ($M_3$)

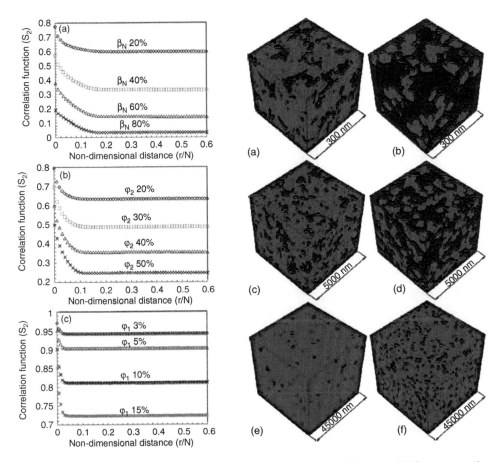

Figure 2.4. Two-point correlation functions $(S_2)$ *versus* a non-dimensional distance $(r/N)$ for representative structures at different used scales.

is 3 nm. Without the scaling technique, the computing domain would need to be generated by $3.375 \times 1012$ control volumes, which would make it impossible to do with the above mentioned computer.

### 2.2.4.1 *Features of reconstructed structures*

Figure 2.4 above, shows (on the left) the two-point correlation functions $(S_2)$ obtained for some representative structures at each studied scale, and the respective reconstructed structures (on the right) as follows: The left curve (a) shows $S_2$ for the carbon phase at the $M_3$ scale, where each curve corresponds to different ionomer loads $(\beta_N)$; also on the left curve, (b) presents the calculated $S_2$ for the agglomerates at the $M_2$ scale; finally, on the left curve, (c) also shows the $S_2$ for the pseudo-solid phase at the $M_1$ scale, that is, in (b) and (c), $S_2$ is calculated for the "pseudo-solid" phase at the $M_2$ and $M_1$ scales, respectively. These curves, (b) and (c), correspond to different porosities $(\Phi_i)$. Notice that when $r = 0$, $S_2$ represents the volume fraction of the characterized phase. In all cases, values of $S_2$ are plotted *versus* a non-dimensional distance $r/N$, where $r$ is the distance between two points and $N$ the domain lattice. It is noticed that all correlation functions $S_2$ decay exponentially down to the squared volume fraction, where its value is independent of $r$.

On the right hand side of the same Figure 2.4, cubic colored images of reconstructed structures are shown. Colored images (a) and (b) represent the $M_3$ scale, with $\beta_N$ equal to 20% and 60%wt. respectively, the blue color representing the carbon phase and the dark blue representing the

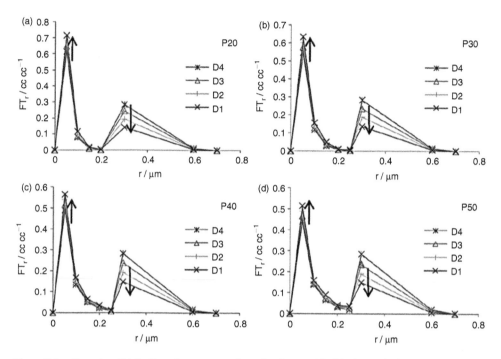

Figure 2.5.   Pore size distribution of reconstructed catalyst layers. (a), (b), (c), and (d) correspond to 20, 30, 40, and 50% of total porosity ($\Phi_T$), respectively. The arrows show how $\Phi_\Phi$ is modified.

ionomer phase. Also on the right of Figure 2.4, (c) and (d) represent the $M_2$ scale, with $\Phi_{M_2}$ equal to 20% and 50%, respectively; the blue color represents the agglomerates and the dark blue represents the empty phase. Finally, figures (e) and (f) on the right show the $M_1$ scale with $\Phi_{M_1}$ equal to 3% and 15%, respectively. Here the blue color represents the pseudo-solid phase and the dark blue represents the empty phase. In general, the reconstructed structures look alike in the three planes for each shown condition. Moreover, along each scale, the effect of extreme values is clear: the ionomer load at the $M_3$ scale, (Fig. 2.4 right (a) and (b)), and the relative porosity at $M_2$ ((c) and (d)) and $M_1$ ((e) and (f)) scales.

The pore size distribution of the CL is obtained by using Equation (2.22). Therefore the resulting expressions, where only two samples contain the empty phase, are:

$$FT_{r_a}(F_{r_a,M_1}) = \frac{F_{r_a,M_1}}{\Phi_R}(\Phi_1) \tag{2.22}$$

$$FT_{r_a}(F_{r_a,M_2}) = \frac{F_{r_a,M_2}}{\Phi_R}(\Phi_2)(1 - \Phi_1) \tag{2.23}$$

These results are expected as $\Phi_\Phi$ is directly proportional to the relative porosity of a sample with a larger scale, therefore, when its value decreases the volume of those pores with a larger size decreases. These results confirm that the pore size distribution of the CL was modified by controlling $\Phi_\Phi$. Such distribution shows two peaks (at 0.05 μm and 0.3 μm), which are strongly influenced by the carbon particle size and the meso-pore size, and both parameters can be considered as input data.

Figure 2.5 shows the distribution of different sizes of pores thus calculated. Figure 2.5a–d correspond to 20, 30, 40, and 50% of total porosity, respectively for a hypothetical CL. In these plots, one can observe the pore size variation by normalizing $\Phi_\Phi$ (total porosity percentage assigned). From Figure 2.5 it can be seen that when $\Phi_\Phi$ decreases, the frequency of the ~50 nm pore size (left peak) increases while the frequency of the $\sim$300 nm pore size (right peak) diminishes. An example of this behavior is seen in Figure 2.4a when $\Phi_\Phi = 0.30$ ($D_4$), $FT_{r_{50}} = 0.61$ and $FT_{r_{300}} = 0.28$, and when $\Phi_\Phi = 0.15$ ($D_1$), $FT_{r_{50}} = 0.71$ and $FT_{r_{300}} = 0.14$. The arrows in those figures point this tendency out.

### 2.2.4.2 *Effective ohmic conductivity*

Ohm's law relates the electric current ($J$) of a conducting material directly to the applied voltage or potential ($\Delta E$), as described by:

$$J = kA\frac{\Delta E}{L} \tag{2.24}$$

where $k$ is the proportionality coefficient so-called "conductivity" which is a property of the material, and $A$ and $L$ are the area and length of the charge transport respectively. By applying the continuity equation in a medium discretized by the FCV, the charge conservation transport equation in a non-reactive system can be expressed by:

$$\nabla \cdot (k_m \nabla \phi_e) = 0 \tag{2.25}$$

where $k_m$ is the material's conductivity and $\phi_e$ the applied potential. By solving the charge transport continuity equation directly in the structural mesh, it is possible to obtain the average of all local current flows ($J_a$) generated by the potential. Taking Equation (2.21) and substituting $J_a$, we can obtain the effective length-to-area ratio of conduction ($L_{meff}/A_{meff} = k\phi_e/J_a$). As it will be shown later, this method will simplify the structure scaling.

The effective resistance ($R_{eff}$) of a heterogeneous material, composed by a conductive phase and one or more insulating phases, is a function of the conductive material's resistivity ($\rho_m$), its effective area ($A_{meff}$), and its effective length ($L_{meff}$) as described in the following equation:

$$R_{eff} = \rho_m \frac{L_{meff}}{A_{meff}} \tag{2.26}$$

From a different angle, the $R_{eff}$ can be calculated as a function of the effective resistivity ($\rho_{meff}$) and the input data: the area of sample ($A_m$) and the length of sample ($L_m$), as described by:

$$R_{eff} = \rho_{meff} \frac{L_m}{A_m} \tag{2.27}$$

Using equations (2.26) and (2.27) we can obtain $\rho_{meff}$, which relates the effective resistivity of the whole sample with the resistivity, effective length, and effective area of conduction of the conductive phase:

$$\rho_{meff} = \left(\rho_m \frac{L_{meff}}{A_{meff}}\right) \frac{A_m}{L_m} \tag{2.28}$$

As the effective resistivity characterizes the material by being an intensive property and being able to be extended to find the resistance of any continuous structure formed by the same material, this equation can be generalized to find the resistivity of an element formed by subdomains of smaller scales:

$$\rho_{meff\_T} = \rho_m * \prod_1^n \frac{L_{meff\_i}}{A_{meff\_i}} \prod_1^n \frac{A_{m\_i}}{L_{m\_i}} \tag{2.29}$$

where $\rho_{meff\_T}$ is the effective resistivity of the global domain formed by various smaller scale subdomains $i$, $L_{meff\_i}$ is the effective length, and $A_{meff\_i}$ is the effective area of the phase under study in every subdomain. $L_{m\_i}$ is the length and $A_{m\_i}$ is the area of the sample of every subdomain.

To normalize and generalize results, in this chapter a calculated resistivity is used to estimate a conduction efficiency ($\varepsilon_k$). As conductivity is the inverse of resistivity, the CLs effective conductivity is the inverse of the effective resistivity value $k_{meff\_T} = \rho_{meff\_T}^{-1}$. $\varepsilon_k$ is calculated by comparing the effective conductivity with the nominal conductivity, as described by Equation

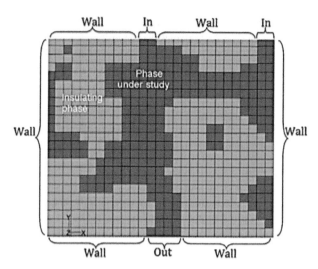

Figure 2.6.   Schematics of the general boundary conditions for the charge transport simulation of the cases studied at the different scales.

(2.30). By substituting Equation (2.29) in Equation (2.30), we can obtain a relation that provides the conduction efficiency of the global domain ($\varepsilon_{k\_T}$) which is formed by several subdomains of smaller scale (Equation (2.31)):

$$\varepsilon_k = \frac{k_{\text{meff}}}{k_m} \tag{2.30}$$

$$\varepsilon_{k\_T} = \prod_1^n \frac{A_{\text{meff}\_i}}{L_{\text{meff}\_i}} \prod_1^n \frac{L_{m\_i}}{A_{m\_i}} \tag{2.31}$$

For example, to study the conduction efficiency with three scaling levels the equation would be:

$$\varepsilon_{k\_T} = \frac{A_{\text{meff}\_1}}{L_{\text{meff}\_1}} * \frac{A_{\text{meff}\_2}}{L_{\text{meff}\_2}} * \frac{A_{\text{meff}\_3}}{L_{\text{meff}\_3}} * \frac{L_{m\_1}}{A_{m\_1}} * \frac{L_{m\_2}}{A_{m\_2}} * \frac{L_{m\_3}}{A_{m\_3}} \tag{2.32}$$

Although as an example the ohmic conduction efficiency will be determined, other transport properties, such as thermal conductivity and diffusion coefficients, just to mention two, could be determined by using the same approach.

### 2.2.4.3   *CL voltage distribution, electric and ionic transport coefficients*

Figure 2.6 shows the general boundary conditions (in 2D) used during simulation of the cases studied at the different scales. As the structures are homogeneous, the side of the cubic domain that is assigned to the "flow current input" is not relevant, but clearly the output should be located at the opposite side. At the input side, a 1.1 V potential is specified exclusively on the FCV of the phase under study, while in the faces of FCV of the isolating phase the boundary condition is a zero-current flow. At the output, a 1.0 V potential is specified only for the phase under study, therefore having a 0.1 V gradient. Other sides in the cubic domain are specified with a zero-current flow condition. It must be mentioned that these limiting values have only the purpose of determining current conducting efficiency, and do not represent actual voltage distribution on the different scales at the same time.

As the material's resistivity does not affect the results of conduction efficiency (Equation (2.31)), the phase under study is specified with an arbitrary high conductivity (1000 S m$^{-1}$) while the isolating phase has an arbitrary low conductivity (0.0001 S m$^{-1}$). At the $M_1$ scale, the phase under study is the pseudo-solid, at the $M_2$ scale, the phase being studied is the agglomerate, and at both $M_1$ and $M_2$ scales, the insulating phase corresponds always to the empty phase.

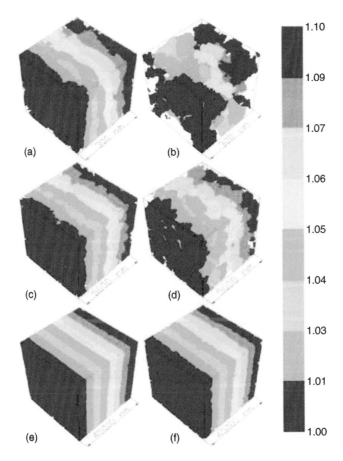

Figure 2.7.    Potential distribution in volts within the electronic conduction phase for reconstructed CLs
shown in the right of Figure 2.4.

To determine the ionic conduction efficiency, the phase under study at the $M_3$ scale is the
ionomer, and the insulating phase corresponds to both carbon and platinum. To determine the
electronic conduction efficiency, the phase under study at the $M_3$ scale corresponds to carbon
and platinum, while the insulating phase is the ionomer.

For a given electronic conduction with the values determined here, one could estimate the
voltage distribution along the CL for a given current. Figure 2.7 shows the potential distribution
[V] of the electronic conduction phase of reconstructed samples shown in Figure 2.4. Figures
2.7a and 2.7b show the potential distribution for the $M_3$ scale, with $\beta_N$ equal to 20 and 60%,
respectively.

Figure 2.7c and 7d show the potential distribution for the $M_2$ scale, with $\Phi_{M2}$ equal to 20
and 50%, respectively. Figures 2.7e and 2.7f show the potential distribution for the M1 scale
with $\Phi_{M1}$ equal to 3 and 15%, respectively. The potential distribution is estimated using the
boundary conditions described above, while the current flows are used exclusively to calculate
the conduction efficiency. The domains of Figures 7a–f are limited by an imaginary line and the
potential distribution is plotted exclusively in the electronic conduction phase.

Curves on the left hand side of Figure 2.8, show the effect of ionomer load ($\beta_N$) on (a) proton
conduction efficiency and (b) electronic conduction efficiency of reconstructed CLs. with 30%
porosity ($\Phi_T$). Each curve represents different pore size distribution. It can be observed that

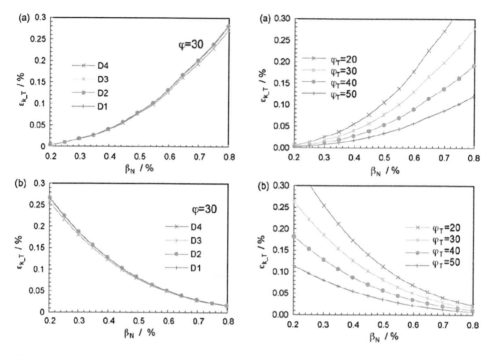

Figure 2.8.   Effect of ionomer load ($\beta_N$) on proton conduction efficiency and the electronic conduction efficiency of reconstructed CLs. See main text for full description.

when $\beta_N$ increases, the proton conduction efficiency increases, while the electronic conduction efficiency diminishes, both in an exponential fashion. This is an expected behavior, as when $\beta_N$ grows what really increases is the volume fraction of the ionomer phase, while the electronic conduction phase decreases inversely. On the other hand, it is relevant to point out that the pore size distribution does not affect the effective ionic or the effective electronic conduction for the same $\beta_N$ value.

Curves on the right of Figure 2.8 show the effect of ionomer load ($\beta_N$) on the proton conduction (a) and the electronic conduction (b) efficiency of the reconstructed CLs. Each curve represents different porosities ($\Phi_T$). As the pore size distribution does not affect these results, conduction efficiency dependent on the pore size distribution, was averaged to be included in this analysis. As expected, one can also observe that the decay rate of the proton conduction efficiency increases, while the electronic conduction efficiency diminishes when $\beta_N$ increases.

In Figure 2.9, the results obtained for proton conduction (a) and electronic conduction (b) efficiency *versus* different total porosity in reconstructed CLs, now with 35% of ionomer loading, are shown. Each curve represents a different pore size distribution and it can be noticed that conduction efficiencies, both electronic and ionic, decrease almost linearly when the CLs porosity increases. This trend is caused by the smaller volumetric proportion of solid components (electronic and ionic) when porosity increases, which clearly affects the effective conductivity of the heterogeneous material. In addition, it should be noted that the slope for the electronic efficiency is different from that for the ionic efficiency. As we shall see later, the decay rate depends on $\beta_N$. Also from these figures, one can see that the pore size distribution does not affect the effective ionic or the effective electronic conduction for the same $\beta_N$ value.

Should the present method be validated experimentally, the following must be considered: (i) The micro-structure reconstruction, at each scale, should be made by a procedure that statistically ensures the correct representation of the real structure (determined experimentally), (ii) The

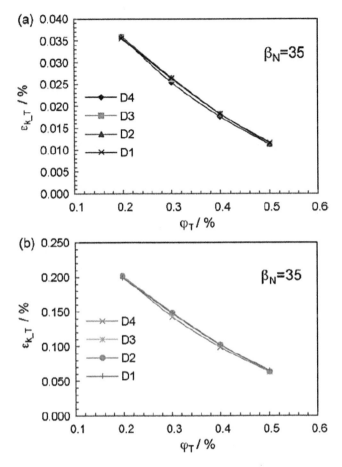

Figure 2.9. Effect of porosity ($\Phi_T$) on proton conduction (a) and electronic conduction (b) efficiency of reconstructed CLs containing 35% of ionomer load ($\beta_N$). Each curve represents different pore size distribution (see main text for more information).

experimentally effective transport coefficients should be exclusive to the CL. This information is not readily available in the literature. Nevertheless, Boyer *et al.* (1998) conducted an experimental study where the proton conductivity of a pseudo-CL is similar to the results presented in this paper. It is noteworthy that experimental validation is a current work in many laboratories, thereby helping to validate all assumptions in reconstruction techniques such as many assumed in this chapter.

### 2.2.5 *Structural reconstruction: annealing route*

In this present section, statistical information based on two-point and linear path correlation functions, obtained from 2D micrographs of real catalyst layers (CL) of proton exchange membrane fuel cells (PEMFC), is used as initial information for the reconstruction of CL structures. As will be seen, with this information, stochastic replicas of CLs' 3D pore networks at two scales can be built (Barbosa *et al.*, 2011b). Once statistical information is available from real samples, this can be used for a scaling strategy in order to determine the effective transport coefficients. In past sections statistical information was determined from reconstructed structures rather than from real samples.

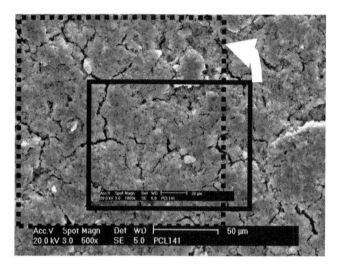

Figure 2.10.    Initial process (P1) to obtain information from micrographs. The internal box is twice the
resolution compared to that in the dotted box. The internal box fixes the superficial fraction
of area in the dotted line box.

When experimental information is used, a proper reconstruction approach can ensure more
realistic information before "recreating" CLs and determining their transport properties. For this,
the continuity equation for charge transport is solved directly on the 3D reconstructed CL to
determine effective electrical conductivities at the "internal-scale" and to simulate the electrical
global performance at the "macro-scale". This electrical performance can be then experimentally
determined. This should enable us to know that the reconstruction technique is valid. To determine
statistical properties an image processing method can be used. One way of doing this is by
determining relative porosities using statistical analysis of SEM (scanning electron microscopy)
micrographs and by verifying this by experimentally measuring the total porosity using, for
example, mercury intrusion porosimetry, or the BET technique (Brunauer Emmett Teller). As
will be seen, in this present approach the method makes use of two immediate scales to improve
the image resolution. In addition, the pore size distribution of the reconstructed scales can be used
to avoid the superposition of equal pore sizes. Such pore size distribution could be contrasted for
example with experimental data.

### 2.2.5.1    Image processing for statistical realistic information
In order to obtain statistical information out of 2D images, the following process was carried out:
the empty phase was selected by the pixel intensity in two stages: (i) first, the superficial fraction
at double resolution of the studied scale was empirically determined by the intensity, (ii) then
the intensity at the studied resolution was automatically selected by the superficial fraction of
the first stage. In this work, intensity at $1000\times$ resolution fixes the superficial fraction of $500\times$
resolution, and $10,000\times$ resolution fixes the superficial fraction of $5000\times$ resolution. This allows
the study of a predetermined area with higher resolution. Figure 2.10 shows schematically this
initial process (P1) on a sample micrograph.

In this annealing technique, the scaling method was applied for two scales. Scales are defined
based on different observable structures at different resolutions of SEM and their statistical
information. The "annealing" stochastic reconstruction and the direct simulation of the charge
continuity equation are performed in the following subsections.

After an "intensity selection process", which can be regarded as an initial process that extracts
information from a micrograph, (P1) the pore size distribution at the two studied scales are
adjusted to avoid the superposition of the same pore size by pore elimination. Figure 2.11 shows

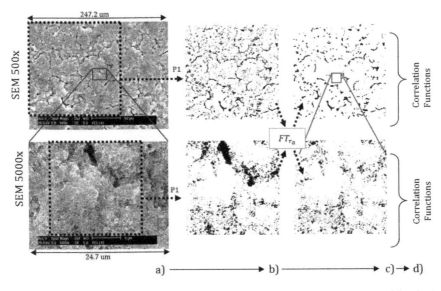

Figure 2.11. Total image process. a)→b) pixel intensity selection; b)→c) proper pore deleted; c)→d) statistical characterization.

a representative example of the total image process: from (a) to (b) the P1 process is applied; from (b) to (c) the pore size distribution is analyzed and the proper size pores are deleted; from (c) to (d) correlation functions (S2F and LPF) are obtained (see below for a description of both functions).

As an example, we will process some images to show the use of this reconstruction technique. In the examples, three loads of pore promoter (ammonium carbonate) were studied: 40, 20, and 0%wt as a base line. The preparation of MEAs containing these structurally different CLs are described elsewhere (Barbosa *et al.*, 2011b). Original and processed images of these real CLs containing the pore promoter additive (PCL) are shown in Figure 2.12. Figures 2.12a and 2.12b correspond to PCL40, 2.12c and 2.12d correspond to PCL20, and 2.12e and 2.12f correspond to PCL00. The left side of the figure corresponds to a 500× resolution, while the right side corresponds to a higher 5000× resolution. 5000× resolution characterizes a micro-pores scale. Statistical information of these samples was used for a 3D isotropic reconstruction. However, the micro-structure at 500× resolution has elements with a longer dimension than the CL thickness. These samples were reconstructed in two stages: (i) representative elements with dimensions greater than 1/3 of the CL thickness, were considered as anisotropic gaps, therefore they were first reconstructed in a 2D isotropic way and then "extruded" to the CL thickness, and (ii) representative elements with a dimension less than 1/3 of the CL thickness were reconstructed in a 3D isotropic way.

### 2.2.5.2 *Structural reconstruction – annealing method*

As mentioned earlier, once experimental information such as the statistical features of CL images presented above, is available, a different path for reconstruction and simulation of transport properties can be implemented using the "annealing" strategy. In this chapter, the simulated annealing method (Torquato, 2002) is applied in the micro-structural reconstruction of a CL. This method generates a system that has the same statistical correlation functions as a specified reference system, i.e. the generated system is then an "image" of the real sample. It involves finding a state of minimum "energy" by interchanging the phases of the pixels in a digitized system. The energy ($E$) is defined in terms of the squared difference of the reference ($f(r)$) and simulated ($f'(r)$) correlation functions:

$$E = \sum \alpha_f \sum [f'(r) - f(r)]^2 \qquad (2.33)$$

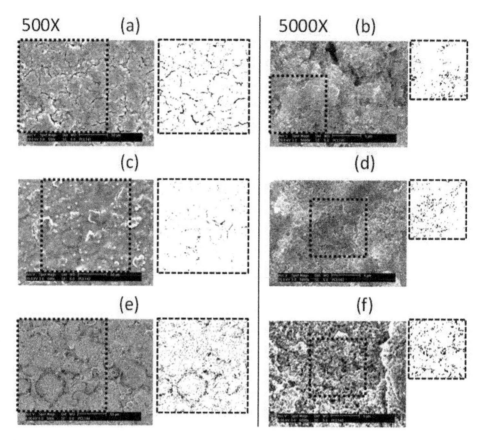

Figure 2.12.   Original and processed images of some real experimental CLs with different pore promoter additive concentrations during manufacture.

where $\alpha_f$ is an arbitrary "weight" that assigns the relative importance of each individual correlation function. First, the algorithm generates a stochastic micro-structure with the same volume fraction of the reference system, as in the initial structure. After this initial step, the states of two random pixels of different phases are interchanged. The energy change ($\Delta E = E'-E$) between the two successive states is then computed. This phase interchange is accepted with a related probability equal to $P(\Delta E)$:

$$P(\Delta E) = 1, \quad \text{if } (\Delta E \leq 0); \quad P(\Delta E) = \exp(-\Delta E/T), \quad \text{if } (\Delta E > 0) \qquad (2.34)$$

where $T$ is a fictitious "temperature". This method causes $f's(r)$ to converge gradually to $fs(r)$. The algorithm ends when the energy $E$ is less than a tolerance value. The concept of finding the lowest error state (lowest energy) by simulated annealing is based on a well known physical fact: if a system is heated to a high temperature $T$ and then slowly cooled down to absolute zero, the system equilibrates to its ground state.

Two correlation functions can then be used: (i) the two-point correlation function ($S2F$) which is obtained by randomly "tossing" a line segment of $r$ length with a specific orientation and counting the number of times that the beginning ($x$) and the end ($x + r$) of the line fall in phase $j$. (ii) The lineal-path function (LPF) which is defined as the probability of finding a line segment with end points at $x$ and $x + r$ entirely in phase $j$. The S2F was applied at the two phases and the

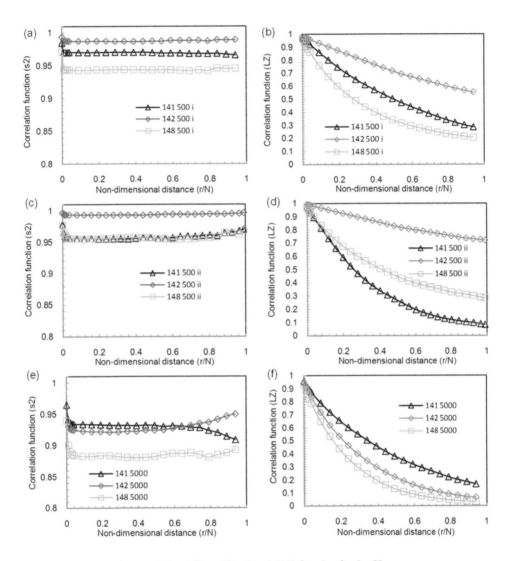

Figure 2.13.   Two-point correlation ($S2$) and lineal-path ($LP$) function for the CLs.

LPF only at the most relevant phase. In this way, the "total energy" was defined by:

$$E = 0.33[S2's - S2s]^2 + 0.33[S2'v - S2v]^2 + 0.33[LPF's - LPFs]^2 \qquad (2.35)$$

where $S2's$ and $S2'v$ are the simulated $S2F$ of the solid phase and the empty phase respectively; $S2s$ and $S2v$ are the reference $S2F$ of the solid phase and the empty phase respectively; $LPF's$ and $LPFs$ are the simulated and the reference $LPF$ of the solid phase; and 0.33 is the assigned function weight.

### 2.2.5.3   *Statistical functions – two scales*

Two-point correlation ($S2F$) and lineal-path ($LPF$) functions for the CLs studied are shown in Figure 2.13. Mean values are plotted *versus* a non-dimensional distance $r/N$, where $r$ is the distance between two points and $N$ is the domain lattice. Each curve corresponds to a different pore promoter additive Load (PCL) and is calculated for the "pseudo-solid" phase. Figures 2.13a

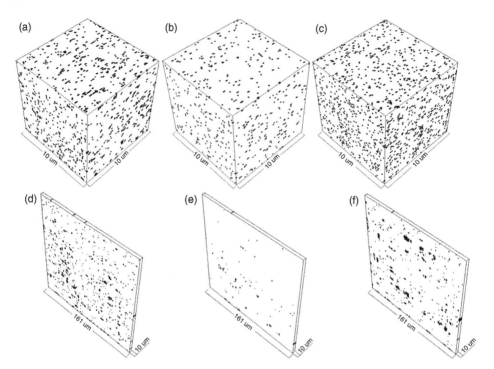

Figure 2.14.    Images of some reconstructed samples. Figures (a) and (d) correspond to PCL40, (b) and (e) correspond to PCL20, and (c) and (f) correspond to PCL00. The white color represents the pseudo-solid phase and black color, the empty phase.

and 2.13b correspond to the 500× resolution with elements of a dimension smaller than 1/3 of the CL thickness. Figures 2.13c and 2.13d correspond to 500× resolution with elements of a dimension greater than 1/3 of the CL thickness. Figures 2.13e and 2.13f correspond to a 5000× resolution. The left side of the figure corresponds to $S2F$ while the right side corresponds to $LPF$.

These statistical correlation functions were used as the reference systems in the annealing micro-structural reconstruction. Each sample was reconstructed and simulated five times with a different random series. In this reconstruction, relative porosities ($\Phi i$) were determined by the statistical analysis of SEM micrographs and verified by the total experimental porosity obtained by mercury intrusion porosimetry. CL thickness ($\varepsilon$) was also determined by SEM. CLs were studied by SEM at different resolutions.

Nevertheless, for the simulation study, two micrometric scales were empirically selected:

- "Superficial scale" at a 500× resolution, where we can obtain samples of ∼120 μm and
- "Micro-pores scale" at a 5000× resolution, where we can obtain samples of ∼12 μm.

It is important to point out that the nano-metric scale ("inside an agglomerate" scale) was not experimentally studied. Nevertheless, the CL internal micro-structure was characterized at 5000× resolution. In this scale we can define stochastic micro-pores but it is not possible to define CL primary components (Pt, C and ionomer). On the other hand, because the DECAL technique (used for the fabrication of CLs) generates very thin CLs (10–20 μm), the 500× resolution characterizes a macro-superficial and isotropic structure.

Figure 2.14 shows representative images of reconstructed samples. Figures 2.14a and 2.14d correspond to PCL40, 2.14b and 2.14e correspond to PCL20, and 2.14c and 2.14f correspond to PCL00. The white color represents the pseudo-solid phase and black color, the empty phase. At

Table 2.3. Electric conduction efficiency on recon-structed CLs with pore promoter.

| Sample | $\eta_{k,500}$ | $\eta_{k,5,000}$ | $\eta_{micro}$ | $\eta_{nano}$ |
|--------|-------|-------|-------|-------|
| PCL40 | 0.90 | 0.93 | 0.84 | 0.12 |
| PCL20 | 0.97 | 0.92 | 0.89 | 0.16 |
| PCL00 | 0.88 | 0.87 | 0.77 | 0.45 |

the 5000× resolution (top of figure), the reconstructed structures look alike in the three planes for each sample. In 500× resolution (bottom of figure) there are two representative structures: (i) elements with a dimension greater than 1/3 of the CL thickness (anisotropic gaps), and (ii) elements with a dimension smaller than 1/3 of the CL thickness (isotropic structure).

### 2.2.5.4 *Effective electric resistivity simulation from a reconstructed structure*
In order to calculate the effective resistivity ($\rho_{eff}$) of a heterogeneous material composed of a conductive phase and one or more insulating phases of an element formed by subdomains of smaller scales, we use the following relation:

$$\rho_{eff} = \rho_m \quad \prod L_{eff\_i}/A_{eff\_i} \quad \prod A_{m\_i}/L_{m\_i} \tag{2.36}$$

where $\rho_{eff}$ is the effective resistivity of the global domain formed by various smaller scale sub-domains, i, $\rho_m$ is the material's resistivity, $L_{eff\_i}$ is the effective length and $A_{eff\_i}$ is the effective area of the phase under study in every subdomain, and $L_{m\_i}$ is the length and $A_{m\_i}$ the area of the scale of every subdomain.

To normalize and generalize the results in this section, calculated resistivities are used to estimate a "conduction efficiency" ($\eta_k$). As conductivity is the inverse of resistivity, the CL's effective conductivity is the inverse of the effective resistivity value $k_{eff} = \rho_{eff}^{-1}$. Then the conduction efficiency $\eta_k$ is calculated by comparing the effective conductivity with a reference or nominal conductivity, as described by Equation (2.37). By substituting Equation (2.36) in Equation (2.37), one can obtain a relation that provides the conduction efficiency of the global domain which is formed by several subdomains (Equation (2.38)):

$$\eta k = k_{eff}/k_m \tag{2.37}$$

$$\eta k = \prod A_{eff\_i}/L_{eff\_i} \prod L_{m\_i}/A_{m\_i} \tag{2.38}$$

The electric conduction efficiency ($\eta_k$), obtained by direct simulation of the charge continuity equation in 500× ($\eta_{k,500}$) and 5000× ($\eta_{k,5000}$) resolutions, can be "generalized" by its product in a single "micrometric efficiency parameter" ($\eta_{micro} = \eta_{k,500}\,\eta_{k,5,000}$). In this way, the electric conduction efficiency of the nano-metric structure can be "estimated" by:

$$\eta_{nano} = k_{eff}/k_m/\eta_{micro} \tag{2.39}$$

where $k_{eff}$ is the effective electric conductivity of the material (electric conductivity of CLs can be experimentally verified by different well-established methods), and $k_m$ is the electric conductivity of the pseudo-solid phase, (for example Vulcan carbon conductivity could be used, $k_m \approx 400\,\text{S}\,\text{m}^{-1}$). Table 2.3 summarizes the electric conduction efficiencies estimated in the studied scales ($\eta_{k,500}$ and $\eta_{k,5,000}$), the generalized micrometric efficiency ($\eta_{micro}$), and the estimated nano-metric efficiency ($\eta_{nano}$).

At the micrometric scale, PCL20 shows higher $\eta_{micro}$ value than the other two manufactured CLs. However, experimental results shows that PCL00 has the highest global conduction efficiency ($\eta_k$). This suggests that the nano-metric scale has a significant influence in the electric global performance. For the examples calculated here, the conduction efficiency increases while the pore promoter load diminishes. On the other hand, the sample without pore promoter (PCL00) is particularly different from other CLs. Such pieces of information can help a PEMFC designer verify or establish adequate mechanisms for better physical overall properties.

## 2.3  NEW MATERIAL SUPPORT AND NEW CATALYST APPROACHES

### 2.3.1  *Carbon nanotubes "decorated" with platinum*

Carbon nanotubes have been around for a few years now and they have been the subject of plenty of research work (Iilima, 1991; Li *et al.*, 2002; Liu *et al.*, 2002; 2005; Lordi *et al.*, 2001; Planeiz *et al.*, 1994; Rajesh *et al.*, 2000; Tae *et al.*, 2012; Terrones, 2009; Vielstich *et al.*, 2009; Xin *et al.*, 2006; Xue *et al.*, 2001; Yu *et al.*, 1998). This explores their use as a support for fuel cell catalysts, as they promote larger available surface areas compared with the traditional support of a PEMFC, normally carbon black particles of various sizes in the micron-scale. Carbon nanotubes can be built from one layer of graphene, a network of carbon atoms with a particular atomic arrangement which can be simplistically seen as a "carbon grid". They can also be manufactured from several layers or "walls" of such graphene, which gives place to the common types of nanotubes, either single-walled nanotubes (SWNTs) or multi-walled nanotubes (MWNTs). Although specific areas have been reported for Vulcan XC-72® and for MWNT (235 $m^2 g^{-1}$ and 100 $m^2 g^{-1}$ respectively), the unavailable pores of the carbon black, which confer such a high value, are less favorable for Pt supporting sites. Meanwhile the large MWNTs' area, from the many small radii tubes, is largely available and this favors the fabrication of catalyst layers with highly dispersed platinum and therefore higher catalyst utilization, as well as higher Pt/C ratios, also favoring thinner and less electrically resistant CLs. MWNTs have shown better corrosion resistance compared with carbon black. Despite that, in practice there have been difficulties in preparing such MWNT-supported CLs as in some cases Pt particles tend to aggregate. From the manufacturing point of view, inks prepared with nanotubes have different rheological properties affecting manufacturing process parameters, which might eventually affect the final CL structure and, therefore, its functionality.

There have been also reports where catalyzers are said to be more stable if supported by CNTs. For example, it has been reported (Lee *et al.*, 2012) the loss of oxygen-related functional groups which are known to accelerate carbon corrosion and Pt agglomeration. In that study, authors observed an effect of catalytic cleavage of hydrogen molecules followed by spillover and a dehydration reaction.

For these reasons, it is of great importance to explore the effects on CL performance not only of variables like Pt content and ionomer content, but also of CNT morphology features, particularly from a structural point of view. As seen in Figures 2.2 and 2.16, different types of carbon support structures (either particles or CNTs) for catalysts, may produce different morphologies which will therefore affect CL properties, particularly structural features, which in turn influence transport phenomena. Such differences in the morphology of supporting material should be taken into consideration during CL design, prediction of properties (simulation), fabrication, and testing.

CNTs, on the other hand, come in various formats, including bundles, straight nanotubes, open nanotubes, or the so-called "unzipped" nanotubes, where the inner areas are more easily available than closed structures. As mentioned earlier, nanotubes are single sheets of graphite rolled up into cylinders, which can be opened along their axis and even flattened in "flake" forms. Handling tubes or such sheets will definitely have different structural effects with operational consequences (namely effective transport properties) and therefore, they need to be studied using computational reconstruction tools just as we did in the past sections.

#### 2.3.1.1  *Substantial differences for CNT structures*

Under the scaling approach used in previous sections of this chapter, it must be recognized that the nano-scale is the only scale that contains more than two solid phases. In addition, it is the only place where primary material structure plays a significant role. At this scale, electrodes' electro-active sites are determined. For all these reasons, this nano-scale can limit or favor to a large degree the fuel cell electrode performance.

From the structural point of view, the selection of a supporting material will be one of the major factors that will ultimately determine the ETCs of a CL. The morphology and the size of such support will be of paramount importance for the formation of specific micro-structures with improved and enhanced effective properties. Nano-structured materials are certainly a focus on many attempts aimed at achieving this. As the reaction taking place in a PEMFC's CL occurs on the surface, clearly the CL's electrochemical activity will be improved by increasing the available catalyst area on a nano-metric structure provided by CNTs. As mentioned above, carbon nanotubes are one of the main R&D targets these days, including their properties (Huang *et al.*, 2006; Lee, 2000; Shunin and Schwartz, 1997). As the electro-cataylzer dispersion on CNTs has shown larger proportions than on conventional carbon blacks, this has promoted a larger surface area available for a reaction, that is, CNTs offer, in principle, excellent conditions at least as supporting material for PEMFCs, by potentially increasing available area for reaction (active sites). Besides, CNTs are relatively inexpensive to produce.

Distribution of CNTs in an RHM micro-structure can be digitally reconstructed in different ways and using different hypothesis, but one must be sure that assumptions are the closest to real conditions, including the possibility of using experimental information. It must be remembered that different CNT structures will include different structural properties (e.g. chirality and length/diameter ratios) and different morphologies, as they can be bundles or well-aligned tubes, which are not so stochastic, therefore they might not be considered as a RHM. For example, one can assume that there exists an initially related "prevailing structure" but not always identical to the primary element. Such structural features can be represented, for example, by a geometrical equation where for CNTs a given diameter and length are known. Nevertheless, during the manufacture of a fuel cell electrode, CNTs could be slightly "bent" by the physicochemical conditions in the manufacturing process. For the computational reconstruction, and based on experimental information, one can use algorithms that take into consideration and simulate such deformations. One could, or maybe should, consider factors such as (i) orientation, that is, a structure with a preferential order or degree of disorder, e.g. aligned CNTs *versus* whiskers of CNTs; (ii) distance between elements and their statistical distribution; (iii) deformation degree, e.g. a deflected angle, a rotational/translational move when another element is encountered, among other aspects.

In an RHM the distribution of the possible structures could be uniform, where uniformity is a systematic parameter defined in the computational domain only. This observation allows both stochastic and deterministic conditions, as well as isotropic or anisotropic features to be considered. By reaching this point, the randomness of the final structure will be affected by initial conditions that introduce some deterministic touches to the final reconstructed heterogeneous material. Some approaches have proposed the use of electrodes with rather deterministic structures for electrochemical processes, for example, parallel and aligned CNTs, "dots" of controlled diameter catalysts supported on flat substrates, etc.

The location of catalysts on supporting structures based on CNTs has recently followed two approaches. Conventional CNTs will provide their surface for the catalyst to be attached. Unzipped CNTs (Long *et al.*, 2011) offer also sites for such attachment on both external and internal surfaces, apparently increasing the electrochemical active area. This spatial location of the catalyst will have to be considered in a subroutine, as well as an openness factor when "building" the CNT structure for the latter case. For this, CNTs' chiral parameters may be used. It is worth remembering that typical single-wall carbon nanotubes are said to be made by wrapping up a graphene monolayer where equivalent hexagonal lattice sites match, to close the newly formed tube. Such a monolayer is rolled up at a particular "chiral" angle and a defined wrapping vector C, describes the relative location of the two matching sites by two integers, $n$ and $m$. Depending on the integer values the nanotube is referred to as having an "armchair" or a "zigzag" chirality. The nature of this feature has a strong effect on the electrical properties of the nanotube, going from metallic for $n = m$

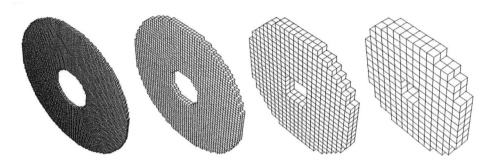

Figure 2.15.   Effect of resolution level for basic reconstruction of a MWCNT (cross section).

Table 2.4.   CV required for the generation of a three dimension domain at 300 nm scale for a CNT of outer diameter of 50 nm and 200 nm length.

| $dX$ nm | NCV domain ($Nx = 300$ nm) | CV CNT, $D = 50$ nm; $L = 200$ nm |
|---|---|---|
| 0.5 | 216,000,000 | 2,140,400 |
| 1.0 | 27,000,000 | 359,200 |
| 2.0 | 3,375,000 | 41,200 |
| 3.0 | 1,000,000 | 12,200 |

(armchair) to semiconducting characteristics. The apparent effect that such a property has, not only when reconstructing CL structures but also, on the final calculated ETCs, is obvious.

For the CL structure's reconstruction, the micro-structure resolution will be directly proportional to the number of volumes required for the digital generation and to the computing requirements for realizing the simulation. As a simplistic exercise, Figure 2.15 shows the cross section of a closed MWCNT with a constant external diameter of 50 nm and a nominal inner diameter of 15 nm. If one assumes that a graphene monolayer is 0.32 nm thick, and assumes that a multi-wall thickness of such structure is simply the sum of the different layers, then the thickness wall for this structure would consist of around 100 graphene layers. Having said that, it can be shown that these structures may be generated at different resolution levels: a) $dx = 0.5$ nm, b) $dx = 1$ nm, c) $dx = 2$ nm y, and d) $dx = 3$ nm. It is also important to notice that at this scale the resolution could be reaching an atomic scale, so again, caution should be practiced as to how simulations will take place regarding scale-related properties of components, as scaling stochastic reconstruction might need particular considerations according to the particular system.

For this example, the number of control volumes necessary to generate a computing domain of 300 nm scale in three dimensions (CV CNT) for an MWCNT structure of 50 nm diameter and 200 nm length. It was mentioned earlier that the diameter-to-length ratio for CTs could be around values of 200,000 or even larger. This gives an idea of the computing domain necessary for the reconstruction with various resolution levels.

Other considerations will arise under the reconstruction approach when it comes to "rebuilding" CNTs for CL structures, these being either aligned, fibrous, whisker-like, flaked, or unzipped features, etc. Such features can be previously defined and designed pixel by pixel in an algorithm, according to previously known morphological characteristics of the element, as well as the resulting distribution, random or not, for the computing domain.

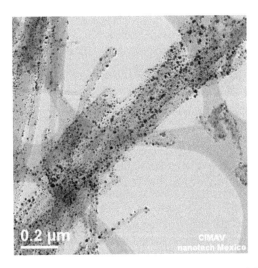

Figure 2.16.   A Pt nano-catalyst on an MWCNT prepared by a colloidal method. Some CNTs seem to have straight geometries while others clearly bend, causing irregular structures.

Despite what has been previously stated regarding potential burdens associated with the initial support material structure, at microscopic levels such materials could tend to form agglomerates, causing micro-structural changes affecting the global performance of the electrodes, especially those related with transport properties. Therefore, the structure of primary elements, i.e. electrical supporting structures (CNTs, CBs, graphene, etc.), the catalyst structure, and even ionomer physical properties like density, viscosity, surface tension, mechanical strength, etc., will greatly influence CL structure and its effective properties, but this will also define the simulation approach.

### 2.3.1.2   *CNT considerations when inputting component properties*

Various tools, like the techniques described in this chapter, make use of macro properties to simulate the effective properties of composite structures. At some scales, those tools will normally give sufficient accuracy to determine effective properties, thanks to the nature of the simulated property at that scale. As seen in the previous section, at the nano-scale, some properties might need the use of quantum mechanics to predict the properties of a composite material as some components show properties such as current transport that are better described by such theories (Lee, 2000; Shunin and Schwartz, 1997), for example, it is well known that graphene may develop a resistivity of $10^{-6}\Omega$ cm (derived from early experiments on electron mobility graphite), but its manufacturing process as well as impurities cause different macroscopic electrical properties.

The length-to-diameter ratio in CNTs may extend to such ranges like 132,000,000:1 while keeping their cross sectional diameter in the nano-metric scale, resulting in propagation of electrons only along the tube's axis with the resulting quantum effects. That is why carbon nanotubes are frequently regarded as one-dimensional conductors. When these considerations are taken into account for the prediction of properties using nano-sized components, it looks like ballistic conduction models might be useful or even needed in order to predict properties like electrical conductivity.

When the same approach is used to simulate thermal conductivity, quantum theory might be needed again to predict the resulting thermal properties of composite structures. On the other hand chirality, diameter, and the thickness of, for example, multi-walled carbon nanotubes, as well as morphology (both non-aligned and random – whiskers or straight and curved nanotubes), may play an important role in how to approach the effective determination of properties. Finally, the conductance at the metal catalyst-nanotube junction in the catalyst layers may be different from what is observed in the two separate components due to synergistic effects. Both models and different theories, like the quantum theory and the electronic density of states (Lee, 2000; Wallace, 1947), can still be included as appropriate methods for the prediction of the effective properties of catalyst layers that make use of new nano-scopic supports like carbon nanotubes. These offer a whole new world of structures with many different structural possibilities and, therefore, with potentially numerous effective properties.

### 2.3.2  *Core-shell-based catalyzers*

#### 2.3.2.1  *General considerations for reconstruction*

The new so-called "core-shell" catalyst approach, offering new options in better performance design catalysts, is generating growing interest for new catalyst layers, because these materials have shown promising paths in replacing expensive noble metals such as platinum (Mazumder *et al.*, 2010; Newsroom: Media and Communications Office of Brookhaven National Laboratory, 2010). It has been claimed that such structures not only increase catalytic activity by four times higher, but researchers have demonstrated that they achieve better resistance to dissolution under conditions encountered in transport applications (Debe, 2012) and this is one of the most appealing fuel cell markets. Core-shell catalysts are usually comprised of a material forming a center while a very thin shell is formed around it, the latter normally being a more active metal. The less expensive core makes it possible to reduce the amount of, for example, platinum, at the exterior of such particles.

From a structural point of view, we are typically talking about sphere-like particles with a known or a designed size distribution. This may facilitate any structural reconstruction, as spheres are amongst the most commonly studied particle shapes and interactions in industrial applications. Nevertheless, it has been observed that there are interactions between both materials (core and shell) that increase the catalytic property of the material. Again, at this nano-scale, classical physics might not be the way forward and interactions at the energy level should be considered in any attempt to predict effective properties in structures made with core-shell catalysts.

Some approaches that extend the range of new properties to core-shell catalysts include the fact that they can be made in different sizes either at the core or with different thicknesses of the shell. Shells can be full "skin" structures or partially covered surfaces and even partially open shell structures. To make things more complex, during the life-cycle of the catalysts, these components suffer from changes due to the partial dissolution of either the shell or their core material (dealloying) introducing new structural properties followed by new core-shell interactions, as well as changes in local stress crystallographic levels.

## 2.4  CONCLUDING REMARKS

Nano-technology seems to be the meeting place for engineering and nano-science. When scientists and engineers discovered the nano-world, they also discovered a new universe where not only isolated materials behave differently at that scale, but they also interact in a new fashion, creating unique synergies. The structures that nano-materials can form have also shown that such "architectural" features need to be taken into consideration before new materials can be directly applied. Then the need for simulation of such structures emerges as one way to understand the creation of new properties through the combinations of nano-structures.

Despite the fact that we already enjoy nano-technology in different applications, many new possibilities are still under study, as we are eager to understand laws and principles of how things work. Nano-technologies have altered the way we conduct scientific research, but they also have affected the way engineers create solutions. In this new realm of knowledge, some tools have been developed to simulate the construction of nano-structures and to discover how their "bricks", "walls", and "pipes", aligned or not, define the functionality of such structures. These tools are also helping us understand how individual properties, compositions, and synergies among individual components, as well as methods of manufacturing nano-technology-based products, can be controlled or designed to improve their performance.

Scale matters. We have seen that scale may be used to facilitate reconstruction of structures with nano-components, but it has also shown that scale is important when simulation takes place. When calculated correctly, properly, or if you like, usefully, transport effective coefficients can be determined and even compared to experimental data. However, in some cases new approaches may need to be considered. Here, approaches like mesoscopic physics, or a model of multiple scattering with effective media approximation (EMA) for condensed matter, based on the approach of atomic cluster, may play important roles. Recently, a review (Debe, 2012) was discussed on the different approaches that scientists and fuel cell developers in general, are using in order to have better and cheaper catalysts. Many have made a great impact on CL structures. Some approaches included supporting material but others considered unsupported catalysts too. The aspect ratio of particles has been recognized as a relevant factor. Metallic membranes, meshes, and bulk materials have also been considered of which the structural features will impact on the final structure and functionality of fuel cell technology. Local structures and at different levels of scale are still subjects of interest in many scientific works (Soboleva *et al.*, 2010).

Nano-engineering aims to make the most out of nano-material's inherent properties by developing and using tools such as those used in the reconstruction of CL structures of PEM fuel cells. In doing so scientists and engineers promote the right effective properties in order to find and promote favorable conditions and interactions, that is, good structures and better synergies, with the right content of primary components and their correct distribution. A CL is a clear example of the complexities that engineers face when it comes to optimizing nano-technologies, as the heart of catalyst layers can perform within a huge spectrum of efficiencies and economies, with the very same basic materials.

## ACKNOWLEDGEMENTS

U. Cano-Castillo wishes to thank Instituto de Investigaciones Eléctricas of Mexico for their support during realization of this work. Romeli Barbosa wishes to thank Universidad de Quintana Roo in Mexico for their support during the writing of this manuscript. Finally, the authors want to thank Dr. Beatriz Escobar from Instituto Tecnológico de Cancún, Mexico for facilitating the TEM image in Figure 2.16, and Dr. Antonino Aricòat CNR-ITAE of Italy for micrographs in Figure 2.2.

## REFERENCES

Arico, A.S., Stassi, A., Modica, E., Ornelas, R., Gatto, I., Passalacqua, E. & Antonucci, V.: Performance and degradation of high temperature polymer electrolyte fuel-cell catalysts. *J. Power Sources* 178 (2008), pp. 525–536.

Barbosa, R., Gatto, I., Squadrito, G., Orozco, G., Arriaga, L.G., Ornelas, R., Antonucci, V. & Passalacqua, E.: Cyclic current profile performance of proton exchange membrane fuel-cells in stationary applications. *ECS Trans.* 11 (2007), pp. 1527–1523.

Barbosa, R., Andaverde, J., Escobar, B. & Cano, U.: Stochastic reconstruction and a scaling method to determine effective transport coefficients of a proton exchange membrane fuel cell catalyst layer. *J. Power Sources* 196 (2011a), pp. 1248–1257.

Barbosa, R., Escobar, B., Cano, U., Pedicini, R., Ornelas, R. & Passalacqua, E.: Stochastic reconstruction at two scales and experimental validation to determine the effective electrical resistivity of a PEMFC catalyst layer. *ECS Trans.* 41:1 (2011b), pp. 2061–2071.

Boyer, C., Gamburzev, S., Velev, O., Srinivasan, S. & Appleby, A.J.: Measurements of proton conductivity in the active layer of PEM fuel-cell gas diffusion electrodes. *Electrochim. Acta* 43 (1998), pp. 3703–3709.

Broka, K. & Ekdunge, P.: Modelling the PEM fuel-cell cathode. *J. Appl. Electrochem.* 27 (1997), pp. 281–289.

Capek, P., Hejtmánek, V., Brabec, L., Zikánová, A. & Kocirík, M.: Stochastic reconstruction of particulate media using simulated annealing: Improving pore connectivity *Transp. Porous Media* 76 (2009), pp. 179–198.

Debe Mark, K.: Electrocatalyst approaches and challenges for automotive fuel-cells. *Nature* 486 (2012), pp. 43–51.

Escobar, B., Gamboa, S.A., Pal, U., Guardián, R., Acosta, D., Magaña, C. & Mathew, X.: Synthesis and characterization of colloidal platinum nanoparticles for electrochemical applications. *Int. J. Hydrogen Energy* 35 (2010), pp. 4215–4221.

Huang, Y., Wu, J. & Hwang, K.C.: Thickness of graphene and single-wall carbon nanotubes. *Phys. Rev.* B 74 (2006), pp. 245,413–245,422.

Iijima, S.: Helical microtubules of graphitic carbon. *Nature* 354, (1991), pp. 56–58.

Jiao, Y., Stillinger, F.H. & Torquato, S.: Modeling heterogeneous materials via two-point correlation functions. II. Algorithmic details and applications. *Phys. Rev.* E 77 (2008), pp. 031135.1–031135.15.

Kamarajugadda, S. & Mazumder, S.: Numerical investigation of the effect of cathode catalyst layer structure and composition on polymer electrolyte membrane fuel-cell performance. *J. Power Sources* 183 (2008), pp. 629–642.

Kim, S.H. & Pitsch, H.: Reconstruction and effective transport properties of the catalyst layer in PEM fuel-cells. *J. Electrochem. Soc.* 156 (2009), pp. B673–B681.

Ledesma-García, J., Barbosa, R., Chapman, T.W., Arriaga, L.G. & Godínez Luis A.: Evaluation of assemblies based on carbon materials modified with dentrimers containing platinum nanoparticles for PEMFC. *Int. J. Hydrogen Energy* 34 (2009), pp. 2008–2014.

Lee, J.-O., Park, C., Kim, J.J., Kim, J., Park, J.W. & Yoo, K.-H.: Formation of low-resistance ohmic contacts between carbon nanotube and metal electrodes by a rapid thermal annealing method. *J. Phys.* D: *Appl. Phys.* 33 (2000), pp. 1953-1956.

Lee T.K., Jung J.H., Kim J.B. & Hur S.H.: Improved durability of Pt/CNT catalysts by the low temperature self-catalyzed reduction for the PEM fuel-cells. *Int. J. Hydrogen Energy* 37, (2012), pp. 17,992–18,000.

Li, W.Z., Liang, C.H., Qiu, J.S., Zhou, W.J., Han, H.M., Wei, Z.B., Sun, G.Q. & Xin Q.: Carbon nanotubes as support for cathode catalyst of a direct methanol fuel cell. *Carbon* 40 (2002), pp. 791–794.

Litster, S. & McLean, G.: PEM fuel-cell electrodes review. *J. Power Sources* 130 (2004), pp. 61–76.

Liu, Z.L., Lin, X., Lee, J.Y., Zhang, W., Hang, M. & Gan L.M.: Preparation and characterization of platinum-based electrocatalysts on multiwalled carbon nanotubes for proton exchange membrane fuel cells. *Langmuir* 18 (2002), pp. 4054–4060.

Liu, Z.L., Gan, L.M., Hong, L., Chen, W.X. & Lee, J.Y.: Carbon-supported Pt nanoparticles as catalysts for proton, exchange membrane fuel-cells. Short communication in *J. Power Sources* 139 (2005), pp. 73–78.

Long, D., Li, W., Qiao, W., Miyawaki, J., Yoon, S.H., Mochida I. & Ling L.: Partially unzipped carbon nanotubes as a superior catalyst support for PEM fuel-cells. Electronic Supplementary Material (ESI) for *Chem. Commun.* 47:33 (2011), The Royal Society of Chemistry, 2011.

Lordi, V., Yao, N. & Wei, J.: Method for supporting platinum on single-walled carbon nanotubes for a selective hydrogenation catalyst. *Chem. Mater.* 13 (2001), pp. 733–737.

Matsumoto, M. & Nishimura T.: Mersenne Twister: A 623-dimensionally equidistributed uniform pseudo-random number generator. *ACM Trans. Modell. Comput. Simul.* 8 (1998), pp. 3–30.

Mazumder, V., Chi, M.F., More, K.L. & Sun, S.H.: Core/shell Pd/FePt nanoparticles as an active and durable catalyst for the oxygen reduction reaction. *J. Am. Chem. Soc.* 132 (2010), pp. 7848–7849.

National Laboratory: New Highly Stable Fuel-Cell Catalyst Gets Strength from its Nano Core, Palladium core protects precious platinum; enhances reactivity/stability, November 10, 2010, Newsroom: Media & Communications Office of Brookhaven National Laboratory, 2010.

Passalacqua, E., Lufrano, F., Squadrito, G., Patti, A. & Giorgi L.: Nafion content in the catalyst layer of polymer electrolyte fuel-cells: effects on structure and performance. *Electrochim. Acta* 46 (2001), pp. 799–805.

Patelli, E. & Schuëller, G.: On optimization techniques to reconstruct microstructures of random heterogeneous media. *Comput. Mater. Sci.* 45 (2009), pp. 536–549.

Planeix, J.M., Coustel, N., Coq, B., Brotons, V., Kumbhar, P.S., Dutartre, R., Geneste, P., Bernier, P. & Ajayan, P.M.: Application of carbon nanotubes as supports in heterogeneous catalysis. *J. Am. Chem. Soc.* 116 (1994), pp. 7395–7936.

Quiblier, J.A.: A new three-dimensional modelling technique for studying porous media. *J. Colloid Interf. Sci.* 98 (1984), pp. 84–102.

Rajesh, B., Thampi, K.R., Bonard, J.M. & Viswanathan, B.: Preparation of a Pt–Ru bimetallic system supported on carbon nanotubes. *Mater J. Chem.* 10 (2000), pp. 1757–1759.

Shah, A.A., Kim, G.-S., Gervais, W., Young, A., Promislow, K., Li, J. & Ye, S.: The effects of water and microstructure on the performance of polymer electrolyte fuel-cells. *J. Power Sources* 160 (2006), pp. 1251–1268.

Scheiba, F., Benker, N., Kunza, U., Rotha, C. & Fuess H.: Electron microscopy techniques for the analysis of the polymer electrolyte distribution in proton exchange membrane fuel-cells. *J. Power Sources* 177 (2008), pp. 273–280.

Schulz, V.P., Becker, J., Wiegmann, A., Mukherjee, P.P. & Wang, C.Y.: Modelling of two-phase behavior in the gas diffusion medium of PEFCs via full morphology approach. *J. Electrochem. Soc.* 154 (2007), B419–B426.

Secanell, M., Karan, K., Suleman, A. & Djilali, N.: Multi-variable optimization of PEMFC cathodes using an agglomerate model. *Electrochim. Acta* 52 (2007), pp. 6318–6337.

Shunin, Yu. N. & Schwartz, K.K.: High resistance of joints CNT/Me, In: R.C. Tennyson, & A.E., Kiv (eds): *Computer modelling of electronic and atomic processes in solids.* Kluwer Acad. Publishers, Dodrecht/Boston/London, 1997, pp. 241–257.

Soboleva, T., Zhao, X.S., Malek, K., Xie, Z. & Navessin, T. & Holdcroft S.: On the micro-, meso-, and macroporous structures of polymer electrolyte membrane fuel-cell catalyst layers. *Appl. Mater. Interfaces* 2:2 (2010), pp. 375–384.

Sun, W., Peppley, B.A. & Karan, K.: An improved two-dimensional agglomerate cathode model to study the influence of catalyst layer structural parameters *Electrochim. Acta* 50 (2005), pp. 3359–3374.

Terrones, M.: Nanotubes unzipped. *Nature* 458 (2009), pp. 845–846.

Torquato, S.: *Random heterogeneous materials: Microstructure and macroscopic properties.* Springer-Verlag, New York, 2002.

Uchida, M., Fukuoka, Y., Sugawara, Y., Eda, N. & Ohta, A.: Effects of microstructure of carbon support in the catalyst layer on the performance of polymer-electrolyte fuel-cells. *J. Electrochem. Soc.* 143 (1996), pp. 2245–2252.

Vielstich, V., Gasteiger, H.A. & Yokokawa, H.: *Handbook of fuel-cells: Advances in electrocatalysis, materials, diagnostics and durability.* John Wiley & Sons, V 5 and 6, 2009.

Wallace, P.R.: The band theory of graphite. *Phys. Rev.* 71:9 (1947), 622W.

Wang, Q.P., Eikerling, M., Song, D.T. & Liu, Z.S.: Structure and performance of different types of agglomerates in cathode catalyst layers of PEM fuel-cells. *J. Electroanal. Chem.* 573 (2004), pp. 61–69.

Wang, X., Li, W.Z., Chen, Z.W., Waje, M. & Yan, Y.S.: Durability investigation of carbon nanotube as catalyst support for proton exchange membrane fuel-cell. Short communication in *J. Power Sources* 158 (2006), pp. 154–159.

Wilson, M.S.: Membrane catalyst layer for fuel-cells. 1993 U.S. Pat. No. 5,234,777.

Xie, J., Wood III, D.L., More, K.L., Atanassov, P. & Borup R.L.: Microstructural changes of membrane electrode assemblies during PEFC durability testing at high humidity conditions. *J. Electrochem. Soc.* 152 (2005), pp. A1011–A1020.

Xue, B., Chen, P., Hong, Q., Lin, J.Y. & Tan, K.L.: Growth of Pd, Pt, Ag and Au nanoparticles on carbon nanotubes, *J. Mater. Chem.* 11 (9), (2001), pp. 2387–2381.

Yan Q. & Wu J.: Modeling of single catalyst particle in cathode of PEM fuel-cells. *Energy Convers. Manage.* 49 (2008), pp. 2425–2433.

Yu, R.Q., Chen, L.W., Liu, Q.P., Lin, J.Y., Tan, K.L., Ng, S.C., Chan, H.S.O., Xu, G.Q. & Hor, T.S.A.: Platinum deposition on carbon nanotubes via chemical modification. *Chem. Mater.* 10 (1998), pp. 718–722.

Zhao X., Yao J. & Yi Y.: A new stochastic method of reconstructing porous media. *Transp. Porous Media* 69 (2007), pp. 1–11.

# CHAPTER 3

## Multi-scale model techniques for PEMFC catalyst layers

Yu Xiao, Jinliang Yuan & Ming Hou

## 3.1 INTRODUCTION

### 3.1.1 *Physical and chemical processes at different length and time scales*

Polymer electrolyte membrane fuel cells (PEMFCs) fueled by hydrogen are considered to be most suitable for automotive applications owing to their fast dynamics and high power densities. However, the performances are limited primarily by processes in PEMFCs appeared in the catalyst layers (CLs), which highly depend on the microscopic structure-related properties. The CLs at both the anode and cathode have complex and multi-phase porous structures, as shown in Figure 3.1. Platinum (Pt) nano-particles are supported by larger carbonaceous substrates, constructing the framework of the agglomerate with the primary pores of 3–10 nm. The Pt clusters range 2–5 nm and the most commonly employed carbon support, Vulcan XC-72, has a diameter of 20–30 nm. The spaces between the agglomerates are the secondary pores with sizes in the range of 10–50 nm (Lim *et al.*, 2008). The ionomer in the CLs serves as both a binder and the pathway for the protons generated/consumed in the electrochemical reactions. In order to reduce the risk of water flooding, polytetrafluoroethylene (PTFE) is commonly used to bind the catalyst particles and form a hydrophobic path (Buryachenko *et al.*, 2005). The typical thickness of a CL ranges from 10 to 30 microns and is a function of the material composition and the fabrication processing.

There is no single, perfect, and all-comprising model for predicting fuel cell properties on all length- and time scales. As shown in Figure 3.2, the density functional theory (DFT) can be applied at the atomistic scale ($10^{-10}$ m) chemical reactions in the three-phase boundary (TPB); the molecular Dynamics (MD) and Monte Carlo (MC) methods, based on classical force fields, can be employed to describe individual atoms or clusters of catalyst materials at the nano-/micro-scale ($10^{-7}-10^{-9}$ m); the particle-based methods (e.g. DPD) or mesh-based methods, for example Lattice-Boltzmann (LB), are used to solve the complex fluid flows in the porous media at the meso-scopic scale ($10^{-6}-10^{-3}$ m); and at the macroscopic scale ($>10^{-3}$ m), continuum models are available for structural and coupled fluid-structural simulation in the optimum design for PEMFC. Thus, due to the limit of computational efficiency, it is important to make sure which phenomena and properties one is primarily interested in. The properties are usually determined by structural hierarchies and processes on very different length- and associated timescales in PEMFC. An efficient modeling technique of the PEMFC requires special consideration of the relevant physicochemical processes at the respective length and time scales.

### 3.1.2 *Needs for multi-scale study in PEMFCs*

The objective of CL modeling is to establish relations between fabrication procedures and conditions, structure, effective properties of transport and reactions, and performance, which evolves over a wide range of scales, from atomistic-dominated processes at supported catalyst nano-particles to the operation in a fuel cell at the macroscopic device level. As discussed above, separate approaches in theory and modeling are considered at different scales, which allow one to focus on essential features under specific conditions. For example, atomistic scale models can predict physical properties of materials with ideal, theoretical or proposed micro-structures under

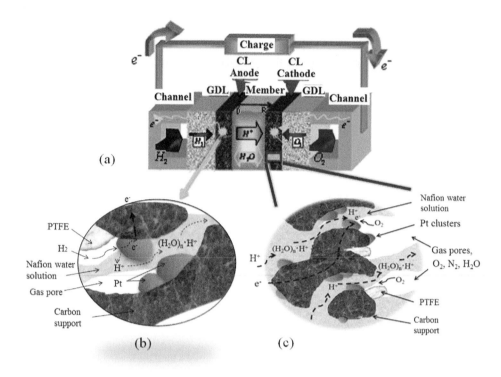

Figure 3.1.    (a) The principle diagram of a working PEMFC and the schematic pictures of the micro-structure and the multi-physics transport phenomena in both (b) anode and (c) cathode catalyst layers.

clearly defined conditions within the size of several nanometers. However, they are not able to probe the random morphology of the complete CLs due to computational limitations. Moreover, it is expected to use the multi-scale simulation strategies to bridge the models, as seamlessly as possible, from one scale to another. Thus, the calculated parameters, properties and numerical information can be efficiently transferred across the scales and passed on to PEMFC designers.

In the following, the key issues and related models at different scales in PEMFCs are outlined in Section 3.2. In Section 3.3, the multi-scale strategies across neighboring scales are summarized and compared. The typical applications based on coarse-grained molecular dynamics (CG-MD) method areal so exhibited to discuss the multi-scale modeling techniques in PEMFC. However, there are still some major challenges and limitations in these modeling and simulations, as summarized in Section 3.4.

## 3.2    MODELS AND SIMULATION METHODS AT DIFFERENT SCALES

### 3.2.1    *Atomistic scale models at the catalyst surface*

At the CLs of PEMFCs, the atomistic scale models are needed to describe the interactions between the atoms and then to predict the lattice structures and chemical reactions due to charge transfer. There are many well-known software packages used in materials science and quantum chemistry, such as ADF, AMPAC, ASE, CASTEP, HyperChem, SIESTA, GAUSSIAN, VASP, etc. Practically all of these codes are based on Density Functional Theory (DFT) and some are employed to analyze the TPB region in the CLs of PEMFC.

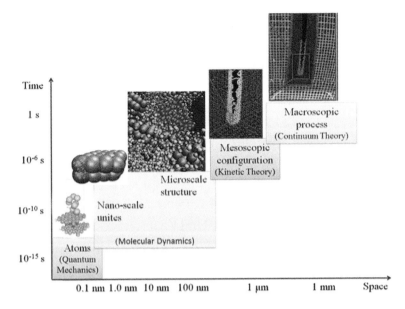

Figure 3.2.   Range of different length and time scales for multi-scale modeling.

### 3.2.1.1   *Dissociation and adsorption processes on the Pt surface*

In the anode, the interactions of hydrogen with Pt surfaces are particularly important and have been investigated in a large number of studies (Poelsema *et al.*, 2005). It was found that the reaction paths were without or with very low barriers leading to dissociation of $H_2$ on the Pt(111) surface (Olsen *et al.*, 1999), and vacant sites increased the surface reactivity of Pt(111) by lowering the activation barriers for the dissociative adsorption of $H_2$ on the substrate (Arboleda and Kasai, 2008). By performing quantum dynamics calculations on *ab initio* potential energy surfaces (PESs), Arboleda *et al.* (2006) also investigated the effects of the initial kinetic and vibrational energies and the orientation of the incident hydrogen molecule ($H_2$) on the dissociative adsorption dynamics of $H_2$ on a Pt(111) surface.

A software package of Automatic Simulation Environmental (ASE) has been developed and applied to investigate the electronic structure of the Pt-H system. The ASE software provides python modules for manipulating atoms, analyzing simulations, visualization etc., because it is easier and more accurate to create the structure without getting lost in the volume of information (Wang *et al.*, 2011). The DFT calculations are performed by the Jacapo calculator, which is an ASE interface for Dacapo and fully compatible with ASE. As observed in Figure 3.3 in Xiao *et al.* (2011), the energy changes look like two humps with the maximum value 0.089 eV (i.e., 8.587 kJ $mol^{-1}$), which represents the diffusion barrier for the H atom on Pt(111) surfaces. It is smaller than that predicted by Blaylock (2011), in which they got 13.4 kJ $mol^{-1}$ for the H atom on the fcc (face-centered cubic) site. The H atom also reached a steady state when it passed through the hcp (hexagonal close-packed) site, where the energy became nearly zero. Thus, the fcc and hcp were both the preferable sites for the H atom diffusion.

### 3.2.1.2   *Reaction thermodynamics*

The thermodynamics of the reactions was established as a function of voltage by calculating the stability of the reaction intermediates and the overpotential of the reaction could be linked directly to the proton and electron transfer (Norskov *et al.*, 2004). A Tafel-Heyrovsky-Volmer three-step reaction model (Wang *et al.*, 2006) for the hydrogen oxidation is considered at the anode, while a Damjanovic three-step reaction model (Malek *et al.*, 2008) for the oxygen reduction mechanism is assumed at the cathode. Assuming non-interaction between adsorbed intermediate species and

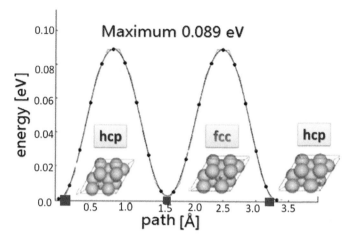

Figure 3.3.    The Minimum Energy Path (MEP) for H atom diffusion on Pt(111) surface and the images under the equilibrium states (the red rectangle dots) (Xiao *et al.*, 2011).

Table 3.1.    Competitive reactions considered for HOR (hydrogen oxidation reaction) and ORR (oxygen reduction reaction) on the Pt surfaces (Malek and Franco, 2011).

| HOR (hydrogen oxidation reaction) | ORR (oxygen reduction reaction) |
|---|---|
| $H_{2(gas)} + 2s \Leftrightarrow 2H_{(ads)}$ | $O_{2(gas)} + H^+ + e^- + s \Leftrightarrow OOH_{(ads)}$ |
| $[v = k_{HOR_1} C_{H_2} \theta_s{}^2 - k_{HOR_{-1}} \theta_H^2]$ | $[v = k_{ORR_1} C_{O_2} C_H + \theta_s \exp(-\alpha_{ORR_1}(F/RT)\eta_F)$ |
| $H_{2(gas)} + s \Leftrightarrow H_{(ads)} + H^+ + e^-$ | $\quad - k_{ORR_{-2}} \theta_{OOH} \exp((1 - \alpha_{ORR_1})(F/RT)\eta_F)]$ |
| $[v = k_{HOR_2} C_{H_2} \theta_s \exp((1 - \alpha_{HOR_2})(F/RT)\eta_F)$ | $OOH_{(ads)} + H_2O_{(gas)} + 2s \Leftrightarrow 3OH_{(ads)}$ |
| $\quad - k_{HOR_{-2}} C_H + \theta_H \exp(-\alpha_{HOR_2}(F/RT)\eta_F)]$ | $[v = k_{ORR_2} \Theta_{H_2O} \theta_s^2 - k_{ORR_{-2}} \theta_{OH}^3]$ |
| $H_{(ads)} \Leftrightarrow s + H^+ + e^-$ | $OH_{(ads)} + H^+ + e^- \Leftrightarrow H_2O_{(gas)} + s$ |
| $[v = k_{HOR_3} \theta_H \exp((1 - \alpha_{HOR_3})(F/RT)\eta_F)$ | $[v = k_{ORR_3} C_H + \theta_{OH_s} \exp(-\alpha_{ORR_3}(F/RT)\eta_F)$ |
| $\quad - k_{HOR_{-3}} C_H + \theta_s \exp(-\alpha_{HOR_3}(F/RT)\eta_F)]$ | $\quad - k_{ORR_{-3}} \theta_s \Theta_{H_2O} \exp((1 - \alpha_{ORR_3})(F/RT)\eta_F)]$ |

between intermediates and water, the rates of these elementary steps are described in Table 3.1. The oxidation of hydrogen occurs readily on Pt-based catalysts (Carrette and Friedrich, 2001). The kinetics of this reaction is very fast on Pt catalysts and in a PEMFC the oxidation of hydrogen at higher current densities is usually controlled by mass-transfer limitations. The process also involves the adsorption of the gas onto the catalyst surface followed by a dissociation of the molecule and the electrochemical reaction to hydrogen ion.

By the ASE software, the $H_2$ dissociative adsorption process has been simulated using the NEB approach (Xiao *et al.*, 2011). In the initial state, the molecule was adsorbed on the same Pt (111) surface as discussed in 3.2.1.1. At the beginning, both of the H atoms moved to the bridge site (shown as the red arrow in Fig. 3.4); after that, they were separated with large energy changes and the dissociation process happened; at last, one H atom went back to the fcc site, and another went to another fcc site to keep the whole system's energy in a higher state. In such cases, the system needed extra energy provided from the environment and the reaction barrier was 4.371 eV.

### 3.2.2    *Modeling methods at nano-/micro-scales*

Modeling performed at the nano-/micro-scale is much more diverse than the typical quantum chemistry, and can be used to calculate wide range properties from thermodynamics to bulk transport properties of components in CLs. For models using semi-empirical or classical force field,

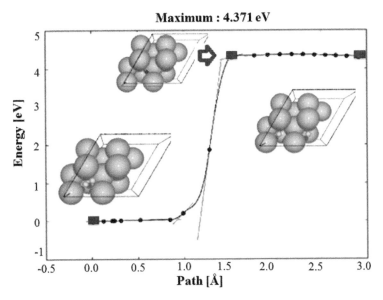

Figure 3.4. Minimum Energy Path (MEP) for $H_2$ dissociative adsorption on the Pt (111) surface (Xiao *et al.*, 2011).

there are several academic software packages available, such as Amber, CHARMM, DL_POLY, GROMOS, Materials Studio, PyMol etc. The phenomena considered at the nano-/micro-scales are mainly determined by their behavior of energy, neglecting the motion of the electrons. Individual atoms or clusters of atoms can be described with the methods based on classical interaction potentials, such as Molecular Dynamics (MD) and Monte Carlo (MC) methods.

### 3.2.2.1 *Molecular dynamics modeling method*
The MD models are developed to predict the time evolution of a system involving interacting particles and provide insights into structural correlations and transport properties of CLs, particularly in three phase boundaries (TPB) of carbon/Pt, ionomer, and gas phases. Furthermore, they provide the information about atomic positions, velocities, and forces. In classical MD models, the system is treated as a set of N interacting particles (Allen, 1989). The atoms are presented by spherical nuclei that attract and repel each other. The forces acting on the particles are derived from a combination of bonding, non-bonding, and electrostatic potentials. The motions of the atoms are calculated using the laws of classical mechanics. The result of an MD simulation is a trajectory in term of positions and velocities of all N particles in the system. The thermodynamic properties, spatial and temporal correlation functions, and transport properties can be exactly calculated, when simulating with an appropriate time step and a sufficient time length.

In MD modeling, the molecular adsorption concept is used to interpret the Pt-C interactions during the fabrication processes. The Pt complexes are mostly attached to the hydrophilic sites on the carbon particles, viz. carbonyl or hydroxyl groups (Hao *et al.*, 2003). The adsorption is based on both the physical and chemical adsorptions. Many efforts have been done on the MD simulations of Pt nano-particles adsorbed on carbon with or without ionomers (Balbuena *et al.*, 2005; Chen and Chan, 2005; Huang and Balbuena, 2002; Lamas and Balbuena, 2003; 2006). The Pt-Pt interactions are modeled with the many-body Sutton-Chen (SC) potential (Rafii-Tabar *et al.*, 2006), whereas a Lennard-Jones (LJ) potential is used to describe the Pt-C interactions. The SC potential for Pt-Pt and Pt-C interactions provides a reasonable description of the properties for small Pt clusters. The diffusion of platinum nano-particles on graphite has also been investigated, with diffusion coefficients in the order of $10^{-5}$ cm$^2$ s$^{-1}$ (Morrow and Striolo, 2007).

Applying MD into polymer composites allows us to evaluate the effects of fillers on polymer structures and dynamics in the vicinity of polymer-filler interfaces, and also to probe the effects of polymer-filler interactions on the materials' properties. As for the properties of hydrated Nafion membrane, the concept of cluster formation for ionomers (Eisenberg and Takahashi, 1970) was suggested for configurational dipole-dipole interactions of water and ions. One widely accepted empirical model for hydrated Nafion is the cluster-network model proposed by Hsu and Gierke (1983) on the basis of small-angle X-ray scattering (SAXS) experiments. In this model, spherical hydrophilic clusters (about 4 nm in diameter) of water are surrounded by sulfonate groups connected through cylindrical channels with ∼1 nm diameter.

### 3.2.2.2  *Monte Carlo methods*

MC methods are a class of computational algorithms that rely on repeated random sampling to calculate the properties of interest. The principle is: choose a site at random, propose a change in the sample, calculate the change in energy, $\Delta E$, and accept or reject the change based on $\Delta E$. When performing such dynamics, it is required that a probability transition function is defined. Different from MD, which gives the non-equilibrium as well as equilibrium properties, MC provides only the information on the equilibrium properties (e.g., free energy, phase equilibrium). In polymer CLs, MC methods have been used to generate a random distribution of three kinds of particles, i.e., Pt/C catalyst, Nafion and poly-tetra-fluoro-ethylene (PTFE) (Wei *et al.*, 2006). Based on such a cluster model, the catalyst utilization was calculated through counting the number of Pt/C clusters and Nafion particle clusters.

In chemistry, dynamic Monte Carlo (DMC) is a method for modeling the dynamic behavior of molecules by comparing the rates of individual steps with random numbers. Unlike the Metropolis MC method, which has been employed to study the systems at equilibrium, the DMC method is used to investigate non-equilibrium systems such as a reaction, diffusion, and so-forth (Meng and Weinberg, 1996). This method is mainly applied to analyze the adsorbates' behavior on the surfaces (Kissel-Osterrieder *et al.*, 2000; Meng and Weinberg, 1996).

### 3.2.3  *Models at meso-scales*

### 3.2.3.1  *Dissipative particle dynamics (DPD)*

The DPD method was introduced by Hoogerbrugge and Koelman (1992) for simulating complex hydrodynamic behavior of isothermal fluids, and further developed by Espanol (1997) who included stochastic differential equations and conservation of energy. It can simulate both Newtonian and non-Newtonian fluids, including polymer melts and blends, on microscopic length and time scales. In DPD models, the particles represent a small cluster of atoms or molecules, and the particle-particle interactions are much softer than the particle-particle interactions used in typical MD models. Therefore, it is feasible to take much larger particle sizes and much larger time steps, and make use of DPD models which are much more efficient than MD models for simulating macroscopic hydrodynamics.

In general, the particles in the DPD method are defined by their mass $M_i$, position $r_i$ and momentum $p_i$. The interaction between two particles can be expressed as the sum of a conservative force $F_{ij}^C$, a dissipative force $F_{ij}^D$, a random force $F_{ij}^R$, and a harmonic spring force $F_{ij}^S$ for the system:

$$f_i = \sum_{j \neq i} (F_{ij}^C + F_{ij}^D + F_{ij}^R + F_{ij}^S) \qquad (3.1)$$

The positions and the velocities of the particles are solved in accordance with the above equations by implementing Newton's equation of motion and a modified version of the velocity. While the interaction potentials in MD models are high-order polynomials of the distance $r_{ij}$ between two particles, in DPD models the potentials are softened in order to approximate the effective potential at microscopic length scales. The form of the conservative force is chosen in particular to decrease linearly with increasing $r_{ij}$. Beyond a certain cut-off separation $r_c$, the weight functions, and thus the forces, are all zero. Because the forces are pair wise and the momentum is conserved, the macroscopic behavior directly incorporates Navier–Stokes hydrodynamics. However, energy is not conserved because of the presence of the dissipative and random force terms.

Recently, DPD methods have been employed to model the morphology evolution of a wide range of copolymer systems, including ionomers, during phase separation (Dongsheng *et al.*, 2009; Dorenbos and Suga, 2009). It could be also used to investigate the micro-scale structure of Nafion membranes at various degrees of hydration (Yamamoto *et al.*, 2002). Because the diffusion of molecules in Nafion depends strongly on the time-scale on which it is measured, water within the pores is very mobile and resembles that of pure water (Pivovar, 2005). Thus, DPD simulations can be used at high water volume fractions for the objective to find a clear relation between the shape of the pore networks and predicted water diffusion constants (Dorenbos and Suga, 2009).

### 3.2.3.2 *Lattice Boltzmann method (LBM)*

LBM is another kind of micro-scale method that is suitable for polymer solution dynamics. The origin of LBM can be traced back to the lattice-gas cellular automata (LGCA), in which the similar kinetic equation is shared (Lim *et al.*, 2002):

$$f_i(x + c_i \Delta t, t + \Delta t) = f_i(x,t) + \Omega_i(f_i(x,t)), \quad i = 0, 1, \ldots, k \qquad (3.2)$$

A typical lattice gas automaton consists of a regular lattice with particles residing on the nodes. A set of Boolean variable $f_i(x,t)$ $(i = 1, \ldots, k)$ describing the particle occupation is defined, where $k$ is the number of directions of the particle velocities at each node; $c_i$ is the particle velocity, the last term $\Omega_i$ in the equation represents the collision operator in accordance with arbitrary collision rules.

Recent lattice Boltzmann models are further simplified by replacing the Boolean algebra with a continuous distribution function and also by linearizing the collision operator by Bhatnagar-Gross-Krook (BGK) approximation (Ramanathan and Koch, 2009), based on the idea that the rate at which collisions drive the distribution function toward the local equilibrium value depends linearly on the deviation from local equilibrium. As an example, LBM provides an effective tool to investigate the transport phenomenon at pore-scale level. In the TPB of fuel cells, several LB models (Gree *et al.*, 2010) have been presented in the literature to describe the diffusion process in complex pore structures; while in the CLs, an interaction-potential based two-phase LB model was developed to study the structure-wettability influence on the underlying two-phase dynamics (Mukherjee *et al.*, 2009). Further, the possible contributions for the water configuration, such as capillary pressure, gravity, vapor condensation, wettability, and micro-structures of the gas diffusion layer (GDL), are discussed using the LBM (Hao and Cheng, 2010; Zhou and Wu, 2010).

### 3.2.3.3 *Smoothed particle hydrodynamics (SPH) method*

SPH was introduced thirty years ago to simulate astrophysical fluid dynamics (Lucy, 1977). It is based on the idea that a continuous field, $A(r)$, can be represented by a superposition of smooth bell-shaped functions, $W(|r-r_i|)$ (usually referred to as the smoothing function or weighting function) centered on a set of the points, $r_i$. The approximation $A(r)$ is expressed by its neighbored particles in a domain $\Omega_r$ (Meakin and Tartakovsky, 2009):

$$< A(r) > = \int_{\Omega} A(r) W(r - r_i, h) d\Omega \qquad (3.3)$$

The SPH method has been also used to simulate pore-scale dissolution of the trapped non-aqueous phase liquids, pore-scale miscible non-reactive flows (Tartakovsky and Meakin, 2005; Zhu *et al.*, 2002), and single- and multi-component reactive transport and precipitation (Tartakovsky *et al.*, 2007). Recently, Jiang *et al.* (2008) developed a meso-scale SPH model for the fluid flow in isotropic porous media. The porous structure was resolved in a meso-scopic level by randomly assigning a certain number of the SPH particles to fixed locations. Particularly, a repulsive force, in a similar form to the 12–6 LJ potential between the atoms, is set in place to mimic the no-penetrating restraint of the interface between the fluid and the solid structures. This force is initiated from the fixed porous material particle and may act on its neighboring moving fluid particles. In this way, the fluid is directed to pass through the porous structure in physically realistic paths.

### 3.2.4   *Simulation methods at macro-scales*

Going yet to larger scales, one encounters typical engineering methods and solves the evolving partial differential equations (PDEs) on a grid. Mesh-based methods such as the finite element method (FEM) are not intrinsically scaled to any length. The function approximation of a solution of a PDE can also be used for the simulation of micro- and sub-microstructures. Different fundamental problems are usually derived from experiments with idealized loading situations (http://tc.dicp.ac.cn/publication_cn.html) and are then extended for other types of loading and more complex geometries. Thus, such constitutive models only show a very low transferability to other situations. If one wants to include some meso-structural features in a mesh-based simulation scheme, one has to make a compromise with respect to the total domain that can be simulated.

## 3.3   MULTI-SCALE MODEL INTEGRATION TECHNIQUE

### 3.3.1   *Integration methods on atomistical scale to nano-scale*

Atomic-scale models can predict physical properties of materials with ideal, theoretical, or proposed micro-structures under clearly defined conditions within the size of several nanometers. However, they are not able to probe the random morphology of the complete CLs due to computational limitations. Therefore, it is expected to use the multi-scale simulation strategies to bridge the models and simulation techniques across the atomic-scale to micro-scale. Several approaches have been proposed, such as the coarse-grained (CG) molecular dynamics (Collins *et al.*, 2010; Izvekov and Violi, 2006). In the CG method, super-atoms are used to represent groups of atoms, and interactions are only defined between these super-atoms. Thus, a reduction in the number of degrees of freedom is achieved. At the same time, it can consolidate the major features of microstructure formation in the CLs of PEMFCs. In this way, the complete description of the CLs microscopic structure can broaden the range of length scales from the chemical bond (at around one angstrom in length) up to chain aggregates, extending for many hundreds of angstroms and beyond, the so-called large scale molecular dynamics method.

In Xiao *et al.* (2012), the CG method is employed to reconstruct the microscopic structure of the CLs in two major steps: (1) Introduce and construct the Nafion chains, water and hydronium molecules, carbon and Pt particles with corresponding spherical beads at a predefined length scale; and (2) Specify the parameters of renormalized interaction energies between the distinct beads as the MARTINI force field (Marrink *et al.*, 2004; 2007), which has been widely used and proved in the applications of lipids (Risselada and Marrink Siewert, 2008), polymers (Hwankyu *et al.*, 2009) and carbohydrates (López *et al.*, 2009). For the defined beads in step (1), four types of interaction sites are mainly considered in step (2): polar (P), nonpolar (N), apolar (C) and charged (Q). The polar sites represent neutral groups of atoms, the apolar sites are hydrophobic moieties, and the nonpolar groups are used for mixed groups, which are partly polar and partly apolar. The charged sites are reserved for ionized groups. For particles of type N and Q, subtypes are further distinguished by a letter denoting the hydrogen bonding capabilities ($d$ = donor, $a$ = acceptor, $0$ = not exist). The defined types of all components are presented in Table 3.2.

As shown in Figure 3.5, the complete CG model of a Nafion copolymer includes a part of the atomistic level configuration of a Nafion chain. According to Zeng *et al.* (2008), a side chain unit in

Table 3.2. Defined CG beads.

| CG beads | Color | Represented atoms | Symbol | Specified type | Radius [nm] |
|---|---|---|---|---|---|
| 1 | Yellow | $-O-CF_2CF(CF_3) - ...$ $-O-CF_2CF_2-SO_3H$ | SO | Qa | 0.43 |
| 2 | Blue | $-(CF_2CF_2)_3-CF_2-$ | C1 | C | 0.43 |
| 3 | Blue | $-(CF_2CF_2)_4-$ | C1 | C | 0.43 |
| 4 | Green | $4 \times H_2O$ | W | P | 0.43 |
| 5 | Magenta | $3 \times H_2O + H_3O^+$ | Wh | Qd | 0.43 |
| 6 | Gray | $36 \times C$ | C | N0 | 0.43 |
| 7 | Golden | $20 \times Pt$ | Pt | Na | 0.43 |

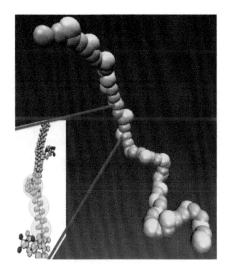

Figure 3.5.   Schematic drawing of the complete Nafion copolymer, which is composed of apolar beads (blue) and charged beads (yellow), and also a part of the atomistic level configuration of a Nafion chain. Four-monomer unit $(-(CF_2CF_2)_4-)$ is represented by one blue bead as the hydrophobic backbone, while the side chain $(-O-CF_2CF(CF_3)-O-CF_2CF_2-SO_3H)$ with the sulfonic acid group is expressed as one yellow bead with hydrophilic property.

the Nafion ionomer has a molecular volume of $0.306 \, \text{nm}^3$, which is comparable to the molecular volume of a four-monomer unit of poly (tetrafluoro-ethylene) (PTFE) $(0.325 \, \text{nm}^3)$. Therefore, one side chain and four-monomer unit $(-(CF_2CF_2)_4-)$ are coarse-grained as spherical beads of volume $0.315 \, \text{nm}^3$ $(r = 0.43 \, \text{nm})$. The Nafion oligomer consists of 20 repeated monomers (Cho *et al.*, 2007); therefore, there are 20 side chains, with the total length of $\sim 30 \, \text{nm}$. The repeated monomers are represented by two blue apolar beads (C) as the hydrophobic backbone and one yellow charged bead (Qa) as the hydrophilic side chain, as shown in the shaded rings in Figure 3.5.

Carbonaceous particles can be coarse-grained in various ways (Zeng *et al.*, 2008); in literature (Xiao *et al.*, 2012), the CG potential parameters are systematically obtained from the atomistic-level interactions. In Figure 3.6a, the carbon slab is firstly constructed by the Atomistic Simulation Environment (ASE) software (https://wiki.fysik.dtu.dk/ase/overview.html), which is accurate to create the structure without getting lost in the volume of information. The carbon slab is modeled by eight layers of the zigzag graphene structure, and each layer has a regular hexagonal net of carbon atoms with the C-C bond length of 1.42 Å. The interlayer spacing is equal to 3.35 Å and the layers are interacted by van der Waals force (Scocchi *et al.*,

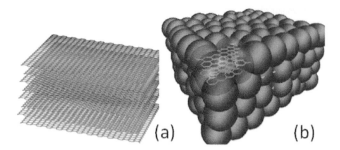

Figure 3.6.    (a) Structural drawing of one carbon slab generated by ASE software, which consists of eight layers of the zigzag graphene structure with a regular hexagonal net of carbon atoms. The interlayer spacing is equal to 3.35 Å. (b) Coarse-graining representations of the carbon slab, in which there are 36 carbon atoms in each bead. The typical carbon slab with the size of $5 \times 5 \times 2.5\,nm^3$ is reconstructed by 224 nonpolar beads (type of N).

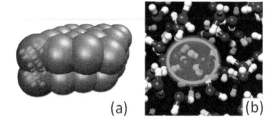

Figure 3.7.    (a) Atomistic (in the shaded rings) and coarse grain representation of one fcc-Pt (111) cluster. The $10 \times 10 \times 6$ Pt cluster is represented by thirty nonpolar beads, and each bead contains 20 Pt atoms. (b) A total of four water molecules (or three water molecules plus a hydronium ion) are represented by one CG bead of radius 0.43 nm.

2007). In the following step, the CG Builder from VMD (Visual Molecular Dynamics) software (http://www.ks.uiuc.edu/Research/vmd) is employed to provide a shape-based coarse graining (SBCG) tool for transforming structures from all-atom representations to CG beads (Arkhipov *et al.*, 2006a; 2006b). Based on a neural network learning algorithm, the shape-based method is used to determine the placement of the CG beads, which also have masses correlated to the clusters of the represented atoms. Neighboring beads are connected by harmonic springs, while separate molecules interact through non-bonded forces (Lennard-Jones and Coulomb potentials) (Arkhipov *et al.*, 2006a). Interactions are parameterized on the basis of all-atom simulations and available experimental data. Therefore, there are thirty-six carbon atoms in one bead of the slab ($r = 0.43$ nm), as shown by the shaded rings in Figure 3.6b. By this method, the typical carbon slab with the size of $5 \times 5 \times 2.5\,nm^3$ is represented by 224 nonpolar beads (type of N).

The Pt nanoclusters supported on carbon slabs are also considered. For the Pt/C sample prepared in Boutaleb *et al.* (2009), the majority of the Pt particles are in the range of 2–3 nm (an average size of 2.4 nm). According to Ray and Okamoto (2003), it is envisaged that the fcc-like clusters will occur at this size. Similar to the carbon slabs, the Pt nanocluster is firstly constructed by the ASE software and then coarse-grained by the VMD software. Thus, the $10 \times 10 \times 6$ fcc-Pt (111) cluster is represented by thirty nonpolar beads (type of Na) in Figure 3.7a. Therefore, each bead consists of 20 Pt atoms with the radius of 0.43 nm, as shown by the shaded rings in Figure 3.7a. In Figure 3.7b, four water molecules are represented as one single polar CG site (type of P) with the radius of 0.43 nm. For the electroneutrality condition, three water molecules plus a hydronium ion are also added into the system and represented by one charged bead (type of Qd) of the radius 0.43 nm.

### 3.3.2 *Microscopic CL structure simulation*

A significant number of micro-scale computational approaches have been employed to understand the phase-segregated morphology and transport properties in membranes and CLs. Based on the self-consistent mean field theory (Galperin and Khokhlov, 2006), CG-MD techniques can describe the system at the microlevel, while still being able to capture the morphology at longer time and larger scales (Komarov *et al.*, 2010; Wescott *et al.*, 2006). The proposed CG method (Xiao *et al.*, 2012) is developed and applied to the step formation process, which follows the preparation of the Catalyst Coated Membranes (CCMs). As shown in Figure 3.8, the numerical reconstruction of a CL is performed through the CG method by the GROMACS package, mimicking the experimental fabrication process (Buryachenko *et al.*, 2005).

In Step I, the Pt/C catalyst (40wt%, BASF) nano-particles, represented by 30 carbon slabs and 16 Pt clusters, are mixed with the water and isopropyl alcohol beads in the $50 \times 25 \times 25 \, nm^3$ box. After MD running for 10 ns, the final formation configuration is obtained for the equilibrium state, as shown in Figure 3.8I. For better clarity, the solvents are not shown and the phenomenon of self-aggregation is easily observed. It is also clear that the lateral part of the Pt clusters adsorbs on the carbon surface. In Step II, add 176 PTFE chains (for the content of 5 wt%, as suggested in Ref. (Song *et al.*, 2010) to the catalyst ink and keep MD running for another 10 ns to reach the equilibrium state, as shown in Figure 3.8II. In Step III, the catalyst ink is sintered to remove all the solvents from the target box. After MD running for 50 ns, the framework structure is formed, in which the PTFE breads embed themselves in the carbon aggregates to cause a more hydrophobic property (see Fig. 3.8III). In Step IV, 100 Nafion chains are introduced to the box, together with three different solvents, respectively, to exploit the solvent effects on the evolvement of the microstructure. There are 200,000 CG beads of EG, IPA or HX in the box to represent a wide range of dielectric permittivity. After MD running for 10 ns, the snapshot (see Fig. 3.8IV) is obtained to show the self-assembled configuration, meanwhile, the solvents are invisible for better clarity. It is easy to find that the ionomer backbones assemble into a separate interconnected phase in the void spaces of the Pt/C aggregates. During the final Step V, excess water and other solvents are evaporated and all the components of a CL can be expressed in the final equilibrium state (see Fig. 3.8V, after MD running for 50 ns). Corresponding to thirteen waters per side chain (Zeng *et al.*, 2008), the number of water beads is 4500 (including 2000 beads with hydronium ions for keeping electroneutrality condition). From the final snapshot in Figure 3.8V, it can be found that the Pt/C nano-particles with the embedding PTFE chains form the aggregates, while the hydrated ionomers do not penetrate into the Pt/C aggregates but form a separate phase that is attached to the surface of the aggregates. The CL thus segregates into hydrophobic and hydrophilic regions. Viewed from the inside of the agglomerate, the hydrophobic region consists of PTFE chains, carbon slabs and the backbones of the Nafion. However, most of the residual solvent beads dwell on the small pores and gas and surround the side chains of the Nafion ionomers to form the hydrophilic regions. It is easy to understand that the hydrophilic beads form a three-dimensional network of irregular water-filled channels, which contains the primary pore structures in the agglomerates of CLs.

### 3.3.3 *Analyses of predicted CLs microscopic structures*

#### 3.3.3.1 *Microscopic parameters evaluation*

As mentioned above, there are lots of the voids and the pore spaces in the agglomerates, which are more intricate and complicated for a big size box (50 nm × 50 nm × 50 nm), as shown by the final image ($t = 50 \, ns$) in Figure 3.9a. The Voss Volume Voxelator (3V) web server provides a technique that can obtain the overall shape of the channels (Fermeglia and Pricl, 2007). It uses the rolling probe method (Sun *et al.*, 2009), which essentially works by rolling a virtual probe or ball of a given radius around the surface of a macromolecule or the agglomerate to calculate the volumes. When a probe of zero-size is used, the van der Waals radius is obtained. For the CG system, the effective bead diameter ($\sigma_{ij}$) is 0.43 nm; therefore, the particle volume ($V$) can be calculated by

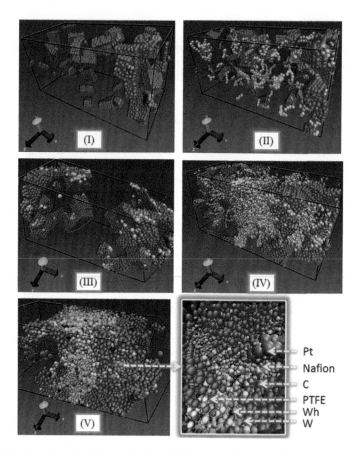

Figure 3.8.   Step formation of a CL based on the CCMs preparation procedure: Step I: randomly disperse Pt/C catalyst in the $50 \times 25 \times$ nm$^3$ box and then the self-aggregation happens in the solvent environment (snapshot at $t = 10$ ns, showing without solvent beads). Step II: add the PTFE to the catalyst ink and reach a well-distributed condition (at $t = 10$ ns, showing without solvent beads). Step III: remove the solvent in the heating and drying process (at $t = 50$ ns). Step IV: add Nafion chains to form the ion transport network (at $t = 10$ ns, showing without solvent beads). Step V: evaporate the residual solvent and exhibit the final snapshot of the equilibrium state at $t = 50$ ns, with the partially enlarged details from the inside. In all these snapshots, Golden: Pt, Yellow and Blue: Side chain and backbone of Nafion ionomer, Grey: Carbon, White: PTFE, Magenta: Hydronium ions and Green: Water.

rolling the probe with the radius of 4.3 Å. As shown in Figure 3.9b, the particle volume ($V$) is 14,650 nm$^3$, and its surface is marked in pink. However, as the probe radius increases, the surface features are filled in. In order to distinguish the interior of the agglomerate from its exterior, the agglomerate volume ($V_0$) is defined here as the space inside the limiting surface of the agglomerate with a larger probe radius. For the probe radius of 4 nm, which is appropriate for the primary pore analysis, the convex hull is obtained and marked in blue (see Fig. 3.9c).

The difference between the agglomerate volume ($V_0$, blue) and the particle volume ($V$, pink) constitutes the primary pore spaces of the CLs and a part of it is shown in Figure 3.9d. Close to the solid surface of the agglomerate, there are mainly two types of internal spaces: cavities and channels. The cavity is the space large enough to accommodate a water molecule, but isolated by the Pt/C aggregates. It is not connected to the agglomerate exterior but regarded as the dead zone for the electrochemical reactions. The second type of internal space is loosely defined as

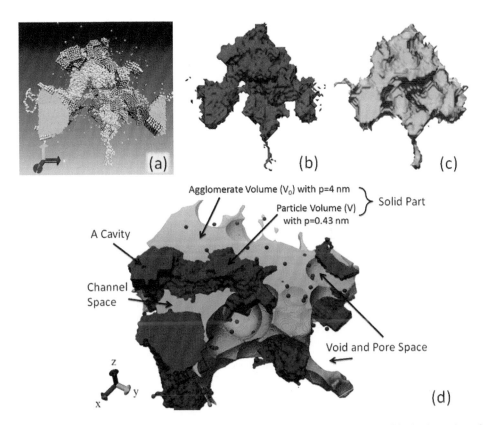

Figure 3.9.   (a) Image of the final equilibrium states after MD running 50 ns with the box size of 50 nm × 50 nm × 50 nm, analyzed by 3V method (Fermeglia and Pricl, 2007): (b) surface of the particles at the probe radius, $p = 4$ Å; (c) surface of the agglomerate at the probe radius, $p = 4$ nm; and (d) a schematic drawing of the space difference between the agglomerate volume ($V_0$, marked in blue) and the particle volume ($V$, marked in pink), which constitutes the primary pore space of the CLs.

the "channel space". For the CLs, the channel space, which contains a major part of the primary pores, acts as a large network within the agglomerates and can be easily filled with liquid water by the capillary force. It is also the place where the ionomer clusters are able to penetrate into, and thus has more influences on the proton conduction. For all the cases in Table 3.3, the channel volume ($V_{channel}$) can be distinguished and calculated by the tools of the 3V web server (Fermeglia and Pricl, 2007), and the results are presented in Table 3.4. In addition, the real density $\rho$, which can be expressed as the mass ($M$) of a unit box divided by the particle volume ($V$), should be almost the same for the cases 1–3, as shown in Table 3.4.

### 3.3.3.2   *Primary pore structure analysis*
In Figure 3.10a, the frame structure of the agglomerate under the equilibrium state ($t = 100$ ns) for the case C (in Table 3.5) is presented by the VMD software. It is easy to find some primary pores of 3–10 nm appearing in the agglomerate. The pores are intricate and connected to form the channels, which can act as large networks within the agglomerates. They are more likely to be filled with liquid water for water and hydronium ions transport, and thus have more influences on the properties of CLs. The original appearance can be obtained by rolling the probe with the radius $p = 0.43$ nm, as shown in Figure 3.10b. However, as the probe radius increases, the surface features are filled in and the calculated volume will increase (see Fig. 3.10c, $p = 5$ nm).

Table 3.3.  Cases studied based on different sizes of the target boxes.

| Cases | 3-D sizes [nm × nm × nm] | Number of component units in different cases | | | |
|---|---|---|---|---|---|
| | | Carbon slabs | Pt clusters* | Nafion chains | Water beads (including $H_3O^+$) |
| 1 | 25 × 25 × 25 | 15 | 8 | 50 | 3250 (1000) |
| 2 | 50 × 25 × 25 | 30 | 16 | 100 | 6500 (2000) |
| 3 | 50 × 50 × 50 | 120 | 65 | 400 | 26000 (8000) |
| 4 | 50 × 50 × 50 | 120 | 25 | 400 | 26000 (8000) |

*The mass percentage of Pt is kept to about 40 wt% for the cases 1–3 and 20 wt% for the case 4 in the Pt/C catalyst.

Table 3.4.  Calculated physical parameters of the CLs.

| Cases | $M$ [$10^{-18}$ g] | $V$ [nm$^3$] | $V_0$ [nm$^3$] | $V_{Channel}$ [nm$^3$] | $\rho$ [g cm$^{-3}$] |
|---|---|---|---|---|---|
| 1 | 6.23 | 1894 | 2574 | 644 | 3.29 |
| 2 | 12.46 | 3717 | 7063 | 3061 | 3.35 |
| 3 | 49.80 | 14650 | 26795 | 10004 | 3.39 |
| 4 | 41.32 | 13060 | 20290 | 6326 | 3.16 |

Table 3.5.  Cases based on solvents with different dielectric permittivities.

| Cases | Solvents | Symbol | Dielectric permittivity | Agglomerate size* [nm$^3$] |
|---|---|---|---|---|
| A | Ethylene glycol | EG | 37.7 | 5430.591 |
| B | Isopropyl alcohol | IPA | 18.3 | 5465.384 |
| C | Hexane | HX | 1.9 | 5844.051 |

*Calculated by the rolling method from the 3V web (Fermeglia and Pricl, 2007) with the probe radius of 4.0 nm.

In addition, the increased probe ball cannot go into some of the primary pores anymore and the calculated volume will increase obviously. Furthermore, the geometric solid obtained by rolling the probe with the radius $p = 2.8$ nm is sliced along with the $x$, $y$ and $z$ axis separately. Thus, the solid part is colored in sky-blue and the primary pores can be observed, as shown in Figure 3.10d. At the same place, the three-dimensional slices of the agglomerate at $p = 3.0$ nm are displayed in yellow. It is clearly found that one pore space, captured at $p = 2.8$ nm, now becomes the solid part, because the probe ball, with the radius $p = 3.0$ nm, is too large to go inside. It means that the size of this primary pore is in the order (the maximum radius) of 2.8 nm.

### 3.3.4  *Model validation*

#### 3.3.4.1  *Pore size distribution*
The primary pore size distributions of the cases A and B in Table 3.5 are experimentally determined by mercury injection porosimetry (MIP) (Fermeglia and Pricl, 2009) on the instrument of PoreMaster GT 60. The MIP enables the measurements of both the pressure required to force mercury into the pores of CLs and the intruded Hg volume at each pressure. The employed equipment operates from 13 kPa to a pressure of 410 MPa, equivalent to the pores with the diameters, $d$, ranging from 100 μm to 0.0036 μm. On the other hand, the 3 V method provides a tool to theoretically calculate the agglomerate volume depending on the probe radius. In analogy to the

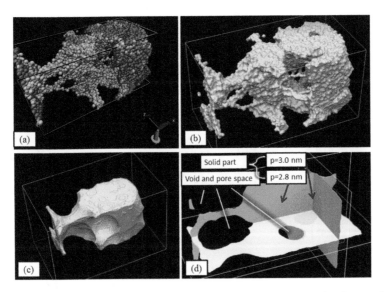

Figure 3.10.    (a) Image of the final equilibrium states after MD running 50 ns for the case C with the box size of 50 nm × 50 nm × 50 nm, analyzed by 3 V method (Fermeglia and Pricl, 2007): (b) surface of the particles at the probe radius, $p = 4$ Å; (c) surface of the agglomerate at the probe radius, $p = 4$ nm; and (d) a schematic drawing of the space difference between the agglomerate volume ($V_0$, marked in blue) and the particle volume ($V$, marked in pink), which constitutes the primary pore space of the CLs.

principle of MIP, the introduced volumes and the corresponding pore diameters can be collected, to identify the primary pore size distribution among the agglomerate, as shown by the colored cures in Figure 3.11.

It is worthwhile to point out that the volume differential distributions (normalized $dV/dD$) are employed to estimate and evaluate the pore size distribution (PSD) of the CLs. In Figure 3.11, the normalized $dV/dD$ values are presented for the two MIP tests and three computational simulations, only in the range of the pore diameters from about 3 to 20 nm. It is clear that the pores are mainly distributed between 3 to 10 nm in both experiments and simulations and the results of MIP tests of cases A and B are nearly the same. In contrast, the simulation results from the three solvents reveal the distinct pore size distributions. For example, with the weakening of the polarity in case C, the PSD becomes more uneven, i.e., more complex pore structures are obtained in the agglomerate. In addition, the agglomerate volume ($V_0$) is also evaluated at the probe radius of 4 nm, which is appropriate for the primary pore analysis. The predicted $V_0$ are also shown in Table 3.5. Obviously, the CCM prepared with nonpolar solvents (such as HX) acquires bigger size agglomerate involving well-developed primary pores, which may produce better cell performance.

### 3.3.4.2    *Pt particle size distribution*
In order to mimic the feature of the CLs, the TEM sample is prepared by mixing the Pt/C electro-catalysts (40 wt%, BASF) in the Isopropanol-Nafion® solution, in which the Nafion® content is 17.2 wt.%. A small volume, approximately 5 μL of the sample, is deposited onto the copper grids in order to ensure that a thin sample is created. The sample is analyzed in the JEOL JEM-2000EX (Japan) microscope at an accelerating voltage of 120 kV. A part of the TEM image with the size of 100 nm × 50 nm (see Fig. 3.12a) is analyzed by the ImageJ software (Cotterell *et al.*, 2007), which is a public domain Java image processing and analysis program. For a comparison purpose, the Pt nano-particles in the reconstructed equilibrium state are displayed in a 2D-image in the same size of 100 nm × 50 nm (repeated by 2 × 2 simulation box), as shown in Figure 3.12b. In Figure 3.12c,

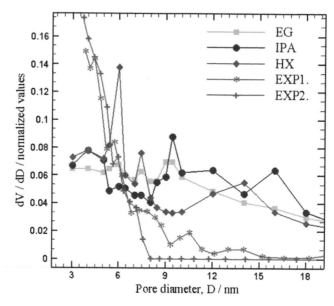

Figure 3.11.    Comparison of the primary pore size distribution (PSD) in the CLs obtained by CG-MD step
formation (colored lines) and experiments (grey lines).

the particle size and dispersion of the Pt nano-particles in Figure 3.12a are evaluated and the
particles are found in the range of 2–16 nm. Due to a high degree of overlap of Pt particles, the
ImageJ software cannot make distinctions of some bigger agglomerates; thus, it may predict a
higher mean size of the Pt particles. Also analyzed by the ImageJ, the number of Pt particles in
each bin (Scott, 1979) in Figure 3.12b is counted and compared in Figure 3.12c. It is found that
most of the Pt nano-particles have the Feret's diameters (which represent the longest dimension
of the particles, independent of their angular rotation at the time that the image is captured) in
the range of 2 to 6 nm. The same finding is also indicated in the experimental measurement, as
shown in Figure 3.12c in terms of the relative abundance [%].

### 3.3.4.3   *The average active Pt surface areas*

The active Pt surface areas of a commercial 20 wt% Pt/C catalyst (E-TEK) are measured by
both hydrogen adsorption and CO-stripping voltammetry (Coker *et al.*, 2007; Lim and Lee,
2008). To compare with these experimental data, the configuration of the corresponding CLs is
reconstructed in the $50 \times 50 \times 50$ nm$^3$ unit volume, as the case 4 in Table 3.3. In the box, there are
120 carbon slabs 25 Pt clusters, 400 Nafion oligomers, and 26000 water beads in the simulation
box with PBC at all the faces. Based on the step formation process, the final equilibrium state is
obtained after MD running for 50 ns. In GROMACS, the information of each component can be
separated and exported individually. Thus, by rolling the probe with the radius of 0.43 nm (Sun
*et al.*, 2009), the surface of the Pt clusters is obtained, as shown in Figure 3.13a; at the same
time, the total surface area ($A_1$) and volume ($V_1$) of the Pt clusters are calculated, as shown in
Table 3.6. Most of these Pt nano-particles are well dispersed among the agglomerates and only
two clusters hold together obviously. With the same method, the surface of the carbon slabs is
plotted in Figure 3.13b; while the overall figure of the Pt/C aggregate is enlarged in Figure 3.13c.
The related geometry parameters ($A_2$, $V_2$ and $A_{12}$, $V_{12}$) can be obtained, as outlined in Table 3.6.
In Figure 3.13c, some parts of the Pt clusters are embed into the carbon aggregate and lose
their electrochemical activities. Therefore, the embedded parts of the surfaces have made no
contribution to the electrochemical active surface areas and should be excluded.

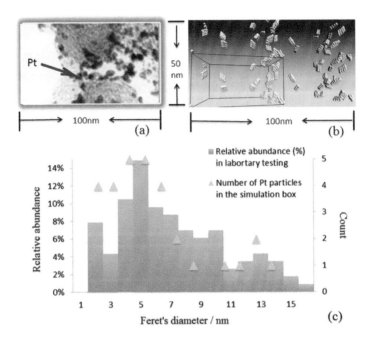

Figure 3.12.  (a) TEM image of the Pt/C catalyst layer prepared with the isopropanol Nafion® solution, (b) Pt nano-particles in the reconstructed equilibrium state shown in the 2 × 2 simulation box (c) Pt particle-size distribution in TEM image, and the counts of Pt particles for each bin (Scott, 1979) in the 2D image of (b).

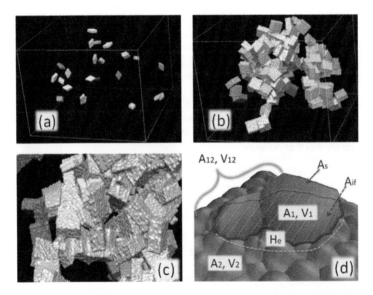

Figure 3.13.  The surface of (a) the Pt clusters, (b) the carbon slabs and (c) the Pt/C aggregate in the same equilibrium state after MD running for 50 ns. (d) a schematic diagram of the Pt and C interface, being applied to calculate the specific active area of the Pt clusters ($A_s$), when the total surface areas and the volumes are available, i.e., known $A_1$, $V_1$ from (a) and $A_2$, $V_2$ from (b), as well as $A_{12}$, $V_{12}$ from (c).

Table 3.6.   Predicted volume and surface area of Pt clusters, carbon slabs and PtC aggregates in an equilibrium state.

| | $j$ | $V_j$ [nm$^3$] | $A_j$ [nm$^2$] |
|---|---|---|---|
| Case 3 | Pt | 595.8 | 1859 |
| | C | 8478 | 10620 |
| | PtC | 9041 | 12060 |
| Case 4 | Pt | 281.2 | 772.7 |
| | C | 8035 | 10120 |
| | PtC | 8247 | 10400 |

Table 3.7.   Active Pt surface area predicted by the current model.

| | Solution | $R$ [nm] | $H$ [nm] | $A_s$ [nm$^{-2}$] | $A_{if}$ [nm$^{-2}$] | Active Pt surface area [m$^2$ g$^{-1}$] |
|---|---|---|---|---|---|---|
| Case 3 | 1 | 16.9 | 0.7 | 1628.1 | 188.1 | 51.04 |
| | 2 | 0.6 | 460.3 | 1602.8 | 162.9 | 50.25 |
| Case 4 | 1 | 10.7 | 0.8 | 503.0 | 223.0 | 20.85 |
| | 2 | 0.7 | 167.5 | 452.7 | 172.7 | 18.76 |

In order to calculate the active Pt surface area ($A_s$), the Pt and C boundary is abstracted as a whole Pt particle ($A_1$, $V_1$), embedding into one big carbon slab ($A_2$, $V_2$), as shown in Figure 3.13d. When the total surface areas and volumes are provided as in Table 3.6, two relations can be deduced from the geometry:

$$A_s + A_2 = A_{12} + A_{if} \tag{3.4}$$

$$V_1 + V_2 - V_{12} = \iint_{A_{if}} H_e dA \tag{3.5}$$

where $A_{if}$ is the boundary interface area and $H_e$ represents the effective height of embedding part. Due to the lack of the function expressions for the real Pt particles, Pt is ideally assumed as a cylinder, which is perpendicular to the carbon slab interface for simplicity. Then, another constrain can be obtained as follows:

$$\frac{H - H_e}{H} = \frac{A_s - \frac{V_1}{H}}{A_1 - \frac{V_1}{H} - A_{if}} \tag{3.6}$$

where $H$ represents the cylinder height and can be calculated from $A_1$ and $V_1$:

$$A_1 = 2\pi R^2 + 2\pi R \cdot H \tag{3.7}$$

$$V_1 = \pi R^2 \cdot H \tag{3.8}$$

The predicted results are listed in Table 3.7. From the assumed perpendicular cylinder model, each case has two real solutions, which have distinct shapes ($H$ and $R$ are quite different), but similar active surface ($A_s$) and interface areas ($A_{if}$). Thus, these will be the approximate results of the active Pt surface areas per gram catalyst, which are little affected by the Pt shapes and the model assumptions.

Table 3.8. Comparison of the active Pt surface area.

| Name | Pt mass percentage | Pt unit or average size | Active Pt surface area/ $[m^2 \, g^{-1}]$ | | Reference |
| | | | Current study | Exp* | |
| --- | --- | --- | --- | --- | --- |
| Case 3 | 40 wt% | $2.8 \times 2.3 \times \sin 60° \times 1.4 \, nm^3$ | 50.7 | – | Xiao *et al.* (2012) |
| TKK | 46.6 wt% | 3.6 nm | – | 41.9 | Cai *et al.* (2006) |
| Case 4 | 20 wt% | $2.8 \times 2.3 \times \sin 60° \times 1.4 \, nm^3$ | 19.8 | – | Xiao *et al.* (2012) |
| E-TEK | 20 wt% | 3.3 nm | – | 37.2 | Lim and Lee (2008), Sun *et al.* (2009) |

*Pt/C catalyst (TKK) by CO-stripping voltammetry; Pt/C catalyst (E-TEK) by $H_2$ desorption cyclic voltammogram

In Table 3.8, the three-dimensional volume of the Pt unit is about $2.8 \times 2.3 \times \sin 60°$ nm$^2$ (crystal plane) $\times 1.4$ nm (height) in the modeling. However, some of these Pt units are very close to each other (see Figure 3.13a) and can be assumed to be the same for all the particles, when observed in the experiments. In fact, the average sizes of the Pt particles in the modeling are bigger than the Pt unit and quite close to the values from the experiments. In this case, the average active Pt surface areas of the cases 3 and 4 (in Table 3.3), which own different Pt mass percentage but similar Pt average sizes, agree reasonably with the experimental ones (Cai *et al.*, 2006; Coker *et al.*, 2007; Lim and Lee, 2008).

### 3.3.5 *Coupling electrochemical reactions in CLs*

Assuming non-interaction between adsorbed intermediate species and between intermediates and water, the rates of these elementary steps were described in Section 3.2.1.2. The value of the surface covering fraction ($\theta_i$) can be obtained by a solution of the conservation equations in Ref. (Chen and Chan, 2005), and specially, the ionomer coverages on Pt and C can be obtained from the CG-MD simulations.

The kinetic parameters (associated with each elementary steps) are given by (Chen and Chan, 2005):

$$k_i = \left(1 - \exp\left(\frac{-hV}{k_B T}\right)\right) \frac{k_B T}{h} \exp\left(-\frac{1}{RT}[\Delta G^0 + f_1(\eta(t)) + f_2(\theta_i(t))]\right) \quad (3.9)$$

where $V$ is the vibrational frequency of the atomic bond on the Pt or C surface. $\Delta G^0$ represents the activation barrier, as determined from ab-initio-calculated data.

According to the superposition principle (Rafii-Tabar *et al.*, 2006), the potential drop $\eta(t)$ is the sum of the ones related to the thickness of the adsorbed water layer ($\Delta \phi_1$) and the drop related to the dipolar surface density ($\Delta \phi_2$):

$$\eta(t) = \Delta \phi_1 + \Delta \phi_2(\sigma) \quad (3.10)$$

In Equation (3.10), the last term is a function of the electronic density $\sigma$, which can be expressed as:

$$-\frac{\partial \sigma}{\partial t} = J(t) - J_{Far} \quad (3.11)$$

where $J_{Far}$ is the Faradaic current density (Morrow and Striolo, 2007). In the case of a volumetric electrode with a total electronic current $I(t)$, the current density $J(t)$ can be calculated by (Rafii-Tabar *et al.*, 2006):

$$J(t) = \frac{I(t)}{S_{electrode} e_{ZA}} \frac{1}{\gamma} \quad (3.12)$$

where $\gamma$ is the specific contact area between metal parts (Pt/C in a PEMFC electrode) and electrolyte fractions (Nafion); $S_{electrode}$ and $e_{ZA}$ are the geometrical surface and thickness of the volumetric electrode, respectively. These physicochemical parameters can be obtained from Eisenberg and Takahashi (1970), and Rafii-Tabar *et al.*, 2006) or deduced from the micro-scale models, such as CG-MD simulations (Chen and Chan, 2005). The boundary conditions, such as the local concentrations of species, can be given as assumptions or provided from the coupling models at a larger scale (Rafii-Tabar *et al.*, 2006).

## 3.4    CHALLENGES IN MULTI-SCALE MODELING FOR PEMFC CLs

One of the main goals of multi-state modeling is to drastically reduce the number of degrees of freedom in engineering problems while maintaining accuracy in the regions of interest. The aim of the CLs modeling is to rapidly and accurately predict the properties and features, which is very difficult to achieve with traditional modeling and simulation methods at a single length and within time scales with the current computer power. However, several challenges, both experimental and theoretical, remain as a road block to successful research and development, as summarized below.

### 3.4.1    *The length scales*

In particular, the smallest and longest scales present in the CLs may span 3–4 orders of magnitude, from about 1 Å (size of an atom) up to hundreds of nanometers (end-to-end distance). This broad range of length scales includes: chemical details at the atomistic level, individual chains, and microscopic features involving aggregates or self-organized chains, up to continuum phenomena at the meso- and macro-scale. A proper study of the CLs requires suitable and simplified models, which allow one to focus on essential features (Zeng *et al.*, 2008). Then, it is expected to use the multi-scale simulation strategies to bridge the models as seamlessly as possible from one scale to another. The calculated parameters, properties, and numerical information, can be efficiently transferred across the scales.

### 3.4.2    *The time scales*

The broad range of length scales brings about a correspondingly wide range of time scales, with chemical bond vibrations occurring over tens of femtoseconds and, at the other extreme, collective motions of many chains taking seconds or much longer. While the time step in the continuum equations is governed by the smallest element in the mesh. Therefore, if the mesh is down to the atomic scale, the simulation will evolve very slowly, and many time-steps will be used to simulate the dynamics in the areas of little interest. This problem will directly affect the computational cost of the simulation. A means of avoiding this issue by completely separating the two time scales would be necessary. By this way, there is no restriction on the mesh to be refined to nano-dimensions; thereby the continuum simulation method can evolve on a larger time step.

### 3.4.3    *The integration algorithms*

Time integration algorithms are based on finite difference methods, where the time is discretized on a finite grid. The time step is the distance between successive points on the grid. These algorithms are only approximate and have two errors associated with them (Ng *et al.*, 2008). First, truncation errors arise because these algorithms involve Taylor expansions, which need to be truncated at a certain order. Secondly, round-off errors arise from the implementation of the algorithms on operating systems, which use a finite number of digits in their arithmetic operations. Both these errors can lead to a divergence of the solution. As very small time steps need to be used in the atomistic domain, the highly iterative nature of these multi-scale simulations can cause error amplification.

3.5   CONCLUSIONS

Typical catalyst layers (CLs) of proton exchange membrane fuel cells (PEMFCs) are fabricated as random and heterogeneous composites to meet multi-functional requirements of transport phenomena and electrochemical activities. The CLs involve various sized particles and pores that span a wide range of length scales, from several to hundred nanometers. Thus, the modeling of the nano-composites in the CLs is a multi-scale problem. Theoretical procedures have to provide quantitative predictions about elementary processes at the nano-scale and also offer the morphology of heterogeneous materials (e.g., ionomers, Pt/C agglomerate) and rationalize their effective properties. In addition, many traditional simulation techniques (e.g. MC, MD, and LB) have been discussed and some novel simulation techniques (e.g., DPD and CG-MD) have also been introduced.

In this chapter, a systematic technique based on CG method is highlighted to provide vivid predictions of self-assembling processes for the CLs in PEMFCs. This technique is based on a step formation process and allows control of the structures while considering the interactions of the Pt/carbon aggregates, Nafion ionomers, and water at a large scale. By the rolling-probe method, the physical parameters of the CLs are evaluated and further used to calculate the real densities of the agglomerates. In addition, the Pt nano-particle size distribution is evaluated and compared with the experimental testing. Furthermore, the primary pore size distributions in the final formations of three cases are predicted and compared with the experiments. The sizes of the reconstructed agglomerates are also considered on the effect of solvent polarity. The active Pt surface areas are also calculated by and compared with the experimental data available in the literature.

Despite the significant achievements in recent years, there still exist major challenges and limitations in the multi-scale modeling and simulations. These methods need new and improved simulation techniques at individual time and length scales. It is also important to integrate the algorithms and explore the contact at a wider range of time and length scales. An ideal coupling method would incorporate all the positive aspects of each method into one sole approach. Therefore, when the computational burden is decreased, it would be possible to reduce the degrees of freedom. Once the efforts are directed to real engineering problems, the structural, dynamic, and mechanical properties, as well as optimizing design, can be explored more effectively.

NOMENCLATURE

| | |
|---|---|
| $\nu$ | reaction rate |
| $k$ | kinetic parameters |
| $c$ | mole concentration |
| $s$ | a free site at the CL surface |
| $\theta$ | the coverage of the sites |
| $\alpha$ | charge transfer coenefficient, the value of which between 0 and 1 |
| $\eta_F$ | electrostatic potential jump between the Pt or carbon surface and the electrolyte |
| F | the Faraday constant |
| R | the universal gas constant |
| $H_{(ads)}$ | an adsorbed H atom on the Pt active site |
| $A_s$ | active Pt surface area |
| $A_{if}$ | boundary interface area |
| $H_e$ | the effective height of embedding part |
| $H$ | the cylinder height |
| $V$ | vibrational frequency |
| $\Delta G^0$ | activation barrier |
| $\eta(t)$ | potential drop |
| $\Delta\phi_1$ | thickness of the adsorbed water layer |

| $\Delta\phi_2$ | dipolar surface density |
| $J_{Far}$ | Faradaic current density |
| $I(t)$ | total electronic current |
| $J(t)$ | current density |
| $\gamma$ | specific contact area |
| $S_{electrode}$ | geometrical surface |
| $e_{ZA}$ | thickness of the volumetric electrode |

## ACKNOWLEDGMENT

The European Research Council and the Swedish Research Council partially support the current research.

## REFERENCES

Allen, M.P. & Tildesley, T.D.: *Computer simulation of liquids*. Oxford University Press, New York, 1989.

Arboleda, J.N.B., Kasai, H., Diño, W.A. & Nakanishi, H.: Quantum dynamics study on the interaction of H2 on a Pt(111) surface. *Thin Solid Films* 509 (2006), pp. 227–229.

Arboleda, N.B. & Kasai, H.: Potential energy surfaces for H2 dissociative adsorption on Pt(111) surface-effects of vacancies. *Surf. Interface Anal.* 40 (2008), pp. 1103–1107.

Arkhipov, A., Freddolino, P.L. & Schulten, K.: Stability and dynamics of virus capsids described by coarse-grained modelling. *Structure* (London), 14 (2006a), pp. 1767–1777.

Arkhipov, A., Freddolino, P.L., Imada, K., Namba, K. & Schulten, K.: Coarse-grained molecular dynamics simulations of a rotating bacterial flagellum. *Biophys. J.* 91 (2006b), pp. 4589–4597.

Balbuena, P.B., Lamas, E.J. & Wang, Y.: Molecular modeling studies of polymer electrolytes for power sources. *Electrochim. Acta* 50 (2005), pp. 3788–3795.

Blaylock, D.W., Teppei, O., William, H.G. & Gregory, J.O.B.: Computational investigation of thermochemistry and kinetics of steam methane reforming on Ni(111) under realistic conditions. *J. Phys. Chem* C 113 (2009), pp. 4898–4908.

Boutaleb, S., Zairi, F., Mesbah, A., Nait-Abdelaziz, M., Gloaguen, J.M., Boukharouba, T. & Lefebvre, J.M.: Micromechanical modelling of the yield stress of polymer-particulate nanocomposites with an inhomogeneous interphase. *Procedia Engineering* 1 (2009), pp. 217–220.

Buryachenko, V.A., Roy, A., Lafdi, K., Anderson, K.L. & Chellapilla, S.: Multi-scale mechanics of nanocomposites including interface: Experimental and numerical investigation. *Compos. Sci. Technol.* 65 (2005), pp. 2435–2465.

Cai, M., Ruthkosky, M.S., Merzougui, B., Swathirajan, S., Balogh, M.P. & Oh, S.H.: Investigation of thermal and electrochemical degradation of fuel cell catalysts. *J. Power Sources* 160 (2006), pp. 977–986.

Carrette, L., Friedrich, K.A. & Stimming, U.: Fuel cells – Fundamentals and applications. *Fuel-Cells* 1 (2001), pp. 5–39.

Chen, J. & Chan, K.Y.: Molecular simulation, size-dependent mobility of platinum cluster on a graphite surface. *Mol. Simulat.* 31 (2005), pp. 527–533.

Cho, J., Luo, J.J. & Daniel, I.M.: Mechanical characterization of graphite/epoxy nanocomposites by multi-scale analysis. *Compos. Sci. Technol.* 67 (2007), pp. 2399–2407.

Coker, E.N., Steen, W.A., Miller, J.T., Kropf, A.J. & Miller, J.E.: Nanostructured Pt/C electrocatalysts with high platinum dispersions through zeolite-templating. *Micropor. Mesopor. Mat.* 101 (2007), pp. 440–444.

Collins, S., Stamatakis, M. & Vlachos, D.: Adaptive coarse-grained Monte Carlo simulation of reaction and diffusion dynamics in heterogeneous plasma membranes. *BMC Bioinformatics* 11 (2010), pp. 218–218.

Cotterell, B., Chia, J.Y.H. & Hbaieb, K.: Fracture mechanisms and fracture toughness in semicrystalline polymer nanocomposites. *Eng. Fract. Mech.* 74 (2007), pp. 1054–1078.

Dongsheng, W., Stephen, J.P. & James, A.E.: Effect of molecular weight on hydrated morphologies of the short-side-chain perfluorosulfonic acid membrane. *Macromolecules* 42 (2009), pp. 3358–3368.

Dorenbos, G. & Suga, Y.: Simulation of equivalent weight dependence of Nafion morphologies and predicted trends regarding water diffusion. *J. Membrane Sci.* 330 (2009), pp. 5–20.

Eisenberg, A. & Takahashi, K.: Viscoelasticity of silicate polymers and its structural implications. *J. Non-Cryst. Solids* 3 (1970), pp. 279–293.

Espanol, P.: Dissipative particle dynamics with energy conservation. *Europhysics Lett.* 40 (1997), pp. 631–636.

Fermeglia, M. & Pricl, S.: Multiscale molecular modeling in nanostructured material design and process system engineering. *Comput. Chem. Eng.* 33 (2009), pp. 1701–1710.

Fermeglia, M. & Pricl, S.: Multiscale modeling for polymer systems of industrial interest. *Prog. Org. Coat.* 58 (2007), pp. 187–199.

Galperin, D.Y. & Khokhlov, A.R.: Mesoscopic morphology of proton-conducting polyelectrolyte membranes of Nafion registered type: A self-consistent mean field simulation. *Macromol. Theor. Simul.* 15 (2006), pp. 137–146.

Grew, K.N., Joshi, A.S. & Peracchio, A.A.: Pore-scale investigation of mass transport and electrochemistry in a solid oxide fuel cell anode. *J. Power Sources* 195 (2010), pp. 2331–2346.

Hao, L. & Cheng, P.: Lattice Boltzmann simulations of water transport in gas diffusion layer of a polymer electrolyte membrane fuel cell. *J. Power Sources* 195 (2010), pp. 3870–3881.

Hao, X., Spieker, W.A. & Regalbuto, J.R.: A further simplification of the revised physical adsorption (RPA) model. *J. Colloid Interf. Sci.* 267 (2003), pp. 259–264.

Hoogebrugge, P.J. & Koelman, J.M.V.A.: Simulating microscopic hydrodynamic phenomena with dissipative particle dynamics. *Europhysics Lett.* 19 (1992), pp. 155–160.

Hsu, W.Y. & Gierke, T.D.: Ion transport and clustering in Nafion perfluorinated membranes. *J. Membrane Sci.* 13 (1983), pp. 307–326.

Huang, S.P. & Balbuena, P.: Platinum nanoclusters on graphite substrates: a molecular dynamics study. *Mol. Phys.* 100 (2002), pp. 2165–2174.

Hwankyu, L., de Vreies, A.H., Siewert-Jan, M. & Richard, W.P.: A coarse-grained model for polyethylene oxide and polyethylene glycol: Conformation and hydrodynamics. *J. Phys. Chem* B 113 (2009), pp. 13,186–13,194.

Izvekov, S. & Violi, A.: A coarse-grained molecular dynamics study of carbon nanoparticle aggregation. *J. Chem. Theor. Comput.* 2 (2006), pp. 504–512.

Jiang, F. & Sousa, A.C.M.: Smoothed particle hydrodynamics modeling of transverse flow in randomly aligned fibrous porous media. *Transport Porous Med.* 75 (2008), pp. 17–33.

Kissel-Osterrieder, R., Behrendt, F. & Warnatz, J.: Dynamic Monte Carlo simulations of catalytic surface reactions. *Symposium (International) on Combustion* 28 (2000), pp. 1323–1330.

Komarov, P.V., Veselov, I.N., Chu, P.P., Khalatur, P.G. & Khokhlov, A.R.: Atomistic and mesoscale simulation of polymer electrolyte membranes based on sulfonated poly(ether ether ketone). *Chem. Phys. Lett.* 487 (2010), pp. 291–296.

Lamas, E.J. & Balbuena, P.B.: Molecular dynamics studies of a model polymer-catalyst-carbon interface. *Electrochim. Acta* 51 (2006), pp. 5904–5911.

Lamas, E.J. & Balbuena, P.B.: Adsorbate effects on structure and shape of supported nanoclusters: A molecular dynamics study. *J. Phys. Chem* B 107 (2003), pp. 11,682–11,689.

Lim, C.Y., Shu, C., Niu, X.D. & Chew, Y.T.: Application of lattice Boltzmann method to simulate microchannel flows. *Phys. Fluids* 14 (2002), pp. 2299–2308.

Lim, D.H., Lee, W.D., Choi, D.H., Park, D.R. & Lee, H.I.: Preparation of platinum nanoparticles on carbon black with mixed binary surfactants: Characterization and evaluation as anode catalyst for low-temperature fuel cell. *J. Power Sources* 185 (2008), pp. 159–165.

Lim, D.H., Lee, W.D. & Lee, H.I.: Highly dispersed and nano-sized Pt-based electrocatalysts for low-temperature fuel cells. *Catal. Surv. Asia* 12 (2008), pp. 310–325.

López, C.A., Rzepiela, A.J., de Vries, A.H., Dijkhuizen, L., Hünenberger, P.H. & Marrink, S.J.: Coarse-grained force field: Extension to carbohydrates. *J. Chem. Theor. Comput.* 5 (2009), pp. 3195–3210.

Lucy, L.B.: A numerical approach to the testing of the fission hypothesis. *Astron. J.* 82 (1977), pp. 1013–1024.

Marrink, S.J., de Vries, A.H. & Mark, A.E.: Coarse grained model for semiquantitative lipid simulations. *J. Phys. Chem* B 108 (2004), pp. 750–760.

Marrink, S.J., Risselada, H.J., Yefimov, S., Tieleman, D.P. & de Vries, A.H.: The MARTINI force field: Coarse grained model for biomolecular simulations. *J. Phys. Chem* B 111 (2007), pp. 7812–7824.

Malek, K., Eikerling, M., Wang, Q., Liu, Z., Otsuka, S., Akizuki, K. & Abe, M.: Nanophase segregation and water dynamics in hydrated Nafion: Molecular modeling and experimental validation. *J. Chem. Phys.* 129 (2008), pp. 204,702–204,710.

Malek, K. & Franco, A.A.: Microstructure-based modeling of aging mechanisms in catalyst layers of polymer electrolyte fuel cells. *J. Phys. Chem.* B 115 (2011), pp. 8088–8101.

Meakin, P. & Tartakovsky, A.: Modeling and simulation of pore-scale multiphase fluid flow and reactive transport in fractured and porous media. *Rev. Geophys.* 47 (2009), RG3002.

Meng, B. & Weinberg, W.H.: Dynamical Monte Carlo studies of molecular beam epitaxial growth models: interfacial scaling and morphology. *Surf. Sci.* 364 (1996), pp. 151–163.

Morrow, B.H. & Striolo, A.: Morphology and diffusion mechanism of platinum nanoparticles on carbon nanotube bundles. *J. Phys. Chem.* C Nanomat. Interf. 111 (2007), pp. 17,905–17,913.

Mukherjee, P.P., Wang, C.Y. & Kang, Q.: Mesoscopic modeling of two-phase behavior and flooding phenomena in polymer electrolyte fuel cells. *Electrochim. Acta* 54 (2009), pp. 6861–6875.

Ng, T.Y., Yeak, S.H. & Liew, K.M.: Coupling of ab initio density functional theory and molecular dynamics for the multiscale modeling of carbon nanotubes. *Nanotechnology* 19 (2008), pp. 055702–055702.

Norskov, J.K., Rossmeisl, J., Logadottir, A., Lindqvist, L., Kitchin, J.R., Bligaard, T. & Jonsson, H.: Origin of the overpotential for oxygen reduction at a fuel-cell cathode. *J. Phys. Chem.* B 108 (2004), 17,886–17,892.

Olsen, R.A., Kroes, G.J. & Baerends, E.J.: Atomic and molecular hydrogen interacting with Pt(111). *J. Chem. Phys.* 111 (1999), pp. 11,155–11,163.

Pivovar, A.M. & Pivovar, B.S.: Dynamic behavior of water within a polymer electrolyte fuel cell membrane at low hydration levels. *J. Phys. Chem.* B 109 (2005), pp. 785–793.

Poelsema, B., Lenz, K. & Comsa, G.: The dissociative adsorption of hydrogen on defect-'free' Pt(111). *J. Phys. Condens. Mat.* 22 (2010), pp. 304,006–304,006.

Rafii-Tabar, H., Shodja, H.M., Darabi, M. & Dahi, A.: Molecular dynamics simulation of crack propagation in fcc materials containing clusters of impurities. *Mech. Mater.* 38 (2006), pp. 243–252.

Rai, V., Aryanpour, M. & Pitsch, H.: First-principles analysis of oxygen-containing adsorbates formed from the electrochemical discharge of water on Pt(111). *J. Phys. Chem.* C Nanomat. Interf. 112 (2008), pp. 9760–9768.

Ramanathan, S. & Koch, D.L.: An efficient direct simulation Monte Carlo method for low Mach number noncontinuum gas flows based on the Bhatnagar-Gross-Krook model. *Phys. Fluids* 21 (2009), pp. 033,103–033,111.

Risselada, H.J. & Marrink, S.J.: The molecular face of lipid rafts in model membranes. *PNAS* 105 (2008), pp. 17,367–17,372.

Scocchi, G., Posocco, P., Danani, A., Pricl, S. & Fermeglia, M.: Multiscale molecular modeling of polymer-clay nanocomposites. *Fluid Phase Equilibr.* 261 (2007), pp. 366–374.

Scott, D.W.: On optimal and data-based histograms. *Biometrika* 66 (1979), pp. 605–610.

Sinha Ray, S. & Okamoto, M.: Polymer/layered silicate nanocomposites: a review from preparation to processing. *Prog. Polym. Sci.* 28 (2003), pp. 1539–1641.

Song, W., Yu, H., Hao, L., Miao, Z., Yi, B. & Shao, Z.: A new hydrophobic thin film catalyst layer for PEMFC. *Solid State Ionics* 181 (2010), pp. 453–458.

Sun, L., Gibson, R.F., Gordaninejad, F. & Suhr, J.: Energy absorption capability of nanocomposites: A review. *Compos. Sci. Technol.* 69 (2009), pp. 2392–2409.

Tartakovsky, A.M. & Meakin, P.: A smoothed particle hydrodynamics model for miscible flow in three-dimensional fractures and the two-dimensional Rayleigh-Taylor instability. *J. Comput. Phys.* 207 (2005), pp. 610–624.

Tartakovsky, A.M., Meakin, P., Scheibe, T.D. & Eichler West, R.M.: Simulations of reactive transport and precipitation with smoothed particle hydrodynamics. *J. Comput. Phys.* 222 (2007), pp. 654–672.

Wang, B., Ma, X., Marco, C., Renald, S. & Li, W.X.: Size-selective carbon nanoclusters as precursors to the growth of epitaxial graphene. *Nano Lett.* 11 (2011), pp. 424–430.

Wang, J.X., Springer, T.E. & Adzic, R.R.: Dual-pathway kinetic equation for the hydrogen oxidation reaction on Pt electrodes. *J. Electrochem. Soc.* 153 (2006), A1732–A1740.

Wei, Z.D., Ran, H.B., Liu, X.A., Liu, Y., Sun, C.X., Chan, S.H. & Shen, P.K.: Numerical analysis of Pt utilization in PEMFC catalyst layer using random cluster model. *Electrochim. Acta* 51 (2006), pp. 3091–3096.

Wescott, J.T., Qi, Y., Subramanian, L. & Weston Capehart, T.: Mesoscale simulation of morphology in hydrated perfluorosulfonic acid membranes. *J. Chem. Phys.* 124 (2006), pp. 134,702–134,714.

Xiao, Y., Dou, M., Yuan, J., Hou, M., Song, W. & Sundén, B.: Fabrication process simulation of a PEM fuel cell catalyst layer and its microscopic structure characteristics. *J. Electrochem. Soc.* 159 (2012), B308–B314.

Xiao, Y., Yuan, J. & Sunden, B.: Review on the properties of nano-/microstructures in the catalyst layer of PEMFC. *ASME J. Fuel-Cell Sci. Technol.* 8 (2011), 034001.

Xiao, Y., Yuan, J. & Sundn, B.: Process based large scale molecular dynamic simulation of a fuel cell catalyst layer. *J. Electrochem. Soc.* 159 (2012), B251–B258.

Yamamoto, S., Maruyama, Y. & Hyodo, S.A.: Dissipative particle dynamics study of spontaneous vesicle formation of amphiphilic molecules. *J. Chem. Phys.* 116 (2002), pp. 5842–5849.

Zeng, Q.H., Yu, A.B. & Lu, G.Q.: Multiscale modeling and simulation of polymer nanocomposites. *Prog. Polym. Sci.* 33 (2008), pp. 191–269.

Zhou, P. & Wu, C.W.: Liquid water transport mechanism in the gas diffusion layer. *J. Power Sources* 195 (2010), pp. 1408–1415.

Zhu, Y. & Fox, P.J.: Simulation of pore-scale dispersion in periodic porous media using smoothed particle hydrodynamics. *J. Comput. Phys.* 182 (2002), pp. 622–645.

# CHAPTER 4

## Fabrication of electro-catalytic nano-particles and applications to proton exchange membrane fuel cells

Maria Victoria Martínez Huerta & Gonzalo García

### 4.1 INTRODUCTION

The proton exchange membrane fuel cell (PEMFC) is considered a very promising system able to provide energy with high efficiency by using pure hydrogen as fuel. However, the performance and durability of the key components, such as membrane electrode assemblies (MEAs), membranes, electrodes, gaskets, or bipolar plates, remain the focus of international research and development (R&D) (Gasteiger and Markovic, 2009; Kunze and Stimming, 2009; Millet *et al.*, 2011; Zhong *et al.*, 2008).

At the heart of a PEMFC is the MEA, which is comprised of two catalytic and two diffusion layers. These layers play a critical role in defining the performance of the MEA. To play its essential function, an electro-catalyst needs to provide high intrinsic activities for the electrochemical oxidation of a fuel at the anode, whether this is hydrogen or alcohol (methanol, ethanol), and for the electrochemical reduction of oxygen at the cathode. Other requirements include high electrical conductivity, appropriate physical and electrical contact with the ionomer, suitable reactant/product gas access/exit, and high stability in the highly corrosive working medium (Thompsett, 2003; Vielstich *et al.*, 2003). To ensure that a fuel cell delivers maximum efficiency, both electrode reactions need to take place as close to their thermodynamic potential as possible.

Research over several decades has found that platinum and platinum-containing catalysts are the most effective catalytic materials for PEMFCs; nonetheless, platinum-based catalysts are expensive and have low durability (Borup *et al.*, 2007; Thompsett, 2003; Rabis *et al.*, 2012). Studies have been intensively conducted to significantly reduce the Pt loading in both electrodes and/or enhance the Pt electro-catalytic efficiency. Alternatively, Pt-free catalysts have been particularly studied to replace Pt (Chen, 2011).

In the past decade, the advances of nano-science and nano-technology have opened new ways to develop novel electro-catalysts for fuel cells (Koenigsmann *et al.*, 2010; Peng and Yang, 2009; Qiao and Li, 2011). The use of nano-materials in fuel cell systems can significantly improve the electro-catalytic performance to achieve high energy density and high power density, while reducing the manufacturing cost. The prominent electro-catalytic behavior of the nano-particles is supplied mainly from their unique physical-chemical properties such as size, shape, pore structure/distribution, surface atomic array, and chemical properties. In comparison to conventional electro-catalysts, the nano-structured counterparts can provide higher catalytic activity and a larger active specific surface area, thereby reducing efficiently the noble metal loading such as Pt material.

It is the aim of this chapter to review recent progress on nano-structured electro-catalysts for application to PEMFCs and direct methanol fuel cells (DMFCs). As one of the most important PEMFCs, DMFCs have been recognized as a potential future power source for portable electronic devices (Aricò *et al.*, 2001; Basri *et al.*, 2010; Guo *et al.*, 2008; Hamnett, 2003; Kamarudin *et al.*, 2009; Zhao *et al.*, 2011). In particular, the relationships between nano-structure and electro-catalysis, and the catalytic mechanisms are discussed. Special attention has been paid in the latest avenues for making low-cost and effective catalysts such as novel nano-structures of Pt, binary and ternary platinum-based catalysts, new cost-effective synthesis routes, and new catalyst supports to replace the generally used carbon black. Besides, ideally the Pt-based catalysts should be replaced with abundant, non-precious materials.

## 4.2   OVERVIEW OF THE ELECTRO-CATALYTIC REACTIONS

### 4.2.1   *Hydrogen oxidation reaction*

The adsorption of hydrogen on metal electrodes such as platinum has been studied extensively in electrochemical systems over the last several decades (Markovic *et al.*, 2002). Non-noble metals also develop high activity for hydrogen oxidation reactions (HORs), but in acidic electrolytes, noble metals show the greatest stability towards corrosion or passivation.

The mechanism for the hydrogen oxidation reaction on a Pt electrode in an acid electrolyte can proceed through two main pathways, Tafel-Volmer and Heyrosky-Volmer, both of which involve the adsorption of molecular hydrogen ($H_{ad}$), followed by a charge transfer step:

$$\text{Tafel:} \quad 2Pt + H_2 \rightarrow 2Pt - H_{ad} \tag{4.1}$$

$$\text{Heyrosky:} \quad Pt + H_2 \rightarrow Pt - H_{ad} + H^+ + e^- \tag{4.2}$$

$$\text{Volmer:} \quad Pt - H_{ad} \rightarrow Pt + H^+ + e^- \tag{4.3}$$

It has been established that $H_{ad}$ is the reaction intermediate during the HOR, and consequently the kinetic is mainly determined by the interactive strength between $H_{ad}$ and the Pt surface atoms (Markovic *et al.*, 2002). There are two different possible states of adsorbed hydrogen. The $H_{upd}$ (underpotential deposition of hydrogen), which is the "strongly" adsorbed state formed on the surface at potentials more positive than the Nernst potential, and the $H_{opd}$ (the overpotential deposition of hydrogen), which is the "weakly" adsorbed state formed at potentials close to, or more negative with respect to the Nernst potential (Markovic *et al.*, 2002; Rojas *et al.*, 2012).

The major effort in HOR electro-catalysis has been focused on understanding the rate dependency on the atomic-scale morphology of a platinum single-crystal surface. Recently, catalysis studies on well-defined Pt single-crystal electrodes clearly demonstrated that the delivered current during the HOR on Pt (hkl) varies with the crystal face symmetry, i.e., it is a "structure sensitivity" reaction (Markovic, 2003; Markovic *et al.*, 2002). The facets of a well formed crystal, or internal planes through a crystal structure, or a lattice, are specified in terms of miller indices, h, k and l. These indices (hkl) represent the set of all parallel planes.

### 4.2.2   *H₂/CO oxidation reaction*

While pure $H_2$ is the ideal fuel for the anode of a PEMFC, the production, storage, and infrastructure necessary for pure hydrogen supply have not been resolved yet, so the hydrogen is produced with CO impurities. The impurities in the hydrogen fuel remain a topic for research, although modern reformers typically produce a hydrogen stream with trace amounts of CO, for instance, 20–50 ppm. However, even low CO concentrations (<5 ppm) can affect the catalytic Pt surface capability of the anode, and reduce the cell performance (Oetjen *et al.*, 1996). Thus, the development of CO tolerant catalysts for PEMFC anodes is still required to improve the performance and durability of PEMFCs operating with hydrogen produced from a reformate process.

Furthermore, CO results from the incomplete oxidation of small organic molecules (methanol, ethanol, etc.) in alcohol fuel cells. The produced CO is adsorbed strongly, poisoning the platinum surface and thereby severely limiting the catalytic oxidation activity (Iwasita, 2003). CO formed along methanol dissociative adsorption on pure Pt, blocks a considerable fraction of the electrode surface in a matter of seconds. At potentials below 0.40 V, the methanol oxidation reaction cannot progress due to the inability of Pt to form enough oxygenated species (i.e. OH) for further oxidation, and the current falls to negligible values. Consequently, an important, although not unique, aspect of the catalysis of methanol oxidation is related to the catalysis of CO oxidation reactions.

Despite its importance for low temperature fuel cells, the exact mechanism of CO oxidation on Pt and the role of cocatalysts (promoters) are still far from being understood (García and Koper, 2011; Koper *et al.*, 2009). It is generally accepted that the oxidation of CO proceeds according to a Langmuir-Hinshelwood mechanism, suggested by Gilman more than 40 years ago (Gilman, 1964). According to Gilman's model, water needs to be activated on a free site on the surface, leading to surface bonded OH:

$$H_2O + * \leftrightarrow OH_{ads} + H^+ + e^- \qquad (4.4)$$

where the * notes a free site on the platinum surface. Next, a chemical reaction between both adsorbed species is supposed to take place:

$$CO_{ad} + OH_{ad} \rightarrow COOH_{ad} \qquad (4.5)$$

followed by a second charge transfer reaction:

$$COOH_{ad} \rightarrow CO_2 + H^+ + e^- + 2* \qquad (4.6)$$

Reaction (4.5) is typically considered to be the rate-determining step, because the Tafel slope observed is close to $60\,mV\,dec^{-1}$. Furthermore, using a series of stepped Pt electrodes, it has been concluded that all CO, both that initially adsorbed at or near the step, and that initially adsorbed on the terrace, reacts to $CO_2$ at the step sites. Thus, CO oxidation on the terrace is essentially negligible, and all CO will diffuse rapidly to the step or defect sites and be oxidized there, resulting in only one anodic peak during the CO stripping experiment. Furthermore, the main anodic CO stripping peak shifts towards lower potentials as the step density is increased, i.e., water dissociation (Equation (4.4)) is enhanced at low coordinated sites (García and Koper, 2011).

For many years it has been well known that CO electro-oxidation on platinum is a structure-sensitive reaction (Chang *et al.*, 1989; Lamy *et al.*, 1983). Studies with single crystal electrodes have shown that the kinetic parameters depend not only on the surface composition of the catalyst but also on the symmetry of the surface, and that the presence of steps and defects alters significantly the reaction rate (Lebedeva *et al.*, 2000). Being so, the surface structure of the nano-particles should also affect its performance for the oxidation of CO. Understanding how the different variables affect CO oxidation on Pt nano-particles dispersed on carbon, requires the control of the platinum surface in a similar way to that which has been achieved for single crystal electrodes. In this sense, the influence of the surface site distribution on CO oxidation using nano-particles of well-defined shapes has been reported (Kinge *et al.*, 2008; Solla-Gullon *et al.*, 2006). For the carbon supported nano-particle electrodes, the shape is generally more complex than that found for single crystal electrodes, since it is affected by the surface structure and nano-particle size and aggregation. López-Cudero *et al.* (2010) have studied the CO oxidation on platinum nano-particles deposited on carbon with different platinum loadings (from 10 to 50%), where the increased loading leads to nano-particle agglomeration. The results demonstrate that CO oxidation takes place at lower overpotentials when agglomeration takes place, since OH and CO species that participate in the reaction are adsorbed on different nano-particles, that is, it is an inter-particle process.

It is well established that binary systems of CO-tolerant electro-catalysts, with Pt as one of the components, can exhibit a substantial resistance to the presence of CO in the fuel stream. It has been found that the use of a second element with Pt, such as Ru, Sn, Co, Ta, Fe, Ni, Au, Mo, W, Ti, etc., in the form of an alloy or a co-deposit, yields significant improvement in the CO-tolerance relative to pure Pt (Abe *et al.*, 2008; Gasteiger *et al.*, 1995a; 1995b; Götz and Wendt, 1998; Grgur *et al.*, 1997; Liu *et al.*, 2009; Papageorgopoulos *et al.*, 2002; Stevens *et al.*, 2007; Watanabe *et al.*, 1999). Among these various Pt-based systems, the most commonly used is PtRu supported on carbon black, whose study started as early as the 1960s as a very promising alloy for the methanol oxidation (Petrii, 2008). To date, this binary catalyst is considered to be the state-of-the-art catalyst for this type of fuel cell and its preparation conditions and intermetallic interactions have been extensively studied. During the 1970s the group of Watanabe-Motoo (Watanabe and

Motoo, 1975a; 1975b; Watanabe *et al.*, 1987), considering a possible nature of high PtRu activity, formulated the mechanism of bifunctional catalysis. According to their theory, ruthenium atoms are responsible for the generation of active oxygen species at potentials more negative than platinum:

$$Ru + H_2O \rightarrow RuOH + H^+ + 1e^- \quad E_0 \approx 0.2 \, V \, NHE \tag{4.7}$$

Therefore, CO adsorbed on platinum sites diffuses and reacts to form carbon dioxide at Ru sites. A second mechanism concerning an electronic effect between the interaction of RU and Pt was also proposed. This mechanism is widely recognized, especially after the *in situ* FTIR work of the Iwasita-Vielstich team in the early 1990s that is associated with an energy shift of the Ptd electronic states caused by the second element, and resulted in a weakening of the Pt-CO bond (Iwasita, 2002; 2003).

### 4.2.3   *Methanol oxidation reaction*

Methanol electro-oxidation is a complex reaction involving the transfer of six electrons and several catalytic steps. Although the thermodynamic potential for the full electro-oxidation of methanol in acid electrolytes is close to that of hydrogen oxidation, the overall reaction is much more demanding due to the multi-electron transfer steps to form carbon dioxide. The half reaction is:

$$CH_3OH + H_2O \rightarrow CO_2 + 6H^+ + 6e^- \quad (E = 0.04V \; vs. \; RHE) \tag{4.8}$$

However, the energy density of the direct methanol fuel cell (DMFC) is still far from that expected due to the methanol crossover and the high overpotentials at the electrodes (Aricò *et al.*, 2001; Basri *et al.*, 2010; Guo *et al.*, 2008; Hamnett, 2003; Kamarudin *et al.*, 2009; Zhao *et al.*, 2011). The anode presents sluggish kinetics of the methanol oxidation reaction and consequently higher metal loading is needed. In fact, the increment of noble metal is about ten times higher than that used in hydrogen-fed PEMFC.

Concerning methanol oxidation on Pt-based electrodes in acidic media, much effort was done during the last decades. For a detailed discussion, the reader is referred to several literature reviews (Iwasita, 2002; Koper *et al.*, 2009; Markovic *et al.*, 2002; Watanabe and Uchida, 2009). To summarize, the mechanism for methanol oxidation involves two main steps:

i) Electro-sorption and dehydrogenation of methanol onto the substrate: methanol needs a suitable surface for its adsorption and the subsequent dehydrogenation steps. In this sense, it is well known that at least three neighboring Pt atoms are necessary to fully complete the dehydrogenation steps yielding adsorbed carbon monoxide ($CO_{ad}$) (Iwasita, 2002). In addition to $CO_{ad}$, several intermediates and by-side products such as formaldehyde and formic acid can be formed (Batista *et al.*, 2003; García *et al.*, 2011; Planes *et al.*, 2007). Recently, Cuesta *et al.* (2006) showed the importance of the ensemble effect during the methanol dehydrogenation on a Pt(111) surface modified with cyanide groups. They demonstrated that methanol could be fully oxidized to carbon dioxide, avoiding the principal catalytic poisoning ($CO_{ad}$), when only two contiguous Pt atoms are present on the surface. It is noticeable that this step becomes important at low (ambient) temperatures (Aricò *et al.*, 2001; García *et al.*, 2007).

ii) Electro-oxidation of $CO_{ad}$: the addition of oxygenated species (O or OH from water) to $CO_{ad}$ to form $CO_2$ is needed, and consequently the catalytic surface is liberated. There is a large body of work of current agreement that the electro-oxidation of $CO_{ad}$ follows the Langmuir-Hinshelwood mechanism (Section 4.2.2).

Pure Pt electrodes initially show very high activity for methanol oxidation, but these very rapidly decay in current on the formation of strongly bound intermediates. Several binary and ternary catalysts have been proposed for methanol oxidation, most of them based on modifications of Pt with some other metal including Ru, Mo, Rh, Os, Sn, Ni, Mo, Ce, W, Ti, and Ir (Abida *et al.*, 2011; Choi *et al.*, 2004; García *et al.*, 2011; Gómez de la Fuente *et al.*, 2009; Huang *et al.*, 2006b; Jian *et al.*, 2009; Martínez-Huerta *et al.*, 2006; 2008; Siwek *et al.*, 2008). At present, there is a general consensus that PtRu offers the most promising results (Petrii, 2008). The catalytic effect has been observed in different kinds of PtRu materials, such as carbon supported and unsupported Pt/Ru alloys (Anumol *et al.*, 2011; Cui *et al.*, 2011; García *et al.*, 2011; Ma *et al.*, 2009; Teng *et al.*, 2007; Wang, H. *et al.*, 2009). The reason for the enhanced rate of methanol oxidation on PtRu is often invoked by the bifunctional mechanism (Section 4.2.2) (Gasteiger *et al.*, 1993). Moreover, the effect of modification in a Pt electronic environment, induced by Ru through an increase in Pt d-band vacancies, has been emphasized (Section 4.2.2) (Iwasita, 2002).

### 4.2.4   *Oxygen reduction reaction*

The oxygen reduction reaction (ORR) is the primary electrochemical reaction occurring at the cathode of a PEMFC, and is central to this promising technology for efficient and clean energy generation. The ORR is a multi-electron reaction that follows the direct four-electron mechanism on platinum-based electro-catalysts. It appears to occur in two pathways in acid electrolytes (Adzic and Lima, 2009):

(i) A "direct" four-electron pathway, wherein four electrons are transferred in concert:

$$O_2 + 4H^+ + 4e^- \rightarrow 2H_2O \tag{4.9}$$

(ii) A "series" pathway, wherein electrons are transferred consecutively:

$$O_2 + 4H^+ + 4e^- \rightarrow H_2O_2 + 2H^+ + 2e^- \tag{4.10}$$

$$H_2O_2 + 2H^+ + 2e^- \rightarrow 2H_2O \tag{4.11}$$

Considerable ongoing research aims to increase the electro-catalytic activity of platinum for the ORR. A particularly difficult problem to resolve is the large loss in potential (0.3–0.4 V) mostly at the cathode, which is the source of major decline in the fuel cell's efficiency. Another drawback of existing electro-catalyst technology is the high Pt loading in cathode electro-catalysts, typically in the range of 0.1–0.5 mg cm$^{-2}$.

Many authors have been discussing the surface structure dependence of Pt toward the ORR under different conditions. Markovic *et al.* (1995) found that the activity for ORR in 0.1 M HClO$_4$ decreases in the sequence of the Pt low-index planes (110) > (111) > (100), whereas the reactivity in H$_2$SO$_4$ increased in the sequence (111) < (100) < (110). These differences in the sequence are ascribed to the strong bisulfate anion adsorption on the highly coordinated surfaces. Therefore, it is important to consider the strength of the anion adsorption and its surface-structure-dependent adsorption properties to understand the oxygen reduction reaction (Rabis *et al.*, 2012).

The catalytic activity of Pt toward the ORR strongly depends on the adsorption energy and dissociation energy of O$_2$, and on the adsorption energy of the OH species on the surface (Mukerjee and Srinivasan, 2003). Alloying Pt with other transition metals has been demonstrated as a successful approach toward advanced ORR electro-catalysts (Greeley *et al.*, 2009; Hernandez *et al.*, 2007; Strasser *et al.*, 2010; Wang, J.X. *et al.*, 2009; 2012b; Zhang *et al.*, 2007). Fundamental studies of well-defined extended surfaces have shown that the enhanced catalytic activity originates from the modified electronic structures of Pt in these alloy catalysts (Wang, C. *et al.*, 2012b), which reduces the adsorption of oxygenated spectator species (e.g., OH$^-$) and thus, increases the number of active sites accessible to molecular oxygen. However, even the best of these catalysts, Pt, is at least 10$^6$ times less active for oxygen reduction than for H$_2$ oxidation.

The effort to increase the active catalyst surface area per unit mass of Pt has centered in recent years on optimization of catalyst *layer* properties, aiming to maximize "catalyst utilization" in fuel cell electrodes based on Pt catalyst particle sizes of 2–5 nm (Gottesfeld, 2009). High catalyst utilization is conditioned on access to the largest possible percentage of the total catalyst surface area embedded in a catalyst layer by the three participants in the electrochemical process— gaseous reactants, protons and electrons—all at the rates called for by the demand current density. Fulfilling the latter condition requires a composite catalyst layer structure in which the electron

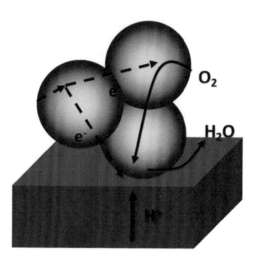

Figure 4.1.    Electrochemical process across the cathodic layer.

mobility, the proton mobility, and the effective gas diffusivity across the thickness dimension of the catalyst layer, are all sufficient at the demand current, to access the maximum fraction of the catalyst particles dispersed uniformly in the (5–20 μm thick) catalyst layer (Fig. 4.1).

## 4.3   NOVEL NANO-STRUCTURES OF PLATINUM

The understanding obtained from fundamental scientific studies can enable us to design and tailor Pt electro-catalysts with a preferential surface structure, in order to enhance the fuel oxidation and oxygen reduction at the anode and cathode of a PEMFC, respectively. However, the activity connection between single crystal and practical catalysts is not so straightforward.

### 4.3.1   *State-of-the-art supported Pt catalysts*

Practical Pt catalysts are carbon supported nano-structured materials that maximize the surface over volume ratio and thus increase the number of catalytic sites per mass. Consequently, with a proper particle arrangement a high fuel turnover rate at low electrode polarization is expected, which may lead to a high power output of the fuel cells. Furthermore, it is predictable that the size, shape, and dispersion uniformity of Pt nano-structures on supporting materials will have significant effects on the catalytic activity and durability. In this sense, it is recognized that electronic and geometrical factors play a fundamental role in order to determine the reactivity and stability (Adzic, 2012; García and Koper, 2011; Hammer and Norskov, 2000). Nowadays, novel catalyst supports based on ceramics, metal oxides (including transition and rare earth metal oxides), carbides and nitrides, are gaining attention, as alternatives to the typically used high surface area carbons. Important factors are their higher stability (less corrosive) and interaction with the Pt particle, which could produce synergetic effects in the activity (Guillén-Villafuerte *et al.*, 2011; 2013; Ho *et al.*, 2011). On the other hand, catalyst synthesis is well developed at present with the possibility to control almost every aspect of Pt nano-catalysts, such as, shape and particle sizes with narrow distribution (Burda *et al.*, 2005; Vidal-Iglesias *et al.*, 2012; Xia *et al.*, 2009). However, some drawbacks still exist and need to be improved, e.g., catalysts surface cleaning, large-scale production with high producibility and stability, and elevated cost. To lower the cost of Pt catalysts, great efforts have focused on the development of novel nano-structures of Pt catalysts, such as nano-particles, nano-wires, or nanotubes, with a high surface area to achieve high catalytic performance and utilization efficiency (Guo *et al.*, 2008; Viswanathan *et al.*, 2012).

### 4.3.2   *Surface structure of Pt catalysts*

The most common metallic crystal structures are body-centered cubic (bcc), face-centered cubic (fcc) and hexagonal close-packed (hcp), with the fcc bulk structure corresponding to Pt, still the most commonly used metal in fuel cells. Nowadays, the shape-controlled synthesis of different nano-structured morphologies is well developed. Thus, it might provide structure-sensitive catalysts such as a tetrahedron and a cube Pt nano-particle, which present (111) and (100) surface facets, respectively. Therefore, different catalytic activities are expected from both nano-crystals. In this sense, Pt nano-structures with a high density of (111) and (100) orientation sites on the surface were tested for the ORR in sulfuric and perchloric acid (Adzic, 2012; Sun *et al.*, 2007; Wang *et al.*, 2008). Higher and lower catalytic activity was observed in sulfuric and perchloric acid at catalysts with (111) surface facets, due to the different adsorption rates of sulfates and perchlorates on this surface, in perfect agreement with those results obtained using Pt single crystals. On the other hand, real catalysts should present the highest quantity of catalytic sites, that is, low-coordinated sites (e.g., edge, corner, step, and kink sites). Fundamental studies on stepped Pt single-crystal surfaces have shown that low-coordinated sites exhibit much higher catalytic activity than that of low-index planes for the CO electro-oxidation (García and Koper, 2011). However, this site exhibits a dual catalytic role, that is, it presents both the highest and lowest catalytic activity, at least towards the CO oxidation reaction (García and Koper, 2009).

### 4.3.3   *Synthesis and performance of Pt catalysts*

The stability of nano-particles is an issue to consider. Usually, the size and shape of active sites, such as low-coordinated atoms, change during the catalytic reaction. Numerous synthetic methods have been developed for the preparation of Pt nano-structures in the literature, each providing varying degrees of size and shape control (Burda *et al.*, 2005; Peng and Yang, 2009; Vidal-Iglesias *et al.*, 2012; Xia *et al.*, 2009). Pt nano-structures are, in all cases, obtained through electron transfer reactions by chemical and electrochemical reduction of Pt salts (precursor) besides thermal decomposition, photo- and sono-chemical reduction (Kundu *et al.*, 2009). The difference between both usually adopted methods are the electrons, which are provided by the cathode and an external power source (electrochemical), or transferred from the parallel proceeding oxidation of a reducing agent (chemical). Among reduction strategies, three important steps delineate the nucleation and growth of Pt nano-structures: the transport of complexed Pt (usually Pt is a complex rather an ion), the expression of the electro-active species (which include not only Pt ion), and the electron transfer. Manipulating the rate of these parameters is the key to developing desired Pt structures. Additionally, to these parameters, energetic factors related to critical nuclei size have to be taken into account. Once the particle size exceeds this critical nucleus size, the particle growth is thermodynamically allowed (Finney and Finke, 2008; Peng and Yang, 2009).

   Therefore, common routes to produce Pt nano-structures can be divided by the electron supplier, e.g., chemical or electrochemical deposition. The foremost is the most popular method to produce well-shaped Pt nano-particles. In this way, catalysts are synthesized by an autocatalytic chemical reduction of the precursor (Pt salt) in an aqueous or organic solution in the presence of a reducing agent, stabilizer or capping agent, and occasionally a buffer, to control the pH (Kloke *et al.*, 2011). The surface of the recently produced metallic particle serves as a catalyst for further Pt ions' reduction (homogeneous nucleation). This methodology presents high control on size, morphology, and exposed crystallographic planes of Pt nano-particles (Burda *et al.*, 2005; Peng and Yang, 2009; Vidal-Iglesias *et al.*, 2012; Xia *et al.*, 2009).

   The size control is essential because by decreasing the particle diameter, a higher percentage of surface atoms are exposed, and thus Pt utilization increases. In this sense, there is huge debate about particle size and catalytic activity (Maillard *et al.*, 2009), in which maximum catalytic activities are found for diameters in the range of 2–5 nm (Sun *et al.*, 2007). A decrease in the particle size leads to an increment of low-coordinated atoms as well as that of the oxophilic character of the surface, which are associated to raise and decrease the catalytic activity, respectively. Additionally, effects associated to changes in the lattice parameter are expected, e.g., lower particle size produces an expansion of the lattice parameter into the surface. Relating to the synthesis of Pt nano-particles, lower sizes will be obtained by increasing the nucleation rate at the expense of the growth rate. Therefore, the most important parameters that affect the particle size are the reaction temperature, the salt concentration, the strength of the reducing agent, and the type of capping agent as well as its concentration (Finney and Finke, 2008; Kloke *et al.*, 2011; Peng and Yang, 2009). For example, the nucleation rate is enhanced by high temperatures, high precursor concentrations, and strong reducing agents. Additionally, a capping agent adsorbs on the metal surface of the growing particle and consequently, the growth rate decreases with the concentration of the capping agent. Another important issue is the size distribution. To obtain homogeneous size distribution, the nuclei must be formed at the same time and grow all together. To avoid agglomeration (Ostwald ripening), once precursor concentration diminishes the particle growth should be stopped (Burda *et al.*, 2005). Again, the use of capping agents prevent the particle agglomeration for the same reasons as explained above. A different method is the use of seeds (small particles obtained before ripening) that enables the separation of nucleation and growth phases. In this sense, Pt salt reduction at seed is preferred (heterogeneous nucleation) over homogeneous nucleation, and an epitaxial overgrowth is obtained, and thus the use of seeds represents a method for controlling the size and shape, as well as producing core-shell nano-particles if the nature of the seed is different from Pt (Qiao and Li, 2011).

It is well known that different atomic surface geometrical arrangements produce diverse catalytic properties (Adzic, 2012; Cuesta, 2006; García and Koper, 2011; Guillén-Villafuerte *et al.*, 2011; Hammer and Norskov, 2000; Koper, 2005). For example, Pt nano-discs that present a bi-dimensional (111) order domain on the surface and a high density of step edge sites with (110) and (100) orientations, were highly active towards CO and methanol oxidation reactions (Fig. 4.2). It was observed that the Pt 2D structures were up to 250% more efficient per catalytic surface area for the electro-oxidation of methanol in acidic media, as compared to a commercial Pt/C catalyst (Guillén-Villafuerte *et al.*, 2011). Consequently, the shape control is as important as particle size regulation.

Pt belongs to the face-centered cubic (fcc) symmetry. Therefore, the thermodynamic equilibrium geometry of Pt is truncated octahedron and its surface energy increases in the following plane symmetry way: (111) < (100) < (110). In this sense, capping agents are organic or inorganic molecules that adsorb onto the surface of nano-particles, and not only prevent agglomeration, but also control their shape. Capping agents adsorb specifically on different crystallographic planes and as a consequence, the choice and concentration can be used to control the particle shape. Thus, the essential requirement to produce the desired particle shape is to control the growth kinetics along the different crystallographic orientations. Up to now, an enormous variety of different particle geometries have been synthesized, the cubic, cub-octahedral, and octahedral being the most common (Burda *et al.*, 2005; Chuan-Jian *et al.*, 2010; Peng and Yang, 2009; Qiao and Li, 2011; Vidal-Iglesias *et al.*, 2012; Xia *et al.*, 2009).

With regards to reducing agents, their strengths promote the preferred effect (in combination with others parameters, see above). Stronger reducing agents will increase the ratio of nucleation against the growth rate and, consequently, a smaller particle size will be obtained. On the other hand, weak reducing agents such as ethylene glycol will produce a narrow size with uniform distribution and high contribution of low-index (111) and (100) planes that present the lowest surface energy (Xia *et al.*, 2009), that is, a longer time will be taken for the adsorbed ions to arrive in position with a higher surface energy (kink or defect sites). Typical chemical reducing agents include hydrogen, hydrazine, sodium borohydride, citrate, ascorbic acid, and alcohols. In most of these cases, the preparation of shape-controlled Pt nano-particles requires the use of a capping layer or stabilizer that must be completely removed before use as electro-catalysts (Koper, 2011). This step seems to be rather difficult and often causes misinterpretations in the analysis of the results. In this sense, physicochemical characterization is regularly performed before the cleanness treatment or electrochemical test. Thus, removing the surface-stabilizing

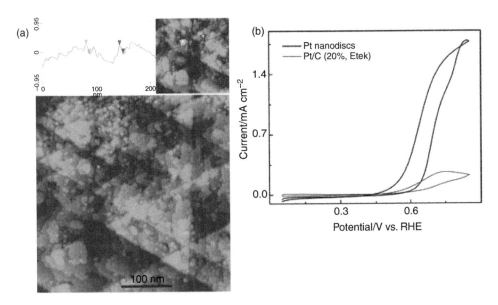

Figure 4.2.   (a) STM image (bottom panel) and the corresponding cross section (upper panel) of Pt nano-discs on Au(111) substrate. (b) Cyclic voltammetry for methanol oxidation on Pt nano-discs (thick solid line) and Pt/C (thin solid line) in 0.5 M $H_2SO_4$, sweep rate 20 mV s$^{-1}$, $E_i = 0.05$ V. (Reprinted with kind permission from Electrocatalysis 2, 231 (2011). Copyright 2011, Springer Science and Business Media).

agent by temperature, electrochemical "activation", or other aggressive treatment regularly leads to different particle shapes and sizes to the original one (García *et al.*, 2007). Consequently, for an fcc metal, such as Pt, the particle will try to have the surface energy as low as possible. Therefore, "clean" Pt nano-particles often contain high contributions of low-index (111) and (100) planes that present the lowest surface energy (Xia *et al.*, 2009).

Finally, electrode formation from nano-particles is usually achieved by the deposition of liquid ink containing Pt nano-particles onto a carbon-supported material, and the surface roughness can be varied by the thickness of the catalyst layer. Besides nano-particle synthesis in solution, the electro-deposition method is widely used to fabricate highly active electro-catalyst structures on top of a conductive electrode (Planes *et al.*, 2007). In this sense, electro-deposited meso-porous Pt electrodes were evaluated as catalysts for CO and methanol electro-oxidation, and the electrochemical analysis showed a mass activity during methanol oxidation at 0.55 V in acidic media, similar to carbon supported catalysts (Planes *et al.*, 2007). This methodology has high controllability using a current source, and the formed structure is covalently bound to the substrate. Therefore, the produced electro-catalyst has a stable structural network and works as a current collector with low electrical resistivity. However, the shape control is not as good as nano-particle synthesis in solution and a significant fraction of the noble metal remains inside the bulk and, consequently, cannot contribute to the catalytic activity. In practice four main approaches with different advantages and disadvantages are used for electro-catalyst synthesis: (i) the galvanostatic method, in which the deposited mass is governable but the electrochemical process can vary during the metal deposition, (ii) the potentio-static mode, in which the electrochemical process is known but the deposited charge is without linear correlation with the time of the potentio-static pulse, (iii) potentio-dynamic and (iv) potentio-static pulses, wherein the information of the growing structure can be extracted and different processes can be varied, respectively (Mard and Faulkner, 2000). In a typical electro-deposition, the nucleation sites are randomly distributed onto the conductive surface and in order to obtain narrow size distribution the double pulse technique is frequently used, i.e., short times applying high overpotential (nucleation) followed by a pulse with low overpotential (growth). Additionally, weak reducing agents such as ethylene glycol that produce narrow sizes with uniform distribution are often utilized (Tian *et al.*, 2007). Also, using ascorbic acid the formation of platinum catalysts rich in low coordinated sites was observed (Tian *et al.*, 2007; 2008). On the other hand, the roughness factor increases drastically from unsupported (20–500) to supported electro-catalysts (3000) (Coutanceau, 2004; Kloke *et al.*, 2011; Planes *et al.*, 2007). Practical electrodes employ support materials with high surface areas such as carbon black or carbon nanotubes, and the ionomer can be incorporated before or after the electro-deposition process.

In addition to the above, the use of templates in the electro-deposition process is regularly utilized (Kloke *et al.*, 2011; Planes *et al.*, 2007; Qiao and Li, 2011). In this case, the template is limited to controlling the size and geometrical shape of the catalyst but does not provide specific crystallographic surface orientations as nano-particle synthesis in solution. Therefore, 3D structures such as nano-wires, nano-brushes, or nanotubes are normally obtained through this technique. After electro-deposition, the template must be completely removed before using the electro-catalyst. Normally, templates are subdivided in two categories, i.e., soft (e.g., liquid crystal and colloidal) and hard templates (e.g., solid membrane structure). Cylindrical metallic structures and thin platinum films with high surface areas (300) were obtained by the electro-depositing of the metal salt dissolved in the aqueous domains of liquid crystals (Kloke *et al.*, 2011; Planes *et al.*, 2007). Typical surfactants used to form liquid crystals are sodium dodecyl sulfonate (SDS), octaethylene glycol monohexadecyl ether, or polyoxyethylene (Brij 78). Then, after the electro-deposition the surfactants are easily removed by washing with water. Polystyrene or silica microspheres are regularly used to produce bi- and tri-dimensional templates on top of a conductive material (Bartlett *et al.*, 2002; Lai and Riley, 2008). Then, platinum is electro-deposited into the voids and interconnected periodic nano-structures can be fabricated, e.g. honeycomb-like structures. After that, the template is removed by immersion in toluene or fluorhydric acid. On the other hand, in the hard template approach the noble metal is electro-deposited inside the cavities of a porous solid structure (e.g., an anodized aluminum oxide membrane with a specific pore diameter) situated onto a conductive substrate. Subsequently, the template is generally dissolved. Besides soft and hard templates, dendrimer is commonly used to fabricate platinum nano-particles of a wide range of sizes, though its removal after metal electro-deposition is quite difficult (Chen and Holt-Hindle, 2010; Kloke *et al.*, 2011; Peng and Yang, 2009; Qiao and Li, 2011).

Besides nano-particle synthesis in solution, chemical reduction can be done on a flat or porous structure, which is termed "electroless deposition". This methodology is widely used to fabricate substrate based materials and in decoration methods (Chen and Holt-Hindle, 2010; Kloke *et al.*, 2011; Peng and Yang, 2009; Qiao and Li, 2011). The autoclave technique or hydrothermal synthesis is an alternative to the electroless method that uses high temperature and pressure (Chen and Holt-Hindle, 2010; Koczkur *et al.*, 2007). Additionally, noble metal nanotubes can be formed by a combination of hard templates (membranes) and electroless depositions, this technique being known as "template wetting" (Peng and Yang, 2009; Steinhart *et al.*, 2004). In addition to these methods, galvanic replacement is widely employed to fabricate platinum nano-structures (e.g., nano-rods, nanotubes, films, hollow nano-particles, and porous materials) from the galvanic replacement of non-Pt structures by its immersion in a platinum salt solution (Peng and Yang, 2009; Sun *et al.*, 2002). Also, formation of platinum nano-columns was reported through the chemical vapor deposition method, and the thermal decomposition technique is normally utilized by heating the precursor solution and depositing it onto a substrate or by just electro-spraying the precursor onto a heated substrate (Lo Nigro *et al.*, 2007).

Other important synthesis techniques are dealloying and decoration (Kloke *et al.*, 2011; Peng and Yang, 2009). The latter is of great magnitude for the reason that nearly all platinum atoms contribute to the catalytic activity and, consequently, the catalyst cost is greatly reduced. This sinthesis can be done onto porous, smooth, nonconductive or conductive material with a thickness ranging from a sub-monolayer to a few monolayers. In this sense, different catalytic activities with enhanced stabilities are observed since atoms of the adlayer have different interatomic distances induced by the substrate lattice. This leads to a shift in d-band filling and thus to diverse catalytic properties from the massive platinum (Adzic, 2012; Hammer and Norskov, 2000). However, homogeneous decoration inside the porous structures is not a trivial issue. Typical methodologies utilized are the electro-deposition, underpotential deposition (UPD), indirect UPD (galvanic replacement of an UPD deposited monolayer of another metal, e.g., Cu), spontaneous deposition by immersion of a non-oxidized metal surface in a solution containing a platinum precursor, and electroless deposition (Xing *et al.*, 2010). In this regard, Adzic recently published an excellent review about properties of Pt monolayer electro-catalysts such as increased activity and stability compared with standard Pt nano-particle catalysts during the ORR (Adzic, 2012). On the other hand, dealloying is an easy technique that offers materials with a high surface roughness (3000) (Kloke *et al.*, 2011). To this end, Liu *et al.* (2006b) synthesized nano-porous platinum films with an enhanced surface area through anodic dissolution of Cu from PtCu alloy. The obtained material exhibited high stability and remarkable catalytic activity for oxygen electro-reduction and methanol electro-oxidation (Liu *et al.*, 2006b). The properties of the obtained porous structure depend mostly on the composition of the initial alloy, and a variety of alloys can be easily found in the market. Briefly, the technique consists of the corrosion and dissolution of the less noble material of the alloy, causing a contiguous nano-porous structure of the more noble material (Pt). The dealloying process can be performed under potential control or by selective etching in alkaline or acidic media.

## 4.4   BINARY AND TERNARY PLATINUM-BASED CATALYSTS

To enhance the mass activity of Pt, the modification of the catalytic platinum surface by the addition of a second or multiple metals has been developed to remarkably reduce Pt loading while enhancing the catalytic performance (Antolini *et al.*, 2006; Anumol *et al.*, 2011; Bing *et al.*, 2010; Ferrando *et al.*, 2008; Gasteiger *et al.*, 2005; Greeley *et al.*, 2009; Huang *et al.*, 2006a; Liu *et al.*, 2009; Peng and Yang, 2009; Stamenkovic *et al.*, 2007; Strasser *et al.*, 2010; Wang, C. *et al.*, 2012a; Yin *et al.*, 2011). One of the major reasons for interest in Pt alloy nano-particles is the fact that their chemical and physical properties may be tuned by varying the composition and atomic order in gas, as well as the size of the clusters. The desire to fabricate electro-catalysts with well-defined, controllable properties and structures on the nanometer scale, coupled with the flexibility afforded by intermetallic materials, has generated considerable interest in bimetallic and trimetallic nano-alloys for the electro-catalytic reaction mentioned above. The understanding of the atomic or molecular engineering of metal or oxide nano-particles is important because such nano-particles could exist in single-phase alloys, phase segregation, or core-shell structures different from their bulk counterparts, which has profound influences on their catalytic properties. Since electro-catalytic reactions occur on the surface of Pt nano-particles, in the last years there has been considerable interest in synthesizing core-shell segregated nano-alloys consisting of a shell of one type of atom (B) surrounding a core of another (A), though there may be some mixing between the shells (Ferrando *et al.*, 2008). A is usually a less expensive metal, which is (generally) less catalytically active, and B is a more expensive, more catalytically active metal (typically Pt). In certain cases, however, formation of core-shells or intermixed particles may also result in synergistic effects upon the catalytic properties of one or both of the component metals. For the requirements of electro-catalytic activity and durability, core-shell-like catalysts seem to be the most promising for practical applications.

### 4.4.1   *Electro-catalysts for CO and methanol oxidation reactions*

PtRu nano-particles are the most widely used electro-catalysts for CO and methanol oxidation reactions (Cui *et al.*, 2011; Dinh *et al.*, 2000; Gasteiger *et al.*, 1993; 1995a; 1995b; Gómez de la Fuente *et al.*, 2005; 2009; Hwang *et al.*, 2006; Kim *et al.*, 2010; Lee *et al.*, 2006; Lewera *et al.*, 2007; Park *et al.*, 2003; 2007; Vidakovic *et al.*, 2007; Wang, H. *et al.*, 2009; Watanabe *et al.*, 1987; Zhou *et al.*, 2007). In general, the current anode catalysts of choice for DMFCs are unsupported PtRu blacks. The use of blacks offers a high concentration of active sites adjacent to the membrane and is considered necessary for high performance. However, unsupported catalysts can suffer from relatively poor surface areas when compared to carbon blacks and are generally used at relatively high electrode loading (Dinh *et al.*, 2000).

A significant amount of work has been carried out and various theoretical and experimental techniques have been brought to bear in order to reveal the details for the PtRu catalytic/co-catalytic effect (Petrii, 2008). An interesting aspect of the Ru effect relates to the effect of the temperature and the optimum Pt:Ru ratio. Gasteiger *et al.* (1994) showed that dissociative methanol adsorption can occur on Ru sites as well, but it is a temperature-activated process. Therefore, at low temperatures (e.g., 25°C) a higher Pt:Ru atomic ratio (above 1:1) is required to facilitate the dissociative adsorption and dehydrogenation of methanol preferentially on Pt, whilst at high temperatures (e.g., 60°C and above) a surface richer in Ru is beneficial (e.g., 1:1 at. ratio) since Ru becomes active for chemisorptions and the rate determining step switches to the reaction between $CO_{ad}$ and $OH_{ad}$ (Gasteiger *et al.*, 1994).

The oxidation state of the Ru component is still a topic of discussion. While some authors refer to the active ruthenium compound mainly as metallic Ru in a bimetallic alloy (Chu and Gilman, 1996; Hamnett, 1997; Wu *et al.*, 2004), early research revealed that hydrous ruthenium oxide as a part of bimetallic PtRu electrodes is the most active catalyst for methanol oxidation (Cao *et al.*, 2006; Chen *et al.*, 2005; Gómez de la Fuente *et al.*, 2009; Jeon *et al.*, 2007; Long *et al.*, 2000; Over, 2012; Rolison *et al.*, 1999). According to this latter point, Rolison *et al.* (Long *et al.*, 2000; Rolison *et al.*, 1999) emphasized the importance of hydrous ruthenium oxides because the $RuO_xH_y$ speciation of Ru in nano-scale PtRu blacks shows both high electron and proton conductivity, which results in a much more active catalyst for methanol oxidation. Cao *et al.* (2006) showed that a new nano-composite $Pt/RuO_xH_y$ supported on carbon nanotubes presented higher activity in the methanol electro-oxidation as compared to that of PtRu commercial catalysts. The superior performance was attributed to the presence of $RuO_xH_y$ species. Gómez de la Fuente *et al.* (2009) revealed that a combination of the electro-catalytic nature of $RuO_xH_y$ species and functionalized carbon black in PtRu/C catalysts greatly improves the performance in a single DMFC. Furthermore, Ma *et al.* (2010) carefully prepared a series of multi-walled carbon nanotube-supported $Pt-(RuO_xH_y)_m$ catalysts without any metallic Ru or anhydrous $RuO_2$. Their data demonstrated that $RuO_xH_y$ alone could act as a very efficient promoter of Pt for the oxidation of CO and methanol oxidation. Meanwhile, it was also pointed out that the promotion of $RuO_xH_y$ in methanol oxidation depended on the proximity and relative amount of $RuO_xH_y$ and Pt.

Pt-based core–shell structured catalysts with Ru as the core and Pt as the shell (Ru@Pt) have been, in recent years, explored (Alayoglu *et al.*, 2008; Kim *et al.*, 2010; Liu *et al.*, 2005; Park *et al.*, 2003; Zhang *et al.*, 2011). These materials show much enhanced CO oxidation activity when compared to PtRu alloy and monometallic Pt and Ru nano-particle catalysts, leading to improved CO tolerance capability. Recent studies also revealed that carbon supported Ru@Pt catalysts could catalyze preferential oxidation of adsorbed CO on the Pt surface in order to quickly release active sites occupied by CO, resulting in an improved hydrogen oxidation reaction in the presence of CO in the fuel feed stream. Due to the fact that Ru metal is confined and kinetically trapped inside a Pt shell, the conventional bifunctional mechanism may not be applied to explain this increased CO removal. Using density functional theory, Alayoglu *et al.* (2008) showed that the enhanced CO oxidation may be achieved through the interactions between the Pt shell and Ru core atoms, which can modify the electronic structure of the Pt surface by the presence of subsurface Ru atoms. This modification may significantly destabilize CO on Pt, leading to a lower CO saturation coverage, thereby providing more free active sites for hydrogen oxidation reactions. Zhang *et al.* (2011) have synthesized a novel PtRu catalyst consisting of a Ru-rich core and a Pt-rich shell using a two-step microwave irradiation technique. For synthesizing this PtRu structure supported on carbon, ruthenium (III) acetylacetonate was first reduced in a tetra (ethylene glycol) solution in a microwave oven, and platinum (II) chloride and carbon black (Vulcan-XC-72R) were added to it (the molar ratio between Pt and Ru was 1:1, and the total metal loading was

20 wt%) under constant stirring overnight to form a slurry mixture. This mixture was then further reduced in the microwave oven. The carbon monoxide (CO) tolerant capability of this electro-catalyst was investigated using the in-house designed electrochemical half-cell with $H_2$ anode gas feedings, which contained five different concentrations of CO. The performances showed that this synthesized PtRu/C catalyst had better electro-catalytic activity than the commercially available 20 wt% Pt/C catalyst in the HOR, and also had CO tolerance in various CO concentrations.

In the study of Shubina and Koper (2002) the binding energies and geometries of CO and OH were computed for PtRu, PtMo, and PtSn catalysts. According to their calculations, the PtMo and PtSn catalysts seemed to be better for the CO oxidation reaction than PtRu materials. Experimentally, bimetallic PtMo catalysts presented high catalytic activity towards $H_2$ oxidation in the presence of CO (Grgur *et al.*, 1998; 1999; Ioroi *et al.*, 2003; Ioroi *et al.*, 2006; Jaksic *et al.*, 2005; Lebedeva and Janssen, 2005; Liu *et al.*, 2009; Mukerjee and Urian, 2002; Mukerjee *et al.*, 2004; Ordoñez *et al.*, 2007; Papageorgopoulos *et al.*, 2002; 2007; Ralph and Hogarth, 2002; Santiago *et al.*, 2004; Urian *et al.*, 2003). Such enhanced performance was ascribed to (i) the lack of CO adsorption on Mo, leaving more adsorption sites for oxygen-containing species that are acting as CO oxidation reagents within the frame of such bifunctional mechanisms, or (ii) changes in Pt-Pt atomic distance which modifies the Pt-CO adsorption energy (Ishikawa *et al.*, 2002). Grgur *et al.* (1997; 1998; 1999) reported the electrochemical oxidation of $H_2$, CO, and $CO/H_2$ on well-characterized PtMo bulk alloys or carbon supported PtMo catalysts in a sulfuric acid solution. Their work suggested a similar bifunctional mechanism with Ru in the role of Mo in PtMo alloy, an increase in free Pt sites by the oxidative removal of adsorbed CO. Mukerjee *et al.* showed a two to three fold enhancement of the CO tolerance of PtMo in a PEM fuel cell compared to that of PtRu, which was ascribed to the onset of CO oxidation at very low potentials ($\sim$100 mV) (Mukerjee *et al.*, 1999; 2004; Urian *et al.*, 2003). Papageorgopoulos *et al.* (2002) related the increased tolerance to the ability of PtMo to promote the CO oxidation process at very low potentials. This was attributed to why oxygen transfer from Mo oxyhydroxide species with only the OH species of the oxyhydroxide states (predominantly $MoO(OH)_2$) being reactive with adsorbed CO (Grgur *et al.*, 1999). In a more recent study, Liu *et al.* (2009) proposes a different mechanism to the bifunctional effects. In this work, PtMo core-shell particles have been prepared, that present a higher CO tolerance and increased activity in the $H_2/CO$ oxidation than commercial Pt/C and PtRu/C catalysts. It is considered likely that the electronic effect of the $MoO_x$ core on the Pt shell weakens the Pt-CO bond, which may dramatically reduce the oxidation overpotential (Fig. 4.3).

With respect to Sn, PtSn nano-composites have been extensively studied as catalysts for the electro-oxidation of hydrogen/carbon monoxide, methanol, and ethanol. In this sense, PtSn catalysts have shown a good CO tolerance during $H_2$ oxidation. Regarding the methanol electro-oxidation, instead, controversial results have been reported (Antolini and Gonzalez, 2010). Mukerjee and McBreen (1999) stated that the electronic effect of Sn on Pt caused partial fill-ing of the Pt d-band vacancies, and an increase in the Pt-Pt bond distance from 2.77 to 2.8 Å, which weakened the adsorption energy of CO but simultaneously inhibited the ability of Pt to adsorb methanol and dissociate C-H bonds. On the other hand, Shubina and Koper (2002), by quantum-chemical calculations on the $Pt_3Sn$ (111) crystal surface, showed that, in contrast to Ru, CO binds only to Pt and not to Sn atoms, whereas OH has an energetic preference for the Sn sites. This implies that PtSn is a good CO oxidation catalyst. According to the study of Kim *et al.* an optimal content of Sn (approximately 25 wt%) was required to balance the electronic and bifunctional effects (Kim *et al.*, 2008). This optimum Sn content provided the pertinent dilation of the lattice parameter assisting the cleavage of C-H bonds, and supplied sufficient surface oxygen-containing species. At low Sn content, however, there might not be sufficient Sn sites to provide surface oxygen derivatives capable of assisting the oxidation of adsorbed reaction intermediates. Recently, Herranz *et al.* (2012) have evaluated the role of two intermetallic phases of PtSn, namely $Pt_3Sn$ (fcc phase) and PtSn (hcp phase) for the electro-oxidation of CO and methanol. The cat-alysts have been studied using electrochemical techniques along with *in situ* techniques such as electrochemical coupled infrared reflection absorption spectroscopy (EC-IRAS) and differential

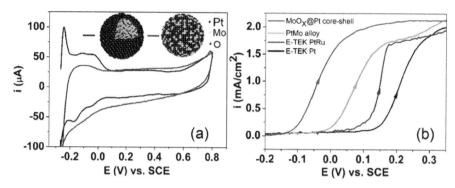

Figure 4.3.    (a) CVs of the $MoO_x$@Pt core-shell and $Pt_{0.8}Mo_{0.2}$ alloy catalysts in 0.5 M $H_2SO_4$ at 25°C, scan rate 100 mV s$^{-1}$. (b) Polarization curves for oxidation of $H_2$ in the presence of 1000 ppm of CO on different catalysts (30% loading) at 25°C. Rotation rate: 1600 rpm. Scan rate: 1 mV s$^{-1}$. Before the potential scan, electrolytes were purged with the CO/$H_2$ mixture for 2 h at −0.20 V electrode potential. All the potential values given in the text are referred to the saturated calomel electrode (SCE). (Reprinted with permission from *J. Am. Chem. Soc.* 131, 6924 (2009). Copyright 2009, American Chemical Society).

electrochemical mass spectrometry (DEMS). Altogether, the results revealed that $Pt_3Sn$ fcc was more active than PtSn hcp for the electro-oxidation of CO and methanol, and that the contribution of the hcp phase in those electro-catalytic processes was negligible.

In order to improve the lifetime of the DMFCs and PEMFCs, fuelled with $H_2$ from steam reforming, without increasing the cost or losing performance, exploring ternary anode catalysts is one of the most interesting and low-cost approaches (Antolini, 2007a; 2007b). However, ternary electro-catalysts must be resistant to changes in morphology and surface properties, and their stability may be a determining factor in the useful lifetime of these systems. The range of compositions that can be studied in such systems is enormous and realistically can only be mapped well through the use of high-throughput material science methods. Several CO-tolerant Pt-containing anodes exist, such as PtRuSn, PtRuW, PtRuMo, PtRuOs, PtRuPd, PtRuIr, PtRuRh, PtRuFe, and PtNiCr (Alcaide *et al.*, 2011; Choi *et al.*, 2004; Franco *et al.*, 2002; Geng *et al.*, 2009; Goetz and Wendt, 2001; Götz and Wendt, 1998; Gurau *et al.*, 1998; He *et al.*, 2003; Jeon and McGinn, 2009; Jusys *et al.*, 2002; Kawaguchi *et al.*, 2006; Lee *et al*, 2009; Ley *et al.*, 1997; Lima *et al.*, 2001; Massong *et al.*, 2001; Roth *et al.*, 2001; Tsiouvaras *et al.*, 2010a; Venkataraman *et al.*, 2003; Ye, 2009; Zhu *et al.*, 2008). Ternary electro-catalysts based on PtRuMo nano-particles have attracted major attention in recent years for PEMFCs fuelled with $H_2$/CO or low molecular weight alcohols. Molybdenum, an inexpensive and widely available metal element, has revealed a remarkable effect for promoting the oxidation of CO and methanol. Experimental studies have suggested that PtRuMo/C could be a better catalyst for CO and methanol oxidation than PtRu/C catalysts (Bauer *et al.*, 2007; Benker *et al.*, 2006; Hou *et al.*, 2003; Lima *et al.*, 2001; Martínez-Huerta *et al.*, 2008; 2010; Morante-Catacora *et al.*, 2008; Oliveira Neto *et al.*, 2003; Papageorgopoulos *et al.*, 2002; Pasupathi and Tricoli, 2008; Tsiouvaras *et al.*, 2009; 2010a; 2010b; Wang, H. *et al.*, 2009). Moreover, there are some reports in literature on the effect of catalyst compositions in the PtRuMo ternary catalysts in direct methanol fuel cells, which are consistent with the finding that relatively low Mo content can improve either CO tolerance or methanol electro-oxidation performance relative to PtRu (Alcaide *et al.*, 2011; Bauer *et al.*, 2007; Lee *et al.*, 2009; Pinheiro *et al.*, 2003; Wang, H. *et al.*, 2009).

### 4.4.2    *Electro-catalysts for the oxygen reduction reaction*

Pt alloys with some transition metals (Co, Ni, Fe, Mn, and Cr), regardless of the ratio between the noble metal and the base metal, displayed better activity toward ORR than the baseline Pt/C

(Bing *et al.*, 2010). However, the stability of Pt alloys was proven to vary between the different base metals. Leaching or dissolution of the base metal from the alloy surface to the acidic solution has been identified as the major cause of catalyst deactivation. Therefore, some treatment of the Pt alloy catalyst is necessary to improve its stability.

The effect of composition has been studied. It was found that PtM alloys with a 3:1 Pt:M ratio exhibited much higher activity and better stability (Bing *et al.*, 2010). Analysis of $Pt_3M$ alloys in a vacuum (M = Ir, Co, Ni and Fe) showed the following trend in stability: $Pt_3Ir$ (111) > $Pt_3Co$ (111) > $Pt_3Ni$(111) > $Pt_3Fe$(111) (Izzo, 2012). The effects of structural arrangements have been also investigated. Stamenkovic *et al.* (2002) carried out extensive investigations on well-defined bulk PtNi and PtCo alloy surfaces, namely the Pt-skeleton surface and the Pt-skin surface, and indicated that the ORR kinetics were dependent on the arrangement of alloying elements on the surface region. The catalytic activity at the different surfaces was in the order of Pt-bulk < Pt-skeleton < Pt-skin. The polycrystalline alloy films of $Pt_3M$ catalysts (M = Ni, Co, Fe and Ti) was also studied to understand the role of 3D metals in the electro-catalytic ORR activity of Pt alloys (Stamenkovic *et al.*, 2006). They found that in these $Pt_3M$ alloys, Pt enriched the first outlayer due to surface segregation. This Pt overlayer structure was found to be fairly stable in 0.1 M $HClO_4$ solution, and the catalytic activity for the ORR on such alloys was dependent on the nature of the 3D metals.

Further study by Stamenkovic *et al.* (2007) led to remarkable progress in new ORR catalyst development. In their study of $Pt_3Ni$(111)-skin surfaces, they reported a tenfold improvement in ORR activity (TOF of 2800 e site$^{-1}$ s$^{-1}$) compared to the corresponding Pt(111) surface, and a ninetyfold greater activity than the current state-of-the-art Pt/C catalysts for PEMFCs. The ORR activity order was identified as: $Pt_3Ni$(100)-skin < $Pt_3Ni$(110)-skin < $Pt_3Ni$(111)-skin. The $Pt_3Ni$(111)-skin surface showed an unusual electronic structure (d-band center position) and arrangement of surface atoms in the near-surface region. The weak interaction between the Pt surface atoms and nonreactive oxygenated species increased the number of active sites for $O_2$ adsorption. In addition to the experimental data on the $Pt_3M$ catalyst, theoretical studies demonstrated that Pt-skin in $Pt_3M$ is a unique surface with special electronic and catalytic properties (Kitchin *et al.*, 2004; Rossmeisl *et al.*, 2009; Stamenkovic *et al.*, 2006).

Another approach reported in recent years is the dealloying concept proposed by Strasser's group (Koh and Strasser, 2007b; Strasser *et al.*, 2010). Structural and compositional analysis showed that the resulting dealloyed catalyst consists of a core-shell structure with a Pt-rich shell and a Pt-poor alloy particle core. The authors report significant activity enhancements for the ORR after Cu or Co dealloying from PtCu/C, PtCo/C, and PtNi/C alloy nano-particle electro-catalysts, respectively. After removal of the second metal atoms from the surface region, the resulting particle catalysts showed improvements of 3–6 fold activity over Pt/C catalyst (Hasché *et al.*, 2011; 2012; Koh and Strasser, 2007a; Mani *et al.*, 2011; Strasser *et al.*, 2008). The activity enhancement in this type of catalyst is attributed to geometric and strain effects. It is suggested that the surface dealloying creates favorable Pt reactive sites for the ORR. In addition to the better ORR kinetics, these systems also show improved durability.

In a recent report, Wang, C. *et al.* (2012a) have investigated well-defined extended surfaces of Pt-ternary alloys ($Pt_3(MN)_1$ with M,N=Fe, Co, or Ni) as electro-catalysts for the ORR. Systematic studies revealed a volcano-type dependence between the measured catalytic activities and DFT-predicted oxygen-binding energies of corresponding model ternary alloy surfaces (Fig. 4.4). The most active system was found to be $Pt_3(CoNi)$. This study demonstrated that the ternary alloy catalysts can be compelling systems for further advancement of ORR electro-catalysis, reaching higher catalytic activities than bimetallic Pt alloys and improvement factors up to 4 *versus* monometallic Pt.

### 4.4.3 *Synthetic methods of binary/ternary catalysts*

Structural characteristics of the nano-particles such as morphology, shape, particle size, chemical state, type of crystal phases, degree of alloying, or distribution of surface metals, depend widely

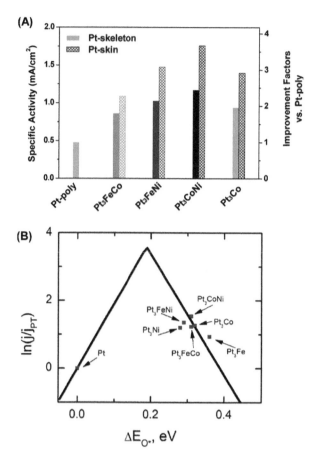

Figure 4.4.    (A) Summary of the ORR catalytic activities for the Pt-bimetallic and Pt-ternary thin films compared to that of Pt-poly. (B) Volcano plot relationship of measured catalyst performance *versus* the DFT-calculated oxygen binding energy. (Reprinted with permission from *J. Phys. Chem. Lett.* 3, 1668 (2012). Copyright 2012. American Chemical Society).

on the synthetic method. Understanding these effects and using appropriate chemical protocols to optimize experimental conditions seem critical in obtaining highly active, stable electro-catalytic nano-particles. Traditional approaches to preparing supported metal nano-particles involve impregnation, co-precipitation, deposition-precipitation, ion exchange, and colloidal methods. While a variety of supported Pt-group binary or ternary catalysts have been prepared by traditional methods (Antolini, 2003; 2007a; Chuan *et al.*, 2010; Deivaraj and Lee, 2005; Dickinson *et al.*, 2004; Esmaeilifar *et al.*, 2010; Gómez de la Fuente *et al.*, 2009; Götz *et al.*, 1998; Thompsett, 2003; Liu *et al.*, 2006a; Martínez-Huerta *et al.*, 2008; Paulus *et al.*, 2002; Rojas *et al.*, 2012; Waszczuk *et al.*, 2001; Yang *et al.*, 2004), the ability to control the size and composition is limited due to the propensity of metals to aggregate at the nano-scale. Aggregation of nano-particles leads to eventual loss of the nano-scale catalytic activity in practical applications.

The current state-of-the-art carbon supported electro-catalyst is made using variants of the colloidal approach. A common approach is to dissolve the metal salt solution in an appropriate solvent followed by reduction, to form a colloidal. A wide variety of recipes using reducing agents, organic stabilizers, or shell-removing approaches have also been developed in recent years. Among colloidal processes, the "alcohol reduction process" by Toshima and Yonezawa (1998) is widely applicable for the preparation of colloidal precious metals of small and uniform metal nano-particles. Refluxing of alcohol solutions of metal ions stabilized by organic polymers such as poly(vinylpyrrolidone) (PVP), poly(vinyl alcohol) (PVA), and poly(methylvinyl ether) (PMVE) gives homogeneous colloidal dispersions of the corresponding metal nano-particles. Alcohols, such as ethanol, methanol, 2-propanol, glycol, or ethoxyethanol, work as solvents as well as reductants, being oxidized into aldehydes or ketones (Deivaraj *et al.*, 2003; Kakade *et al.*, 2012; Wang *et al*, 2012; Xia *et al.*, 2006). The intrinsic problem underlying this process is that the stabilizing organic material remains on the surface of metal colloids. This should be removed prior to the application of metal particles for electro-catalysis. Removal of the organic material is important as it hinders the access of fuel to the catalyst sites. In general, the removal of stabilizer involves heat treatment. Consequently, due to the sintering effect, the phase separation and the distribution of metal particles are affected, resulting in lowered catalytic performance. In this respect, preparation via the "polyol process" is preferred due to several advantages. The polyol process is a technique in which a polyol such as ethylene glycol is used as both solvent and reducing agent (Bock *et al.*, 2004; Boehm, 1994; He *et al.*, 2010; Kadirgan *et al.*, 2009; Lee *et al.*, 2010; Li *et al.*, 2004; Sau *et al.*, 2009). A unique property of the polyol process is that it does not require any type of polymer stabilizer. In the polyol process using ethylene glycol, metal ions are reduced to form a metal colloid by receiving the electrons from the oxidation of ethylene glycol to glycolic acid. Glycolic acid is present in its deprotonated form as a glycolate anion in alkaline solution. It is believed that the glycolate anion acts as a stabilizer by adsorbing the metal colloids (Bock *et al.*, 2004). Furthermore, removal of these organics on the metal surface by heat treatment below 160°C has been reported, which is low enough to avoid the deleterious effects associated with heat treatment. However, no information is available on metal loading as a function of solution pH and different gas environments (Esmaeilifar *et al.*, 2010).

The micro-emulsion method was used by Boutonnet *et al.* (1982) for the first time to prepare monodisperse particles (in the size range of 3–5 nm) of Pt, Pd, Rh, and Ir by reduction of metal salts which are dissolved in the water pools of micro-emulsions with hydrogen or hydrazine. The method needs the following conditions to be generally applicable: (i) the solubility of the salts should not be limited by specific interactions with the solvent or the surfactant, and (ii) the reducing agent should react only with the salt. Under properly chosen conditions the particles can be transferred to supports without agglomeration. This consists of incorporating metal salts in the aqueous core of the small aggregates that are formed by surfactant molecules in a non-polar solvent at certain concentrations. The reduction of the metal salts with hydrazine at room temperature engenders the formation of metal and metal oxide particles. On the other hand, since the micelles have the same composition, i.e., metal precursors are distributed homogeneously, the nucleation of metallic particles renders particles of the same composition. This latter feature is very important for the synthesis of bimetallic (or ternary) catalysts (Esmaeilifar *et al.*, 2010; Xiong *et al.*, 2005; Zhang *et al.*, 2003; 2004). The main drawback of the micro-emulsion, or any other approach using surfactants, is surfactant removal. Severe thermal treatments are in order to achieve complete removal of the surfactant, although they promote the particle aggregation and/or surface enrichment, or complete phase segregation of the components of the bimetallic samples.

Among many emerging approaches to the preparation of nano-particles or nano-structures, one particular class of nano-particles with core–shell type structures is attracting great interest for addressing some of the challenges in nano-scale catalyst preparation (Fig. 4.5) (Luo *et al.*, 2008; Shao *et al.*, 2007; Templeton *et al.*, 1999; Wang, J.X. *et al.*, 2009).

Core/shell type nano-particles can be broadly defined as being comprised of a core and shell of different materials in close interaction, including inorganic/organic, inorganic/inorganic, organic/organic, or inorganic/biological combinations. There has been an increasing number of studies in recent years aimed at synthesizing metal nano-particles in the presence of organic cap-ping agents (Duan *et al.*, 2013; Gan *et al.*, 2013; Gao *et al.*, 2011; Huang *et al.*, 2012; Karan *et al.*, 2012; Kuai *et al*, 2012; Liu *et al.*, 2009; Luo *et al.*, 2008; Xia *et al.*, 2012; Zhang *et al.*, 2013). In contrast to traditional approaches to the preparation of supported catalysts, the molec-ular encapsulation based synthesis and processing strategy involves a sequence of three steps

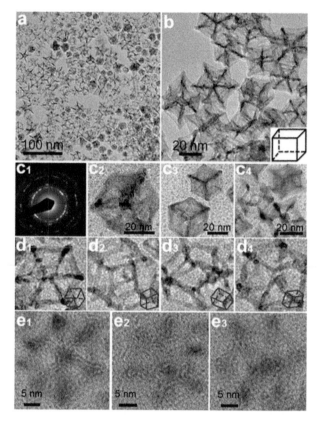

Figure 4.5.    TEM (a, b, c2–c4, d1–d4) and HRTEM (e1–e3) images of PtCu₃ nano-cages: (a) low magnification, (b) high magnification, inset of (b) is the model of a cubic nano-cage. (c1) Selected-area
electron-diffraction (SAED) pattern; (c2–c4) nano-cages. (d1–d4) Nano-cages with different
orientations; the insets are the corresponding crystal model. (Reprinted with permission from
*J. Am. Chem. Soc.* 134, 13934 (2012). Copyright 2012. American Chemical Society).

for the preparation of multi-metallic catalysts: (i) chemical synthesis of the metal nano-crystal
cores with molecular encapsulation, (ii) assembly of the encapsulated nano-particles on support
materials (e.g., carbon powders or carbon nanotubes), and (iii) thermal treatment of the supported
nano-particles.

The nano-technology guided design and fabrication of catalysts for enhancing catalytic activity
and reducing the cost of the catalyst will have enormous impacts on bettering catalyst preparation
for fuel cell applications (Chuan *et al.*, 2010; Zhong *et al.*, 2008). Addressing both fundamental
and engineering issues will certainly lead to new insights for rational design and better control of
nano-structured catalysts, leading to optimal performance and stability of fuel cells.

## 4.5    NEW ELECTRO-CATALYST SUPPORTS

The performance and durability of low temperature fuel cells seriously depend on electro-catalyst
support materials. The main requirements of suitable supports for fuel cell catalysts are (i) good
electrical conductivity, (ii) a high surface area to obtain high metal dispersion, (iii) suitable
porosity to boost gas flow, (iv) corrosion resistance at high voltages and low pH, and (v) platinum
bound strongly enough to shift the dissolution potential to higher values to prevent platinum

corrosion during stop-and-go runs (Antolini, 2010; Rabis *et al.*, 2012; Sharma and Pollet, 2012). Conventionally, highly conductive carbon blacks of turbostratic structures with high surface areas, such as Vulcan XC-72R (Cabot Corp, $237\,m^2\,g^{-1}$), Black Pearls 2000 (BP2000, Cabot Corp., $1500\,m^2\,g^{-1}$), or Ketjen Black (KB EC600JD & KB EC600J, Ketjen International, $1270\,m^2\,g^{-1}$ and $800\,m^2\,g^{-1}$, respectively) are currently used as fuel cell electro-catalyst supports for PEMFCs and DMFCs, to ensure large electrochemical reaction surfaces (Antolini, 2009; Sharma and Pollet, 2012). The high availability and low cost make carbon blacks the most used support for fuel cell catalysts. The disadvantages of these carbons are the presence of a high amount of micro-pore, which can hinder the reactant flow, and their low resistance to corrosion caused by electrochemical oxidation of the carbon surface, especially in the cathode.

Durability work has demonstrated the detrimental effects of potential cycling on carbon corrosion (Borup *et al.*, 2007) and Pt particle growth and movement of the support within the catalyst layer are two identified mechanisms of deterioration (Ferreira *et al.*, 2005; Hara *et al.*, 2012; Meier *et al.*, 2012a; 2012b; Wilson *et al.*, 1993). In particular, the corrosion of the carbon black support for cathode catalysts may be a critical problem for cell durability under certain PEMFC operating conditions. The standard electrode potential $E_0$ for the oxidation of carbon to carbon dioxide is low:

$$C + 2H_2O \rightarrow CO_2 + 4H^+ + 4e^-, \quad E_0 = +0.206V \text{ } vs. \text{ NHE at } 25°C \qquad (4.12)$$

Once the carbon support degrades and Pt particle growth initiates, the electrochemical surface area and thus performance, are significantly reduced. Moreover, despite the fact that the dissolution potential of bulk platinum is high (1.2 V at standard conditions), as compared with the operational potential of low-temperature fuel cells (0.6–0.7 V), the ramping of the potential to the open circuit voltage potential (1.1–1.2 V) during the stop-and-go drive, can speed up platinum leaching, which would severely degrade catalyst performance (Tripković *et al.*, 2012). Therefore, various alternatives for electro-catalyst supports are being researched. Over the last decade, the focus has shifted towards nano-structured supports as they enable faster electron transfer and high electro-catalytic activity. These supports may be classified in carbon-based/carbonaceous supports (typically by graphitization), and conductive ceramic-based support materials. However, the use of these support materials is not always completely satisfactory. Thus, in the last few years composite materials such as polymer-carbon, ceramic-carbon, and polymer-ceramic materials have been proposed as fuel cell catalyst supports (Antolini, 2010; Armstrong *et al.*, 2012; Bae *et al.*, 2012; Baoli *et al.*, 2010; Huang *et al.*, 2012a; 2012b; Orilall *et al.*, 2009).

The first category consists of carbon nano-structures like meso-porous carbon, carbon nanotubes (CNTs), nano-diamonds, carbon nano-fibers (CNF), and graphenes. CNTs are the most well-known and by far the most widely explored carbon nano-structures for application as catalyst support in fuel cells (Che *et al.*, 1998; Noked *et al.*, 2011; Tans *et al.*, 1998; Zhang *et al.*, 2011). CNTs are 2D nano-structures, typically tubes formed by rolled up single sheets of hexagonally arranged carbon atoms. They may be single-walled (SWCNT) or multi-walled (MWCNT). Depending on the structure, SWCNTs can be conducting, i.e., metallic as well as semi-conducting in nature. Both SWCNTs and MWNTs have been extensively studied for PEMFC and DMFC catalyst support applications (Girishkumar *et al.*, 2004; Li *et al.*, 2003; Wang, X. *et al.*, 2006; Zhang *et al.*, 2011). As innovative catalyst supports, CNTs have drawn a great deal of attention due to their unique geometric shape, excellent mechanical and thermal properties, high electric conductivity, large surface areas, and fascinating chemical stability (De Volder *et al.*, 2013; Zhang *et al.*, 2011). In spite of these advantages, the uniform dispersion of Pt nano-particles onto CNTs still remains a formidable challenge because of the inert surface of CNTs. Therefore, harsh chemical or electrochemical oxidations applied with concentrated strong acid are typically employed to create oxygen-containing surface functionalities, which facilitate the attachment of Pt nano-particles (Ebbesen, 1996; Lee *et al.*, 2006; Sharma and Pollet, 2012). Surface modified CNTs have been used to support a wide variety of mono, binary (e.g., Pt–Ru, Pt–Co, Pt–Fe), as well as ternary catalyst (e.g., Pt–Ru–Pd, Pt–Ru–Ni, Pt–Ru–Os) systems using both noble and non-precious metals (Baglio *et al.*, 2008; Esmaeilifar *et al.*, 2010; Liu *et al.*, 2002; Oh *et al.* 2005; Wang, C.H. *et al.*, 2007). However, these severe treatments also significantly deteriorate the preferred structure and electrical properties of CNTs due to the destruction of the graphitic structure. Alternatively, doping with foreign atoms (e.g., nitrogen) represents a feasible path to increase reactivity toward the deposition of metal nano-particles. The doping of nitrogen in CNTs creates defects that break out the chemical inertness of pure CNTs, yet preserves the electrical conductivity. Nitrogen-doped carbon nanotubes contain nitrogenated sites that are chemically active and have a certain activity toward ORR (Chen *et al.*, 2011; Lefevre *et al.*, 2009; Shao *et al.*, 2008). Nevertheless, a problem for the commercialization of carbon nanotubes is their higher cost compared to that of carbon blacks.

Recently, graphene nano-sheets (GNSs) have been investigated as a support for low temperature fuel cell catalysts (Antolini, 2012). Graphene is a two-dimensional one atom thick planar sheet of $sp^2$ bonded carbon atoms, having a thickness of 0.34 nm, which is considered as the fundamental foundation for all fullerene allotropic dimensionalities (Geim and Novoselov, 2007; Chen *et al.*, 2010). The combination of the high surface area (theoretical value of $2630 \, m^2 \, g^{-1}$) (Stankovich *et al.*, 2006), high conductivity (Geim and Novoselov, 2007), unique graphitized basal plane structure, and potential low manufacturing cost (Stankovich *et al.*, 2006; Xu *et al.*, 2008), makes the graphene sheet a promising candidate as a low temperature fuel cell catalyst support. In comparison with CNTs, graphene not only possesses similar stable physical properties but also a larger surface area. Additionally, production cost of GNSs in large quantities is much lower than that of CNTs. Thus, fuel cell catalysts supported on GNSs have been synthesized and characterized, their electro-catalytic activity and durability have been evaluated by half-cell measurements, and tests in single fuel cells have also been performed. Graphene's oxidized counterpart, i.e., GO, has also drawn a lot of interest and attention. Although GO has lower conductivity (a difference of two to three orders of magnitude compared to graphene), it offers a different set of properties (hydrophilicity, high mechanical strength, and chemical tunability) compared to graphene, which makes it suitable for a wide range of applications. The use of GO as catalyst support material in PEMFCs and DMFCs is one of the latest applications of GO which have shown promising results (Antolini, 2012; Brownson *et al.*, 2011; Chen *et al.*, 2010; Li *et al.*, 2010; Qiu *et al.*, 2011; Sharma and Pollet, 2012; Sharma *et al.*, 2010; Wang, S. *et al.*, 2011; Wu *et al.*, 2011; Zhou *et al.*, 2010). Oxygen groups introduced into the graphene structure during the preparation of GO create defect sites on surface planes as well as edge planes. These defect sites act as nucleation centers and anchoring sites for growth of metal nano-particles. Nitrogendoped graphene has been also investigated as a support for fuel cell catalysts. It has been shown using DFT that N-doping of graphene increases the binding energy of a Pt atom to the substrate (Groves *et al.*, 2009b). In general, the more N atoms and the closer they are to the C atom which bonds directly to the Pt, the stronger the binding energy. This can be attributed to how N atoms want to form pentagonal structures. This disrupts the delocalized double bond typical of graphene sheets, and causes the C–Pt bond to focus on their 2s/6s orbitals respectively, instead of their 2p/5d orbitals. This demonstrates the usefulness of using N-doped carbon structures as Pt catalyst supports since a twofold increase in the binding energy was observed, resulting in an improvement of Pt catalyst durability and activity for oxygen reduction (Groves *et al.*, 2009a).

During the past ten years, several research groups started to look for corrosion-resistant (carbon-free) support materials for Pt-based electro-catalysts in PEMFCs (Rabis *et al.*, 2012; Wang, Y.J. *et al.*, 2011). For this reason, several conducting and semiconducting materials have been studied as electro-catalyst supports and co-catalyst materials for PEMFCs. Conductive oxides especially gained much interest during that time because of their potentially high stability in acidic environments and unique chemical and physical properties related to the multi-valence of transition metal oxides, but also, carbides and nitrides have promising properties as electro-catalyst support material (Rabis *et al.*, 2012; Tripković *et al.*, 2012).

According to the properties of conductive metal oxides, Sasaki *et al.* conducted thermochemical calculations to determine stable substances under fuel cell relevant conditions (potential of $1.0 \, V_{SHE}$, pH $= 0$ and 80°C) (Sasaki *et al.*, 2010). Results are shown in Figure 4.6, where W, Ti,

Sn, Nb, Ta, Sn, and Sb are stable substances in the form of oxides, hydroxides, or metals under these conditions.

Titanium oxide is the most commonly studied metal oxide support material but also other conducting metal oxides such as tungsten oxide, zirconium oxide, niobium oxide, tantalum oxide, and tin oxide have been investigated as catalyst support (Abida *et al.*, 2011; Bauer *et al.*, 2010; Ho *et al.*, 2011; Huang *et al.*, 2012; Kim *et al.*, 2007; Krishnan *et al.*, 2012; Liu *et al.*, 2010; Masao *et al.*, 2009; Masud *et al.*, 2012; Matsui *et al.*, 2006; Park *et al.*, 2007; Sasaki *et al.*, 2008; 2010; Wesselmark *et al.*, 2010; Wickman *et al.*, 2011; Wu *et al.*, 2010). The greatest progress has been made by improving the electrical conductivity by doping or modifications of the oxide surface, increasing the surface area through nano-structures, meso-spheres, etc., and mainly by a deeper understanding of metal-support interaction and particle size influences. Recently, Tripković *et al.* (2012) investigated the activity and stability of $n = (1, 2, 3)$ platinum layers supported on a number of rutile metal oxides ($MO_2$; M = Ti, Sn, Ta, Nb, Hf, and Zr). They found that both the activity and the stability depend on the number of platinum layers and, as expected, both converge toward platinum values as the number of layers is increased. Futhermore, they established a correlation between stability and activity for supported platinum monolayers, which suggests that activity can be increased at the expense of stability, and vice versa. Finally, it was concluded that the most promising systems are the three platinum layers supported on $NbO_2$ and $TaO_2$.

Recently, different transition metal nitrides and carbides have received considerable attention as new electro-catalyst support, because of their unique physical and chemical properties, which combine the characteristic properties of three different classes of materials: covalent solids, ionic crystals, and transition metals (Chen, 1996; Ham and Lee, 2009; Oyama, 1996; Rabis *et al.*, 2012; Stottlemyer *et al.*, 2012). They demonstrate the extreme hardness and brittleness of covalent solids, they possess the high melting temperature and simple crystal structures typical of ionic crystals, and they have electronic and magnetic properties similar to transition metals. At the same time, values of their electric conductivity, magnetic susceptibility, and heat capacity are in the metallic range. Therefore, the stability and conducting nature of carbides and nitrides makes them good candidates for PEMFCs. Among the metal carbides, tungsten carbides (WC, $W_2C$) have been widely applied in the electrochemical systems including the electro-catalyst for low-temperature fuel cells. This is due to their surface reactivity resembling that of platinum and stability in an acidic medium, and their unique resistance to CO poisoning (Cui *et al.*, 2011; Hwu *et al.*, 2003a; 2003b; Zellner and Chen, 2005). Further, the electrochemical stability of WC could be improved in acidic electrolytes by adding Ta (Lee *et al.*, 2004). According to transition metal nitrides, they display the similar formation and physicochemical properties, including catalysis, to those of transition metal carbides (Ham and Lee, 2009). Titanium nitride (TiN) is known to be exceptionally stable under acidic conditions, and therefore, it has been used as a corrosion-resistant coating on stainless steel. TiN also has more than a factor of ten greater electronic conductivity compared to those of typical oxides or carbons. Avasarala *et al.* (2009) and Avasarala and Haldar (2010) initially reported promising results for Pt catalysts supported on TiN (Pt/TiN), but later reported surface passivation under more stringent conditions, leading to a lower Electrochemically Active Surface Area (ECSA) and durability compared to commercial Pt/Carbon black. In contrast, Kakinuma *et al.* (2011) reported that a Pt catalyst supported on highly crystallized TiN nano-particles exhibited superior ECSA and durability in the high potential region compared to those of commercial Pt/carbon black. Moreover, they reported the synthesis of a Pt catalyst with both controlled morphology and a highly preferred orientation of Pt nano-particles on the TiN nano-particle support, with resulting improved activity for the oxygen reduction reaction (ORR) with high durability in high potential step cycling (Kakinuma *et al.*, 2012).

## 4.6   CONCLUSIONS

In this chapter, we have presented a selected survey of the literature devoted to the investigation of Pt-based nano-structures for PEMFCs and DMFCs. Moreover, the relationships between Pt nano-structures and the electro-catalytic reactions, and common methods of synthesizing Pt and Pt-based nano-structured materials, including chemical and electrochemical deposition, have been discussed.

To enhance the mass activity of Pt, the modification of the catalytic platinum surface by the addition of a second or multiple metals has been developed to remarkably reduce Pt loading while enhancing the catalytic performance. For CO and methanol oxidation reactions, while PtRu is the practical electro-catalyst, PtMo (core-shell) and other ternary catalysts show very high activity in the oxidation of CO and/or methanol. Well-defined $Pt_3Ni$ surfaces and $Pt_3(CoNi)$ are two good examples to show how judicious design can be beneficial to dramatically enhanced activity in catalyzing ORRs.

Electro-catalyst supports play a vital role in ascertaining the performance, durability, and cost of PEMFC and DMFC systems. A myriad of nano-structured materials including carbon nano-structures, metal oxides, conducting polymers, transition metals nitrides and carbides, and many hybrid conjugates, have been exhaustively researched to improve the existing support and also to develop novel PEMFC/DMFC catalyst support. One of the main challenges in the immediate future is to develop new catalyst supports that improve the durability of the catalyst layer and, in a best-case scenario, also impact the electronic properties of the active phase to leapfrog to improve catalyst kinetics.

Through efforts made in the development of novel platinum nano-structures with high catalytic performance, utilization efficiency, and stability, enormous progress has been made in nano-particle synthesis techniques, and understanding the effective parameters which enhance the catalytic activity and stability. Choosing a proper synthesis method, the inclusion of appropriate content of suitable promoters in Pt-based catalysts and reducing agents, and using a proper support material, seem to be the major requirements of an effective catalyst. Optimizing the synthesis conditions, the amount of each component in the catalyst, and the metal-support interactions, in order to reach the highest catalytic activity and stability should be the most crucial efforts for future research.

## ACKNOWLEDGEMENTS

This work has been supported by the Spanish Science and Innovation Ministry under project CTQ2011-28913-C02-02. G. García acknowledges to the JAE program (CSIC) for financial support.

## REFERENCES

Abe, H., Matsumoto, F., Alden, L.R., Warren, S.C., Abruña, H.D. & DiSalvo, F.J.: Electrocatalytic performance of fuel oxidation by $Pt_3Ti$ nanoparticles. *J. Am. Chem. Soc.* 130 (2008), pp. 5452–5458.

Abida, B., Chirchi, L., Baranton, S.V., Napporn, T.W., Kochkar, H., Léger, J.M. & Ghorbel, A.: Preparation and characterization of $Pt/TiO_2$ nanotubes catalyst for methanol electro-oxidation. *Appl. Catal.* B *Environ.* 106 (2011), pp. 609–615.

Adzic, R.: Platinum monolayer electrocatalysts: Tunable activity, stability, and self-healing properties. *Electrocatalysis* 3 (2012), pp. 163–169.

Adzic, R. & Lima, F.H.: Platinum monolayer oxygen reduction electrocatalysts. In: *Handbook of fuel-cells: fundamentals, technology and applications.* Vol. 5: *Advances in electrocatalysis, materials, diagnostics and durability.* Wiley, NJ, 2009, pp. 5–15.

Alayoglu, S., Nilekar, A.U., Mavrikakis, M. & Eichhorn, B.: Ru-Pt core-shell nanoparticles for preferential oxidation of carbon monoxide in hydrogen. *Nat. Mater.* 7 (2008), pp. 333–338.

Alcaide, F., Álvarez, G., Tsiouvaras, N., Peña, M.A., Fierro, J.L. & Martínez-Huerta, M.V.: Electrooxidation of $H_2/CO$ on carbon-supported PtRu-MoOx nanoparticles for polymer electrolyte fuel-cells. *Int. J. Hydrogen Energy* 36 (2011), pp. 14590–14598.

Antolini, E.: Formation of carbon-supported PtM alloys for low temperature fuel-cells: A review. *Mater. Chem. Phys.* 78 (2003), pp. 563–573.

Antolini, E.: Platinum-based ternary catalysts for low temperature fuel-cells: Part I. Preparation methods and structural characteristics. *Appl. Catal. B Environ.* 74 (2007a), pp. 324–336.

Antolini, E.: Platinum-based ternary catalysts for low temperature fuel-cells: Part II. Electrochemical properties. *Appl. Catal. B Environ.* 74 (2007b), pp. 337–350.

Antolini, E.: Carbon supports for low temperature fuel-cell catalysts. *Appl. Catal. B Environ.* 88 (2009), pp. 1–24.

Antolini, E.: Composite materials: An emerging class of fuel-cell catalyst supports. *Appl. Catal. B Environ.* 100 (2010), pp. 413–426.

Antolini, E.: Graphene as a new carbon support for low-temperature fuel-cell catalysts. *Appl. Catal. B Environ.* 123–124 (2012), pp. 52–68.

Antolini, E. & Gonzalez, E.R.: The electro-oxidation of carbon monoxide, hydrogen/carbon monoxide and methanol in acid medium on Pt-Sn catalysts for low-temperature fuel-cells: A comparative review of the effect of Pt-Sn structural characteristics. *Electrochim. Acta* 56 (2010), pp. 1–14.

Antolini, E., Salgado, J.R.C. & Gonzalez, E.R.: The methanol oxidation reaction on platinum alloys with the first row transition metals: The case of Pt-Co and -Ni alloy electrocatalysts for DMFCs: A short review. *Appl. Catal. B Environ.* 63 (2006), pp. 137–149.

Anumol, E.A., Halder, A., Nethravathi, C., Viswanath, B. & Ravishankar, N.: Nanoporous alloy aggregates: synthesis and electrocatalytic activity. *J. Mater. Chem.* 21 (2011), pp. 8721–8726.

Aricò, A.S., Srinivasan, R. & Antonucci, V.: DMFCs: From fundamental aspects to technology development. *Fuel-cells* 1 (2001), pp. 133–161.

Armstrong, K.J., Elbaz, L., Bauer, E., Burrell, A.K., McCleskey, T.M. & Brosha, E.L.: Nanoscale titania ceramic composite supports for PEM fuel-cells. *J. Mater. Res.* 27 (2012), pp. 2046–2054.

Avasarala, B. & Haldar, P.: Electrochemical oxidation behavior of titanium nitride based electrocatalysts under PEM fuel-cell conditions. *Electrochim. Acta* 55 (2010), pp. 9024–9034.

Avasarala, B., Murray, T., Li, W. & Haldar, P.: Titanium nitride nanoparticles based electrocatalysts for proton exchange membrane fuel-cells. *J. Mater. Chem.* 19 (2009), pp. 1803–1805.

Bae, S.J., Nahm, K.S. & Kim, P.: Electroreduction of oxygen on Pd catalysts supported on Ti-modified carbon. *Curr. Appl. Phys.* 12 (2012), pp. 1476–1480.

Baglio, V., Di Blasi, A., D'Urso, C., Antonucci, V., Aricò, A.S., Ornelas, R., Morales-Acosta, D., Ledesma-Garcia, J., Godinez, L.A., Arriaga, L.G. & Varez-Contreras, L.: Development of Pt and PtFe catalysts supported on multiwalled carbon nanotubes for oxygen reduction in direct methanol fuel-cells. *J. Electrochem. Soc.* 155 (2008), pp. B829–B833.

Bartlett, P.N., Baumberg, J.J., Birkin, P.R., Ghanem, M.A. & Netti, M.C.: Highly ordered macroporous gold and platinum films formed by electrochemical deposition through templates assembled from submicron diameter monodisperse polystyrene spheres. *Chem. Mater.* 14 (2002), pp. 2199–2208.

Basri, S., Kamarudin, S.K., Daud, W.R.W. & Yaakub, Z.: Nanocatalyst for direct methanol fuel-cell (DMFC). *Int. J. Hydrogen Energy* 35 (2010), pp. 7957–7970.

Batista, E.A., Malpass, G.R.P., Motheo, A.J. & Iwasita, T.: New insight into the pathways of methanol oxidation. *Electrochem. Commun.* 5 (2003), pp. 843–846.

Bauer, A., Gyenge, E.L. & Oloman, C.W.: Direct methanol fuel-cell with extended reaction zone anode: PtRu and PtRuMo supported on graphite felt. *J. Power Sources* 167 (2007), pp. 281–287.

Bauer, A., Lee, K., Song, C., Xie, Y., Zhang, J. & Hui, R.: Pt nanoparticles deposited on $TiO_2$ based nanofibers: Electrochemical stability and oxygen reduction activity. *J. Power Sources* 195 (2010), pp. 3105–3110.

Benker, N., Roth, C., Mazurek, M. & Fuess, H.: Synthesis and characterization of ternary Pt/Ru/Mo catalysts for the anode of the PEM fuel-cell. *J. New Mater. Electrochem. Syst.* 9 (2006), pp. 121–126.

Bing, Y., Liu, H., Zhang, L., Ghosh, D. & Zhang, J.: Nanostructured Pt-alloy electrocatalysts for PEM fuel-cell oxygen reduction reaction. *Chem. Soc. Rev.* 39 (2010), pp. 2184–2202.

Bock, C., Paquet, C., Couillard, M., Botton, G.A. & MacDougall, B.R.: Size-selected synthesis of PtRu nano-catalysts: Reaction and size control mechanism. *J. Am. Chem. Soc.* 126 (2004), pp. 8028–8037.

Boehm, H.P.: Some aspects of the surface chemistry of carbon blacks and other carbons. *Carbon* 32 (1994), pp. 759–769.

Borup, R., Meyers, J., Pivovar, B., Kim, Y.S., Mukundan, R., Garland, N., Myers, D., Wilson, M., Garzon, F., Wood, D., Zelenay, P., More, K., Stroh, K., Zawodzinski, T., Boncella, J., McGrath, J.E., Inaba, M., Miyatake, K., Hori, M., Ota, K., Ogumi, Z., Miyata, S., Nishikata, A., Siroma, Z., Uchimoto, Y., Yasuda, K., Kimijima, K. & Iwashita, N.: Scientific aspects of polymer electrolyte fuel-cell durability and degradation. *Chem. Rev.* 107 (2007), pp. 3904–3951.

Boutonnet, M., Kizling, J., Stenius, P. & Maire, G.: The preparation of monodisperse colloidal metal particles from microemulsions. *Colloid. Surface.* 5 (1982), pp. 209–225.

Brownson, D.A., Kampouris, D.K. & Banks, C.E.: An overview of graphene in energy production and storage applications. *J. Power Sources* 196 (2011), pp. 4873–4885.

Burda, C., Chen, X., Narayanan, R. & El-Sayed, M.A.: Chemistry and properties of nanocrystals of different shapes. *Chem. Rev.* 105 (2005), pp. 1025–1102.

Cao, L., Scheiba, F., Roth, C., Schweiger, F., Cremers, C., Stimming, U., Fuess, H., Chen, L.Q., Zhu, W.T. & Qiu, X.P.: Novel nanocomposite $Pt/RuO_2$ center dot $xH_{(2)}O$/carbon nanotube catalysts for direct methanol fuel-cells. *Angew. Chem.* Int. Ed. 45 (2006), pp. 5315–5319.

Chang, S.C., Leung, L.W. & Weaver, M.J.: Comparisons between coverage-dependent infrared frequencies for carbon monoxide adsorbed on ordered platinum (111), (100), and (110) in electrochemical and ultrahigh-vacuum environments. *J. Phys. Chem.* 93 (1989), pp. 5341–5345.

Che, G., Lakshmi, B.B., Fisher, E.R. & Martin, C.R.: Carbon nanotubule membranes for electrochemical energy storage and production. *Nature* 393 (1998), pp. 346–349.

Chen, A. & Holt-Hindle, P.: Platinum-based nanostructured materials: Synthesis, properties, and applications. *Chem. Rev.* 110 (2010), pp. 3767–3804.

Chen, D., Tang, L. & Li, J.: Graphene-based materials in electrochemistry. *Chem. Soc. Rev.* 39 (2010), pp. 3157–3180.

Chen, J.G.: Carbide and nitride overlayers on early transition metal surfaces: Preparation, characterization, and reactivities. *Chem. Rev.* 96 (1996), pp. 1477–1498.

Chen, Y., Wang, J., Liu, H., Banis, M.N., Li, R., Sun, X., Sham, T.K., Ye, S. & Knights, S.: Nitrogen doping effects on carbon nanotubes and the origin of the enhanced electrocatalytic activity of supported Pt for proton-exchange membrane fuel-cells. *J. Phys. Chem.* C 115 (2011), pp. 3769–3776.

Chen, Z.G., Qiu, X.P., Lu, B., Zhang, S.C., Zhu, W.T. & Chen, L.Q.: Synthesis of hydrous ruthenium oxide supported platinum catalysts for direct methanol fuel-cells. *Electrochem. Commun.* 7 (2005), pp. 593–596.

Chen, Z., Higgins, D., Yu, A., Zhang, L. & Zhang, J.: A review on non-precious metal electrocatalysts for PEM fuel-cells. *Energy Environ. Sci.* 4 (2011), pp. 3167–3192.

Choi, J.H., Park, K.W., Park, I.S., Nam, W.H. & Sung, Y.E.: Methanol electro-oxidation and direct methanol fuel-cell using Pt/Rh and Pt/Ru/Rh alloy catalysts. *Electrochim. Acta* 50 (2004), pp. 787–790.

Chu, D. & Gilman, S.: Methanol electro-oxidation on unsupported Pt-Ru alloys at different temperatures. *J. Electrochem. Soc.* 143 (1996), pp. 1685–1690.

Coutanceau, C., Rakotondrainibé, A.F., Lima, A., Garnier, E., Pronier, S., Léger, J.M. & Lamy, C.: Preparation of Pt-Ru bimetallic anodes by galvanostatic pulse electrodeposition: characterization and application to the direct methanol fuel-cell. *J. Appl. Electrochem.* 34 (2004), pp. 61–66.

Cuesta, A.: At least three contiguous atoms are necessary for CO formation during methanol electrooxidation on platinum. *J. Am. Chem. Soc.* 128 (2006), pp. 13,332–13,333.

Cui, G., Shen, P.K., Meng, H., Zhao, J. & Wu, G.: Tungsten carbide as supports for Pt electrocatalysts with improved CO tolerance in methanol oxidation. *J. Power Sources* 196 (2011), pp. 6125–6130.

Cui, Z., Li, C.M. & Jiang, S.P.: PtRu catalysts supported on heteropolyacid and chitosan functionalized carbon nanotubes for methanol oxidation reaction of fuel-cells. *Phys. Chem. Chem. Phys.* 13 (2011), pp. 16,349–16,357.

De Volder, M.F., Tawfick, S.H., Baughman, R.H. & Hart, A.J.: Carbon nanotubes: Present and future commercial applications. *Science* 339 (2013), pp. 535–539.

Deivaraj, T.C. & Lee, J.Y.: Preparation of carbon-supported PtRu nanoparticles for direct methanol fuel-cell applications – a comparative study. *J. Power Sources* 142 (2005), pp. 43–49.

Deivaraj, T.C., Chen, W.X. & Lee, J.Y.: Preparation of PtNi nanoparticles for the electrocatalytic oxidation of methanol. *J. Mater. Chem.* 13 (2003), pp. 2555–2560.

Dickinson, A.J., Carrette, L.P.L., Collins, J.A., Friedrich, K.A. & Stimming, U.: Performance of methanol oxidation catalysts with varying Pt:Ru ratio as a function of temperature. *J. Appl. Electrochem.* 34 (2004), pp. 975–980.

Dinh, H.N., Ren, X., Garzon, F.H., Zelenay, P. & Gottesfeld, S.: Electrocatalysis in direct methanol fuel-cells: in-situ probing of PtRu anode catalyst surfaces. *J. Electroanal. Chem.* 491 (2000), pp. 223–233.

Duan, M.Y., Liang, R., Tian, N. & Li, Y.J.: Self-assembly of Au-Pt core-shell nanoparticles for effective enhancement of methanol electrooxidation. *Electrochim. Acta* 87 (2013), pp. 432–437.

Ebbesen, T.W.: Decoration of carbon nanotubes. *Adv. Mater.* 8 (1996), pp. 155–157.

Esmaeilifar, A., Rowshanzamir, S., Eikani, M.H. & Ghazanfari, E.: Synthesis methods of low-Pt-loading electrocatalysts for proton exchange membrane fuel-cell systems. *Energy* 35 (2010), pp. 3941–3957.

Ferrando, R., Jellinek, J. & Johnston, R.L.: Nanoalloys: From theory to applications of alloy clusters and nanoparticles. *Chem. Rev.* 108 (2008), pp. 845–910.

Ferreira, P.J., la O', G.J., Shao-Horn, Y., Morgan, D., Makharia, R., Kocha, S. & Gasteiger, H.A.: Instability of Pt/C electrocatalysts in proton exchange membrane fuel-cells: A mechanistic investigation. *J. Electrochem. Soc.* 152 (2005), pp. A2256–A2271.

Finney, E.E. & Finke, R.G.: Nanocluster nucleation and growth kinetic and mechanistic studies: A review emphasizing transition-metal nanoclusters. *J. Colloid Interf. Sci.* 317 (2008), pp. 351–374.

Franco, E.G., Neto, A.O., Linardi, M. & Arico, E.: Synthesis of electrocatalysts by the Bonnemann method for the oxidation of methanol and the mixture $H_2/CO$ in a proton exchange membrane fuel-cell. *J. Braz. Chem. Soc.* 13 (2002), pp. 516–521.

Gan, L., Heggen, M., O'Malley, R., Theobald, B. & Strasser, P.: Understanding and controlling nanoporosity formation for improving the stability of bimetallic fuel-cell catalysts. *Nano Lett.* 13 (2013), pp. 1131–1138.

Gao, H., Liao, S., Zeng, J., Xie, Y. & Dang, D.: Preparation and characterization of core-shell structured catalysts using $Pt_xPd_y$ as active shell and nano-sized Ru as core for potential direct formic acid fuel-cell application. *Electrochim. Acta* 56 (2011), pp. 2024–2030.

García, G. & Koper, M.T.: Dual reactivity of step-bound carbon monoxide during oxidation on a stepped platinum electrode in alkaline media. *J. Am. Chem. Soc.* 131 (2009), pp. 5384–5385.

García, G. & Koper, M.T.: Carbon monoxide oxidation on Pt single crystal electrodes: Understanding the catalysis for low temperature fuel-cells. *Chemphyschem.* 12 (2011), pp. 2064–2072.

García, G., Baglio, V., Stassi, A., Pastor, E., Antonucci, V. & Aricò, A.S.: Investigation of Pt-Ru nanoparticle catalysts for low temperature methanol electro-oxidation. *J. Solid State Electrochem.* 11 (2007), pp. 1229–1238.

García, G., Florez-Montaño, J., Hernandez-Creus, A., Pastor, E. & Planes, G.A.: Methanol electrooxidation at mesoporous Pt and Pt-Ru electrodes: A comparative study with carbon supported materials. *J. Power Sources* 196 (2011), pp. 2979–2986.

Gasteiger, H.A. & Markovic, N.M.: Just a dream-or future reality? *Science* 324 (2009), pp. 48–49.

Gasteiger, H.A., Markovic, N., Ross, P.N. & Cairns, E.J.: Methanol electrooxidation on well-characterized platinum-ruthenium bulk alloys. *J. Phys. Chem.* 97 (1993), pp. 12,020–12,029.

Gasteiger, H.A., Markovic, N., Ross, J. & Cairns, E.J.: Temperature-dependent methanol electro-oxidation on well-characterized Pt-Ru alloys. *J. Electrochem. Soc.* 141 (1994), pp. 1795–1803.

Gasteiger, H.A., Markovic, N.M. & Ross Jr., P.N.: $H_2$ and CO electrooxidation on well-characterized Pt, Ru and PtRu. 2. Rotating disk electrode studies of $CO/H_2$ mixtures at 62°C. *J. Phys. Chem.* 99 (1995a), pp. 16,757–16,767.

Gasteiger, H.A., Markovic, N.M. & Ross Jr., P.N.: $H_2$ and CO electroxidation on well-characterized Pt, Ru and PtRu. 1. Rotating disk electrode studies of the pure gases including temperature effects. *J. Phys. Chem.* 99 (1995b), pp. 8290–8301.

Gasteiger, H.A., Kocha, S.S., Sompalli, B. & Wagner, F.T.: Activity benchmarks and requirements for Pt, Pt-alloy, and non-Pt oxygen reduction catalysts for PEMFCs. *Appl. Catal. B Environ.* 56 (2005), pp. 9–35.

Geim, A.K. & Novoselov, K.S.: The rise of graphene. *Nat. Mater.* 6 (2007), pp. 183–191.

Geng, D., Matsuki, D., Wang, J., Kawaguchi, T., Sugimoto, W. & Takasu, Y.: Activity and durability of ternary PtRuIr/C for methanol electro-oxidation. *J. Electrochem. Soc.* 156 (2009), pp. B397–B402.

Gilman, S.: The mechanism of electrochemical oxidation of carbon monoxide and methanol on platinum. II. The reactant-pair mechanism for electrochemical oxidation of carbon monoxide and methanol. *J. Phys. Chem.* 68 (1964), pp. 70–80.

Girishkumar, G., Vinodgopal, K. & Kamat, P.V.: Carbon nanostructures in portable fuel-cells: Single-walled carbon nanotube electrodes for methanol oxidation and oxygen reduction. *J. Phys. Chem.* B 108 (2004), pp. 19,960–19,966.

Goetz, M. & Wendt, H.: Composite electrocatalysts for anodic methanol and methanol-reformate oxidation. *J. Appl. Electrochem.* 31 (2001), pp. 811–817.

Gómez de la Fuente, J.L., Martínez-Huerta, M.V., Rojas, S., Terreros, P., Fierro, J.L.G. & Peña, M.A.: Enhanced methanol electrooxidation activity of PtRu nanoparticles supported on $H_2O_2$-functionalized carbon black. *Carbon* 43 (2005), pp. 3002–3005.

Gómez de la Fuente, J.L., Martínez-Huerta, M.V., Rojas, S., Hernández-Fernández, P., Terreros, P., Fierro, J.L.G. & Peña, M.A.: Tailoring and structure of PtRu nanoparticles supported on functionalized carbon for DMFC applications: New evidence of the hydrous ruthenium oxide phase. *Appl. Catal.* B *Environ.* 88 (2009), pp. 505–514.

Gottesfeld, S.: Electrocatalysis of oxygen reduction in polymer electrolyte fuel-cells: A brief history and a critical examination of present theory and diagnostics. Chapter 1 in M. Koper (ed): *Fuel-cell catalysis, a surface science approach*. Wiley, NJ, 2009, pp. 1–30.

Götz, M. & Wendt, H.: Bynary and ternary anode catalyst formulations including the elements W, Sn and Mo for PEMFCs operated on methanol or reformate gas. *Electrochim. Acta* 43 (1998), pp. 3637–3644.

Greeley, J., Stephens, I.E.L., Bondarenko, A.S., Johansson, T.P., Hansen, H.A., Jaramillo, T.F., Rossmeisl, J., Chorkendorff, I. & Norskov, J.K.: Alloys of platinum and early transition metals as oxygen reduction electrocatalysts. *Nature Chem.* 1 (2009), pp. 552–556.

Grgur, B.N., Zhuang, G., Markovic, N.M. & Ross, J.P.N.: Electrooxidation of $H_2/CO$ mixtures on a well-characterized Pt75Mo25 alloy surface. *J. Phys. Chem.* B 101 (1997), pp. 3910–3913.

Grgur, B.N., Markovic, N.M. & Ross Jr., P.N.: Electrooxidation of $H_2$, CO and $H_2/CO$ mixtures on a well-characterized Pt70Mo30 bulk alloy electrode. *J. Phys. Chem.* B 102 (1998), pp. 2494–2501.

Grgur, B.N., Markovic, N.M. & Ross, P.N.: The electrooxidation of $H_2$ and $H_2/CO$ mixtures on carbon supported PtxMoy alloy catalysts. *J. Electrochem. Soc.* 146 (1999), pp. 1613–1619.

Groves, M.N., Chan, A., Malardier-Jugroot, C. & Jugroot, M.: Improving Pt catalyst durability and activity for $H_2$ and $O_2$ reactions by N-doping of graphene and SWCNTs. *Technical Proceedings of the 2009 NSTI Nanotechnology Conference and Expo, NSTI-Nanotech* 3, 2009a, pp. 465–468.

Groves, M.N., Chan, A.S.W., Malardier-Jugroot, C. & Jugroot, M.: Improving platinum catalyst binding energy to graphene through nitrogen doping. *Chem. Phys. Lett.* 481 (2009b), pp. 214–219.

Guillén-Villafuerte, O., García, G., Orive, A.G., Anula, B., Creus, A.H. & Pastor, E.: Electrochemical characterization of 2D Pt nanoislands. *Electrocatal.* 2 (2011), pp. 231–241.

Guillén-Villafuerte, O., García, G., Guil-López, R., Nieto, E., Rodríguez, J.L., Fierro, J.L.G. & Pastor, E.: Carbon monoxide and methanol oxidations on Pt/X@$MoO_3$/C (X = $Mo_2$C, $MoO_2$, $Mo^0$) electrodes at different temperatures. *J. Power Sources* 231 (2013), pp. 163–172.

Guo, Y.-G., Hu, J.-S. & Wan, L.-J.: Nanostructured materials for electrochemical energy conversion and storage devices. *Adv. Mater.* 20 (2008), pp. 2878–2887.

Gurau, B., Viswanathan, R., Liu, R., Lafrenz, T., Ley, K., Smotkin, E., Reddington, E., Sapienza, A., Chan, B., Mallouk, T. & Sarangapani, S.: Structural and electrochemical characterization of binary, ternary, and quaternary platinum alloy catalysts for methanol electro-oxidation. *J. Phys. Chem.* B 102 (1998), pp. 9997–10,003.

Ham, D.J. & Lee, J.S.: Transition metal carbides and nitrides as electrode materials for low temperature fuel-cells. *Energies* 2 (2009), pp. 873–899.

Hammer, B. & Norskov, J.K.: Theoretical surface science and catalysis-calculations and concepts. In: H.K. Bruce & C. Gates (eds): *Advances in catalysis*. Academic Press, San Diego, CA, 2000, pp. 71–129.

Hamnett, A.: Mechanism and electrocatalysis in the direct methanol fuel-cell. *Catal. Today* 38 (1997), pp. 445–457.

Hamnett, A.: Direct methanol fuel-cells (DMFC). In: *Handbook of fuel-cells–Fundamental, technology and applications, Vol. 1: Fundamentals and survey of systems*. Wiley, NJ, 2003, pp. 305–322.

Hara, M., Lee, M., Liu, C.H., Chen, B.H., Yamashita, Y., Uchida, M., Uchida, H. & Watanabe, M.: Electrochemical and Raman spectroscopic evaluation of Pt/graphitized carbon black catalyst durability for the start/stop operating condition of polymer electrolyte fuel-cells. *Electrochim. Acta* 70 (2012), pp. 171–181.

Hasché, F., Oezaslan, M. & Strasser, P.: Activity, stability, and degradation mechanisms of dealloyed PtCu3 and PtCo3 nanoparticle fuel-cell catalysts. *ChemCatChem* 3 (2011), pp. 1805–1813.

Hasché, F., Oezaslan, M. & Strasser, P.: Activity, structure and degradation of dealloyed PtNi3 nanoparticle electrocatalyst for the oxygen reduction reaction in PEMFC. *J. Electrochem. Soc.* 159 (2012), pp. B25–B34.

He, C., Kunz, H.R. & Fenton, J.M.: Electro-oxidation of hydrogen with carbon monoxide on Pt/Ru-based ternary catalysts. *J. Electrochem. Soc.* 150 (2003), pp. A1017–A1024.

He, W., Liu, J., Qiao, Y., Zou, Z., Zhang, X., Akins, D.L. & Yang, H.: Simple preparation of PdPt nanoalloy catalysts for methanol-tolerant oxygen reduction. *J. Power Sources* 195 (2010), pp. 1046–1050.

Hernandez, J., Solla-Gullon, J., Herrero, E., Aldaz, A. & Feliu, J.M.: Electrochemistry of shape-controlled catalysts: Oxygen reduction reaction on cubic gold nanoparticles. *J. Phys. Chem.* C 111 (2007), pp. 14,078–14,083.

Herranz, T., García, S., Martínez-Huerta, M.V., Peña, M.A., Fierro, J.L.G., Somodi, F., Borbáth, I., Majrik, K., Tompos, A. & Rojas, S.: Electrooxidation of CO and methanol on well-characterized carbon supported $Pt_x$Sn electrodes. Effect of crystal structure. *Int. J. Hydrogen Energy* 37 (2012), pp. 7109–7118.

Ho, V.T.T., Pan, C.J., Rick, J., Su, W.N. & Hwang, B.J.: Nanostructured $Ti_{0.7}Mo_{0.3}O_2$ support enhances electron transfer to Pt: High-performance catalyst for oxygen reduction reaction. *J. Am. Chem. Soc.* 133 (2011), pp. 11,716–11,724.

Hou, Z., Yi, B., Yu, H., Lin, Z. & Zhang, H.: CO tolerance electrocatalyst of PtRu-$H_x$MeO3/C (Me=W, Mo) made by composite support method. *J. Power Sources* 123 (2003), pp. 116–125.

Huang, K., Sasaki, K., Adzic, R.R. & Xing, Y.: Increasing Pt oxygen reduction reaction activity and durability with a carbon-doped $TiO_2$ nanocoating catalyst support. *J. Mater. Chem.* 22 (2012a) pp. 16,824–16,832.

Huang, R., Wen, Y.H., Zhu, Z.Z. & Sun, S.G.: Two-stage melting in core-shell nanoparticles: An atomic-scale perspective. *J. Phys. Chem.* C 116 (2012b), pp. 11,837–11,841.

Huang, S.Y., Chang, S.M. & Yeh, C.: Characterization of surface composition of platinum and ruthenium nanoalloys dispersed on active carbon. *J. Phys. Chem.* B 110 (2006a), pp. 234–239.

Huang, S.Y., Chang, C.M. & Yeh, C.T.: Promotion of platinum-ruthenium catalyst for electro-oxidation of methanol by ceria. *J. Catal.* 241 (2006b), pp. 400–406.

Huang, S.Y., Ganesan, P. & Popov, B.N.: Electrocatalytic activity and stability of titania-supported platinum-palladium electrocatalysts for polymer electrolyte membrane fuel-cell. *ACS Catal.* 2 (2012), pp. 825–831.

Huang, X., Qi, X., Boey, F. & Zhang, H.: Graphene-based composites. *Chem. Soc. Rev.* 41 (2012), pp. 525–944.

Hwang, B.J., Chen, C.H., Sarma, L.S., Chen, J.M., Wang, G.R., Tang, M.T., Liu, D.G. & Lee, J.F.: Probing the formation mechanism and chemical states of carbon-supported Pt-Ru nanoparticles by in situ X-ray absorption spectroscopy. *J. Phys. Chem.* B 110 (2006), pp. 6475–6482.

Hwu, H.H. & Chen, J.G.: Potential application of tungsten carbides as electrocatalysts: 4. Reactions of methanol, water, and carbon monoxide over carbide-modified W(110). *J. Phys. Chem.* B 107 (2003a), pp. 2029–2039.

Hwu, H.H. & Chen, J.G.G.: Potential application of tungsten carbides as electrocatalysts. *J. Vac. Sci. Technol.* A 21 (2003b), pp. 1488–1493.

Ioroi, T., Yasuda, K., Siroma, Z., Fujiwara, N. & Miyazaki, Y.: Enhanced CO-tolerance of carbon-supported platinum and molybdenum oxide anode catalyst. *J. Electrochem. Soc.* 150 (2003), pp. A1225–A1230.

Ioroi, T., Akita, T., Yamazaki, S.I., Siroma, Z., Fujiwara, N. & Yasuda, K.: Comparative study of carbon-supported Pt/Mo-oxide and PtRu for use as CO-tolerant anode catalysts. *Electrochim. Acta* 52 (2006), pp. 491–498.

Ishikawa, Y., Liao, M.S. & Cabrera, C.R.: Energetics of $H_2O$ dissociation and $CO_{ads}$+$OH_{ads}$ reaction on a series of Pt-M mixed metal clusters: a relativistic density-functional study. *Surface Sci.* 513 (2002), pp. 98–110.

Iwasita, T.: Electrocatalysis of methanol oxidation. *Electrochim. Acta* 47 (2002), pp. 3663–3674.

Iwasita, T.: Methanol and CO electrooxidation. In: *Handbook of fuel-cells–Fundamentals, technology and applications*, Vol. 2: *Fuel cell electrocatalysis*. Wiley, NJ, 2003, pp. 603–624.

Izzo, A.E.: Highly dispersed alloy cathode catalyst for durability 2008. *Progress Report for the DOE hydrogen program focuses on fuel-cells*, Washington, DC, 2008.

Jaksic, J.M., Vracar, Lj., Neophytides, S.G., Zafeiratos, S., Papakonstantinou, G., Krstajic, N.V. & Jaksic, M.M.: Structural effects on kinetic properties for hydrogen electrode reactions and CO tolerance along Mo-Pt phase diagram. *Surface Sci.* 598 (2005), pp. 156–173.

Jeon, M.K. & McGinn, P.J.: Composition dependence of ternary PtNiCr catalyst activity for the methanol electro-oxidation reaction. *J. Power Sources* 194 (2009), pp. 737–745.

Jeon, M.K., Won, J.Y. & Woo, S.I.: Improved performance of direct methanol fuel-cells by anodic treatment. *Electrochem. Solid State Lett.* 10 (2007), pp. B23–B25.

Jian, X.H., Tsai, D.S., Chung, W.H., Huang, Y.S. & Liu, F.J.: Pt-Ru and Pt-Mo electrodeposited onto Ir-IrO2 nanorods and their catalytic activities in methanol and ethanol oxidation. *J. Mater. Chem.* 19 (2009), pp. 1601–1607.

Jusys, Z., Schmidt, T.J., Dubau, L., Lasch, K., Jörissen, L., Garche, J. & Behm, R.J.: Activity of PtRuMeO$_x$ (Me=W, Mo or V) catalysts towards methanol oxidation and their characterization. *J. Power Sources* 105 (2002), pp. 297–304.

Kadirgan, F., Beyhan, S. & Atilan, T.: Preparation and characterization of nano-sized PtPd/C catalysts and comparison of their electro-activity toward methanol and ethanol oxidation. *Int. J. Hydrogen Energy* 34 (2009), pp. 4312–4320.

Kakade, B.A., Tamaki, T., Ohashi, H. & Yamaguchi, T.: Highly active bimetallic PdPt and CoPt nanocrystals for methanol electro-oxidation. *J. Phys. Chem.* C 116 (2012), pp. 7464–7470.

Kakinuma, K., Wakasugi, Y., Uchida, M., Kamino, T., Uchida, H. & Watanabe, M.: Electrochemical activity and durability of platinum catalysts supported on nanometer-size titanium nitride particles for polymer electrolyte fuel-cells. *Electrochem.* 79 (2011), pp. 399–403.

Kakinuma, K., Wakasugi, Y., Uchida, M., Kamino, T., Uchida, H., Deki, S. & Watanabe, M.: Preparation of titanium nitride-supported platinum catalysts with well controlled morphology and their properties relevant to polymer electrolyte fuel-cells. *Electrochim. Acta* 77 (2012), pp. 279–284.

Kamarudin, S.K., Achmad, F. & Daud, W.R.: Overview on the application of direct methanol fuel-cell (DMFC) for portable electronic devices. *Int. J. Hydrogen Energy* 34 (2009), pp. 6902–6916.

Karan, H.I, Sasaki, K. Kuttiyiel, K., Farberow, C.A., Mavrikakis, M. & Adzic, R.: Catalytic activity of platinum mono layer on iridium and rhenium alloy nanoparticles for the oxygen reduction reaction. *ACS Catal.* 2 (2012), pp. 817–824.

Kawaguchi, T., Rachi, Y., Sugimoto, W., Murakami, Y. & Takasu, Y.: Performance of ternary PtRuRh/C electrocatalyst with varying Pt:Ru:Rh ratio for methanol electro-oxidation. *J. Appl. Electrochem.* 36 (2006), pp. 1117–1125.

Kim, J.H., Ishihara, A., Mitsushima, S., Kamiya, N. & Ota, K.I.: Catalytic activity of titanium oxide for oxygen reduction reaction as a non-platinum catalyst for PEFC. *Electrochim. Acta* 52 (2007), pp. 2492–2497.

Kim, J.H., Choi, S.M., Nam, S.H., Seo, M.H., Choi, S.H. & Kim, W.B.: Influence of Sn content on PtSn/C catalysts for electrooxidation of C1-C3 alcohols: Synthesis, characterization, and electrocatalytic activity. *Appl. Catal.* B *Environ.* 82 (2008), pp. 89–102.

Kim, Y.T., Lee, H., Kim, H.J. & Lim, T.H.: PtRu nano-dandelions on thiolated carbon nanotubes: a new synthetic strategy for supported bimetallic core-shell clusters on the atomic scale. *Chem. Commun.* 46 (2010), pp. 2085–2087.

Kinge, S., Urgeghe, C., Battisti, A. & Bonnemann, H.: Dependence of CO oxidation on Pt nanoparticle shape: A shape-selective approach to the synthesis of PEMFC catalysts. *Appl. Organometal. Chem.* 22 (2008), pp. 49–54.

Kitchin, J.R., Norskov, J.K., Barteau, M.A. & Chen, J.G.: Modification of the surface electronic and chemical properties of Pt(111) by subsurface 3d transition metals. *J. Chem. Phys.* 120 (2004), pp. 10,240–10,246.

Kloke, A., von Stetten, F., Zengerle, R. & Kerzenmacher, S.: Strategies for the fabrication of porous platinum electrodes. *Adv. Mater.* 23 (2011), pp. 4976–5008.

Koczkur, K., Yi, Q. & Chen, A.: Nanoporous Pt-Ru networks and their electrocatalytical properties. *Adv. Mater.* 19 (2007), pp. 2648–2652.

Koenigsmann, C., Zhou, W.P., Adzic, R.R., Sutter, E. & Wong, S.S.: Size-dependent enhancement of electrocatalytic performance in relatively defect-free, processed ultrathin platinum nanowires. *Nano Lett.* 10 (2010), pp. 2806–2811.

Koh, S. & Strasser, P.: Electrocatalysis on bimetallic surfaces: Modifying catalytic reactivity for oxygen reduction by voltammetric surface dealloying. *J. Am. Chem. Soc.* 129 (2007a), pp. 12,624–12,625.

Koper, M.T.: Combining experiment and theory for understanding electrocatalysis. *J. Electroanal. Chem.* 574 (2005), pp. 375–386.

Koper, M.T.: Structure sensitivity and nanoscale effects in electrocatalysis. *Nanoscale* 3 (2011), pp. 2054–2073.

Koper, M.T., Lai, S.C. & Herrero, E.: Mechanism of the oxidation of carbon monoxide and small organic molecules at metal electrodes. In M. Koper (ed): *Fuel-cell catalysis, a surface science approach.* Wiley, NJ, 2009, pp. 159–207.

Krishnan, P., Advani, S.G. & Prasad, A.: Magneli phase $Ti_nO_{2n-1}$ as corrosion-resistant PEM fuel-cell catalyst support. *J. Solid State Electrochem.* 16 (2012), pp. 2515–2521.

Kuai. L., Yu, X., Wang, S., Sang, Y. & Geng, B.: Au-Pd alloy and core-shell nanostructures: One-pot coreduction preparation, formation mechanism, and electrochemical properties. *Langmuir* 28 (2012), pp. 7168–7173.

Kundu, S., Wang, K. & Liang, H.: Photochemical generation of catalytically active shape selective rhodium nanocubes. *J. Phys. Chem.* C 113 (2009), pp. 18,570–18,577.

Kunze, J. & Stimming, U.: Electrochemical *versus* heat-engine energy technology: A tribute to Wilhelm Ostwalds visionary statements. *Angew. Chem.* Int. Ed. 48 (2009), pp. 9230–9237.

Lai, M. & Riley, D.J.: Templated electrosynthesis of nanomaterials and porous structures. *J. Colloid Interf. Sci.* 323 (2008), pp. 203–212.

Lamy, C., Leger, J.M., Clavilier, J. & Parsons, R.: Structural effects in electrocatalysis: A comparative study of the oxidation of CO, HCOOH and CH$_3$OH on single crystal Pt electrodes. *J. Electroanal. Chem.* 150 (1983), pp. 71–77.

Lebedeva, N.P. & Janssen, G.J.: On the preparation and stability of bimetallic PtMo/C anodes for proton-exchange membrane fuel-cells. *Electrochim. Acta* 51 (2005), pp. 29–40.

Lebedeva, N.P., Koper, M.T., Herrero, E., Feliu, J.M. & van Santen, R.A.: Cooxidation on stepped Pt(n(111)Î(111)) electrodes. *J. Electroanal. Chem.* 487 (2000), pp. 37–44.

Lee, K., Ishihara, A., Mitsushima, S., Kamiya, N. & Ota, K.I.: Stability and electrocatalytic activity for oxygen reduction in WC+Ta catalyst. *Electrochim. Acta* 49 (2004), pp. 3479–3485.

Lee, K., Zhang, J., Wang, H. & Wilkinson, D.P.: Progress in the synthesis of carbon nanotube- and nanofiber-supported Pt electrocatalysts for PEM fuel-cell catalysis. *J. Appl. Electrochem.* 36 (2006), pp. 507–522.

Lee, K.R., Jeon, M.K. & Woo, S.I.: Composition optimization of PtRuM/C (M=Fe and Mo) catalysts for methanol electro-oxidation via combinatorial method. *Appl. Catal. B Environ.* 91 (2009), pp. 428–433.

Lee, K.S., Park, H.Y., Cho, Y.H., Park, I.S., Yoo, S.J. & Sung, Y.E.: Modified polyol synthesis of PtRu/C for high metal loading and effect of post-treatment. *J. Power Sources* 195 (2010), pp. 1031–1037.

Lee, Y.H., Lee, G., Shim, J.H., Hwang, S., Kwak, J., Lee, K., Song, H. & Park, J.T.: Monodisperse PtRu nanoalloy on carbon as a high-performance DMFC catalyst. *Chem. Mater.* 18 (2006), pp. 4209–4211.

Lefevre, M., Proietti, E., Jaouen, F. & Dodelet, J.P.: Iron-based catalysts with improved oxygen reduction activity in polymer electrolyte fuel-cells. *Science* 324 (2009), pp. 71–74.

Lewera, A., Zhou, W.P., Hunger, R., Jaegermann, W., Wieckowski, A., Yockel, S. & Bagus, P.S.: Core-level binding energy shifts in Pt-Ru nanoparticles: A puzzle resolved. *Chem. Phys. Lett.* 447 (2007), pp. 39–43.

Ley, K.L., Liu, R., Pu, C., Fan, Q., Leyarovska, N., Segre, C. & Smotkin, E.S.: Methanol oxidation on single-phase Pt-Ru-Os ternary alloys. *J. Electrochem. Soc.* 144 (1997), pp. 1543–1548.

Li, W., Liang, C., Zhou, W., Qiu, J., Zhou, Z., Sun, G. & Xin, Q.: Preparation and characterization of multiwalled carbon nanotube-supported platinum for cathode catalysts of direct methanol fuel-cells. *J. Phys. Chem.* B 107 (2003), pp. 6292–6299.

Li, W.Z., Zhou, W.J., Li, H.Q., Zhou, Z.H., Zhou, B., Sun, G.Q. & Xin, Q.: Nano-structured Pt-Fe/C as cathode catalyst in direct methanol fuel-cell. *Electrochim. Acta* 49 (2004), pp. 1045–1055.

Li, Y., Gao, W., Ci, L., Wang, C. & Ajayan, P.M.: Catalytic performance of Pt nanoparticles on reduced graphene oxide for methanol electro-oxidation. *Carbon* 48 (2010), pp. 1124–1130.

Lima, A., Coutanceau, C., Leger, J.M. & Lamy, C.: Investigation of ternary catalysts for methanol electrooxidation. *J. Appl. Electrochem.* 31 (2001), pp. 379–386.

Liu, D.-G., Lee, J.-F. & Tsang, M.-T.: Characterization of Pt-Ru/C catalysts by X-ray absorption spectroscopyand temperature-programmed surface reaction. *J. Mol. Catal.* A *Chem.* 240 (2005), pp. 97–206.

Liu, H., Song, C., Zhang, L., Zhang, J., Wang, H. & Wilkinson, D.P.: A review of anode catalysis in the direct methanol fuel-cell. *J. Power Sources* 155 (2006a), pp. 95–110.

Liu, H., He, P., Li, Z. & Li, J.: High surface area nanoporous platinum: facile fabrication and electrocatalytic activity. *Nanotechnology* 17 (2006b), pp. 2167–2173.

Liu, Y., Ishihara, A., Mitsushima, S. & Ota, K.i.: Influence of sputtering power on oxygen reduction reaction activity of zirconium oxides prepared by radio frequency reactive sputtering. *Electrochim. Acta* 55 (2010), pp. 1239–1244.

Liu, Z., Hu, J.E., Wang, Q., Gaskell, K., Frenkel, A.I., Jackson, G.S. & Eichhorn, B.: PtMo alloy and MoOx@Pt core-shell nanoparticles as highly CO-tolerant electrocatalysts. *J. Am. Chem. Soc.* 131 (2009), pp. 6924–6925.

Liu, Z.L., Lin, X.H., Lee, J.Y., Zhang, W., Han, M. & Gan, L.M.: Preparation and characterization of platinum-based electrocatalysts on multiwalled carbon nanotubes for proton exchange membrane fuel-cells. *Langmuir* 18 (2002), pp. 4054–4060.

Lo Nigro, R., Malandrino, G., Fiorenza, P. & Fragalá, I.L.: Template-free and seedless growth of Pt nanocolumns: Imaging and probing their nanoelectrical properties. *ACS Nano* 1 (2007), pp. 83–190.

Long, J.W., Stroud, R.M., Swider-Lyons, K.E. & Rolison, R.: How to make electrocatalysts more active for direct methanol oxidation-avoid PtRu bimetallic alloys. *J. Phys. Chem.* B 104 (2000), pp. 9772–9776.

Lopez-Cudero, A., Solla-Gullon, J., Herrero, E., Aldaz, A. & Feliu, J.M.: CO electrooxidation on carbon supported platinum nanoparticles: Effect of aggregation. *J. Electroanal. Chem.* 644 (2010), pp. 117–126.

Luo, J., Wang, L., Mott, D., Njoki, P., Lin, Y., He, T., Xu, Z., Wanjana, B.N., Lim, I.-I.S. & Zhong, C.J.: Core/shell nanoparticles as electrocatalysts for fuel-cell reactions. *Adv. Mater.* 20 (2008), pp. 4342–4347.

Ma, J.H., Feng, Y.Y., Yu, J., Zhao, D., Wang, A.J. & Xu, B.Q.: Promotion by hydrous ruthenium oxide of platinum for methanol electro-oxidation. *J. Catal.* 275 (2010), pp. 34–44.

Ma, L., Liu, C., Liao, J., Lu, T., Xing, W. & Zhang, J.: High activity PtRu/C catalysts synthesized by a modified impregnation method for methanol electro-oxidation. *Electrochim. Acta* 54 (2009), pp. 7274–7279.

Maillard, F., Pronkin, S.N. & Savinova, E.: Size effects in electrocatalysis of fuel-cell reactions on supported metal nanoparticles. In: *Fuel-cell catalysis: A surface science approach*. Wiley, NJ, 2009, pp. 567–592.

Mani, P., Srivastava, R. & Strasser, P.: Dealloyed binary PtM$_3$ (M=Cu, Co, Ni) and ternary PtNi$_3$M (M=Cu, Co, Fe, Cr) electrocatalysts for the oxygen reduction reaction: Performance in polymer electrolyte membrane fuel-cells. *J. Power Sources* 196 (2011), pp. 666–673.

Mard, A.J. & Faulkner, L.R.: Electrochemical methods: fundamentals and applications. Wiley, NJ, 2000.

Markovic, N.: The hydrogen electrode reaction and the electrooxidation of CO and H$_2$/CO mixtures on well-characterized Pt and Pt-bimetallic surfaces. In: *Handbook of fuel cells: fundamentals technology and applications*. Vol. 2: *Electrocatalysis*. Wiley, NJ, 2003, pp. 368–393.

Markovic, N.M. & Ross, P.N.: Surface science studies of model fuel-cell electrocatalysts. *Surf. Sci. Rep.* 45 (2002), pp. 117–229.

Markovic, N.M., Gasteiger, H.A. & Ross, J.P.N.: Oxygen reduction on platinum lo-index single-crystal surfaces in sulfuric acid solution: rotating ring-Pt(hkl) disk studies. *J. Phys. Chem.* 99 (1995), pp. 3411–3415.

Martínez-Huerta, M.V., Rojas, S., Gomez de la Fuente, J.L., Terreros, P., Pena, M.A. & Fierro, J.L.G.: Effect of Ni addition over PtRu/C based electrocatalysts for fuel-cell applications. *Appl. Catal. B Environ.* 69 (2006), pp. 75–84.

Martínez-Huerta, M.V., Rodríguez, J.L., Tsiouvaras, N., Peña, M.A., Fierro, J.L.G. & Pastor, E.: Novel synthesis method of CO-tolerant PtRu-MoO$_x$ nanoparticles: structural characteristics and performance for methanol electrooxidation. *Chem. Mater.* 20 (2008), pp. 4249–4259.

Martínez-Huerta, M.V., Tsiouvaras, N., Peña, M.A., Fierro, J.L.G., Rodriguez, J.L. & Pastor, E.: Electrochemical activation of nanostructured carbon-supported PtRuMo electrocatalyst for methanol oxidation. *Electrochim. Acta* 55 (2010), pp. 7634–7642.

Masao, A., Noda, S., Takasaki, F., Ito, K. & Sasaki, K.: Carbon-free Pt electrocatalysts supported on SnO$_2$ for polymer electrolyte fuel-cells. *Electrochem. Solid-State Lett.* 12 (2009), pp. B119–B122.

Massong, H., Wang, H., Samjeske, G. & Baltruschat, H.: The co-catalytic effect of Sn, Ru and Mo decorating steps of Pt(111) vicinal electrode surfaces on the oxidation of CO. *Electrochim. Acta* 46 (2001), pp. 701–707.

Masud, J., Alam, M.T., Awaludin, Z., El-Deab, M.S., Okajima, T. & Ohsaka, T.: Electrocatalytic oxidation of methanol at tantalum oxide-modified Pt electrodes. *J. Power Sources* 220 (2012), pp. 399–404.

Matsui, T., Fujiwara, K., Okanishi, T., Kikuchi, R., Takeguchi, T. & Eguchi, K.: Electrochemical oxidation of CO over tin oxide supported platinum catalysts. *J. Power Sources* 155 (2006), pp. 152–156.

Meier, J.C., Galeano, C., Katsounaros, I., Topalov, A.A., Kostka, A., Schüth, F. & Mayrhofer, K.J.J.: Degradation mechanisms of Pt/C fuel-cell catalysts under simulated start-stop conditions. *ACS Catal.* 2 (2012a), pp. 832–843.

Meier, J.C., Katsounaros, I., Galeano, C., Bongard, H.J., Topalov, A.A., Kostka, A., Karschin, A., Schuth, F. & Mayrhofer, K.J.J.: Stability investigations of electrocatalysts on the nanoscale. *Energy Environ. Sci.* 5 (2012b), pp. 9319–9330.

Millet P., Ngameni, R., Grigoriev, S.A. & Fateev, V.N.: Scientific and engineering issues related to PEM technology: Water electrolysers, fuel-cells and unitized regenerative systems. *Int. J. Hydrogen Energy* 36 (2011), pp. 4156–4163.

Morante-Catacora, T.Y., Ishikawa, Y. & Cabrera, C.R.: Sequential electrodeposition of Mo at Pt and PtRu methanol oxidation catalyst particles on HOPG surfaces. *J. Electroanal. Chem.* 621 (2008), pp. 103–112.

Mukerjee, S. & McBreen, J.: An in situ X-ray absorption spectroscopy investigation of the effect of Sn additions to carbon-supported Pt electrocatalysts: Part I. *J. Electrochem. Soc.* 146 (1999), pp. 600–606.

Mukerjee, S. & Srinivasan, S.: O$_2$ reduction and structure-related parameters for supported catalysts. In: *Handbook of fuel cells*. Wiley, NJ, 2003, pp. 502–519.

Mukerjee, S. & Urian, R.C.: Bifunctionality in Pt alloy nanocluster electrocatalysts for enhanced methanol oxidation and CO tolerance in PEM fuel-cells: electrochemical and in situ synchrotron spectroscopy. *Electrochim. Acta* 47 (2002), pp. 3219–3231.

Mukerjee, S., Lee, S.J., Ticianelli, E.A., McBreen, J., Grgur, B.N., Markovic, N.M., Ross, P.N., Giallombardo, J.R. & De Castro, E.S.: Investigation of enhanced CO tolerance in proton exchange membrane fuel-cells by carbon supported PtMo alloy catalyst. *Electrochem. Solid State Lett.* 2 (1999), pp. 12–15.

Mukerjee, S., Urian, R.C., Lee, S.J., Ticianelli, E.A. & McBreen, J.: Electrocatalysis of CO tolerance by carbon-supported PtMo electrocatalysts in PEMFCs. *J. Electrochem. Soc.* 151 (2004), pp. A1094–A1103.

Noked, M., Soffer, A. & Aurbach, D.: The electrochemistry of activated carbonaceous materials: Past, present, and future. *J. Solid State Electrochem.* 15 (2011), pp. 1563–1578.

Oetjen, H.F., Schmidt, V.M., Stimming, U. & Trila, F.: Performance data of a proton exchange membrane fuel-cell using $H_2/CO$ as fuel gas. *J. Electrochem. Soc.* 143 (1996), pp. 3838–3842.

Oh, S.M., Li, J.G. & Ishigakia, T.: Nanocrystalline $TiO_2$ powders synthesized by in-flight oxidation of TiN in thermal plasma: Mechanisms of phase selection and particle morphology evolution. *J. Mater. Res.* 20 (2005), pp. 529–537.

Oliveira, N.A., Franco, E.G., Arico, E., Linardi, M. & Gonzalez, E.R.: Electro-oxidation of methanol and ethanol on Pt-Ru/C and Pt-Ru-Mo/C electrocatalysts prepared by Bönnemann's method. *J. Eur. Ceram. Soc.* 23 (2003), pp. 2987–2992.

Ordoñez, L.C., Roquero, P., Sebastian, P.J. & Ramirez, J.: CO oxidation on carbon-supported PtMo electrocatalysts: Effect of the platinum particle size. *Int. J. Hydrogen Energy* 32 (2007), pp. 3147–3153.

Orilall, M.C., Matsumoto, F., Zhou, Q., Sai, H., Abruña, H.D., DiSalvo, F.J. & Wiesner, U.: One-pot synthesis of platinum-based nanoparticles incorporated into mesoporous niobium oxide-carbon composites for fuel-cell electrodes. *J. Am. Chem. Soc.* 131 (2009), pp. 9389–9395.

Over, H.: Surface chemistry of ruthenium dioxide in heterogeneous catalysis and electrocatalysis: From fundamental to applied research. *Chem. Rev.* 112 (2012), pp. 3356–3426.

Oyama, S.T.: *Chemistry of transition metal carbides and nitrides.* Chapman & Hall, UK, 1996.

Papageorgopoulos, D.C., Keijzer, M. & de Bruijn, F.A.: The inclusion of Mo, Nb and Ta in Pt and PtRu carbon supported 3 electrocatalysts in the quest for improved CO tolerant PEMFC anodes. *Electrochim. Acta* 48 (2002), pp. 197–204.

Papakonstantinou, G., Paloukis, F., Siokou, A. & Neophytides, S.G.: The electrokinetics of CO oxidation on $Pt_4Mo(20\ wt\%)/C$ interfaced with Nafion membrane. *J. Electrochem. Soc.* 154 (2007), pp. B989–B997.

Park, I.S., Park, K.W., Choi, J.H., Park, C.R. & Sung, Y.E.: Electrocatalytic enhancement of methanol oxidation by graphite nanofibers with a high loading of PtRu alloy nanoparticles. *Carbon* 45 (2007), pp. 28–33.

Park, K.W. & Seol, K.S.: $Nb-TiO_2$ supported Pt cathode catalyst for polymer electrolyte membrane fuel-cells. *Electrochem. Commun.* 9 (2007), pp. 2256–2260.

Park, S., Wieckowski, A. & Weaver, M.J.: Electrochemical infrared characterization of CO domains on ruthenium-decorated platinum nanoparticles. *J. Am. Chem. Soc.* 125 (2003), pp. 2282–2290.

Pasupathi, S. & Tricoli, V.: Effect of third metal on the electrocatalytic activity of PtRu/Vulcan for methanol electro-oxidation. *J. Solid State Electrochem.* 12 (2008), pp. 1093–1100.

Paulus, U.A., Wokaun, A., Scherer, G.G., Schmidt, T.J., Stamenkovic, V., Radmilovic, V., Markovic, N.M. & Ross, P.N.: Oxygen reduction on carbon-supported Pt-Ni and Pt-Co alloy catalysts. *J. Phys. Chem.* B 106 (2002), pp. 4181–4191.

Peng, Z. & Yang, H.: Designer platinum nanoparticles: Control of shape, composition in alloy, nanostructure and electrocatalytic property. *Nano Today* 4 (2009), pp. 143–164.

Petrii, O.: PtRu electrocatalysts for fuel-cells: a representative review. *J. Solid State Electrochem.* 12 (2008), pp. 609–642.

Pinheiro, A.L.N., Oliveira Neto, A., de Souza, E.C., Perez, J., Paganin, V.A., Ticianelli, E. & Gonzalez, E.R.: Electrocatalysis on noble metal and noble metal alloys dispersed on high surface area carbon. *J. New Mater. Electrochem. Syst.* 6 (2003), pp. 1–8.

Planes, G.A., García, G. & Pastor, E.: High performance mesoporous Pt electrode for methanol electrooxidation. A DEMS study. *Electrochem. Commun.* 9 (2007), pp. 839–844.

Qiao, Y. & Li, C.M.: Nanostructured catalysts in fuel-cells. *J. Mater. Chem.* 21 (2011), pp. 4027–4036.

Qiu, J.D., Wang, G.C., Liang, R.P., Xia, X.H. & Yu, H.W.: Controllable deposition of platinum nanoparticles on graphene as an electrocatalyst for direct methanol fuel-cells. *J. Phys. Chem.* C 115 (2011), pp. 15,639–15,645.

Rabis, A., Rodriguez, P. & Schmidt, T.J.: Electrocatalysis for polymer electrolyte fuel-cells: Recent achievements and future challenges. *ACS Catal.* 2 (2012), pp. 864–890.

Ralph, T.R. & Hogarth, M.: Catalysis for low temperature fuel-cells. Part II: the anode challenges. *Platinum Metals Rev.* 46 (2002), pp. 117–135.

Rojas, S., Martínez-Huerta, M.V. & Peña, M.A.: Supported metals for application in fuel-cells. In: J.A. Anderson and M. Fernandez Garcia (eds): *Supported metals in catalysis.* Imperial College Press, London, UK, 2012, pp. 407–491.

Rolison, D.R., Hagans, P.L., Swider, K.E. & Long, J.W.: Role of hydrous ruthenium oxide in Pt-Ru direct methanol fuel-cell anode electrocatalysts: The importance of mixed electro/proton conductivity. *Langmuir* 15 (1999), pp. 774–779.

Rossmeisl, J., Karlberg, G.S., Jaramillo, T. & Norskov, J.K.: Steady state oxygen reduction and cyclic voltammetry. *Faraday Discuss.* 140 (2009), pp. 337–346.

Roth, C., Goetz, M. & Fuess, H.: Synthesis and characterization of carbon-supported Pt-Ru-WO$_x$ catalysts by spectroscopic and diffraction methods. *J. Appl. Electrochem.* 31 (2001), pp. 793–798.

Santiago, E.I., Batista, M.S., Assaf, E.M. & Ticianelli, E.A.: Mechanism of CO tolerance on molybdenum-based electrocatalysts for PEMFC. *J. Electrochem. Soc.* 151 (2004), pp. A944–A949.

Sasaki, K., Zhang, L. & Adzic, R.R.: Niobium oxide-supported platinum ultra-low amount electrocatalysts for oxygen reduction. *Phys.* 10 (2008), pp. 159–167.

Sasaki, K., Takasaki, F., Noda, Z., Hayashi, S., Shiratori, Y. & Ito, K.: Alternative electrocatalyst support materials for polymer electrolyte fuel-cells. *ECS Trans.* 33 (2010), pp. 473–482.

Sau, T.K., Lopez, M. & Goia, D.V.: Method for preparing carbon supported Pt-Ru nanoparticles with controlled internal structure. *Chem. Mater.* 21 (2009), pp. 3649–3654.

Shao, M., Sasaki, K., Marinkovic, N.S., Zhang, L. & Adzic, R.R.: Synthesis and characterization of platinum monolayer oxygen-reduction electrocatalysts with Co-Pd core-shell nanoparticle supports. *Electrochem. Commun.* 9 (2007), pp. 2848–2853.

Shao, Y., Sui, J., Yin, G. & Gao, Y.: Nitrogen-doped carbon nanostructures and their composites as catalytic materials for proton exchange membrane fuel-cell. *Appl. Catal. B Environ.* 79 (2008), pp. 89–99.

Sharma, S. & Pollet, B.G.: Support materials for PEMFC and DMFC electrocatalysts: A review. *J. Power Sources* 208 (2012), pp. 96–119.

Sharma, S., Ganguly, A., Papakonstantinou, P., Miao, X., Li, M., Hutchison, J.L., Delichatsios, M. & Ukleja, S.: Rapid microwave synthesis of CO tolerant reduced graphene oxide-supported platinum electrocatalysts for oxidation of methanol. *J. Phys. Chem.* C 114 (2010), pp. 19,459–19,466.

Shubina, T.E. & Koper, M.T.M.: Quantum-chemical calculations of CO and OH interacting with bimetallic surfaces. *Electrochim. Acta* 47 (2002), pp. 3621–3628.

Siwek, H., Tokarz, W., Piela, P. & Czerwinski, A.: Electrochemical behavior of CO, CO$_2$ and methanol adsorption products formed on Pt-Rh alloys of various surface compositions. *J. Power Sources* 181 (2008), pp. 24–30.

Solla-Gullon, J., Vidal-Iglesias, F.J., Herrero, E., Feliu, J.M. & Aldaz, A.: CO monolayer oxidation on semi-spherical and preferentially oriented (100) and (111) platinum nanoparticles. *Electrochem. Commun.* 8 (2006), pp. 189–194.

Stamenkovic, V., Schmidt, T.J., Ross, P.N. & Markoviç, N.M.: Surface composition effects in electrocatalysis: Kinetics of oxygen reduction on well-defined Pt$_3$Ni and Pt$_3$Co alloy surfaces. *J. Phys. Chem.* B 106 (2002), pp. 11,970–11,979.

Stamenkovic, V., Mun, B.S., Mayrhofer, K.J., Ross, P.N., Markovic, N.M., Rossmeisl, J., Greeley, J. & Noskov, J.K.: Changing the activity of electrocatalysts for oxygen reduction by tuning the surface electronic structure. *Angew. Chem.* Int. Ed. 45 (2006), pp. 2897–2901.

Stamenkovic, V., Fowler, B., Mun, B.S., Wang, G., Ross, P.N., Lucas, C.A. & Markovic, N.M.: Improved oxygen reduction activity on Pt$_3$Ni(111) via increased surface site availability. *Science* 315 (2007), pp. 493–497.

Stamenkovic, V., Mun, B.S., Arenz, M., Mayrhofer, K.J.J., Lucas, C.A., Wang, G., Ross, P.N. & Markovic, N.M.: Trends in electrocatalysis on extended and nanoscale Pt-bimetallic alloy surfaces. *Nat. Mater.* 6 (2009), pp. 241–247.

Stankovich, S., Dikin, D.A., Dommett, G.H.B., Kohlhaas, K.M., Zimney, E.J., Stach, E.A., Piner, R.D., Nguyen, S.T. & Ruoff, R.S.: Graphene-based composite materials. *Nature* 442 (2006), pp. 282–286.

Steinhart, M., Wehrspohn, R.B., Gösele, U. & Wendorff, J.H.: Nanotubes by template wetting: A modular assembly system. *Angew. Chem.* Int. Ed. 43 (2004), pp. 1334–1344.

Stevens, D.A., Rouleau, J.M., Mar, R.E., Bonakdarpour, A., Atanasoski, R.T., Schmoeckel, A.K., Debe, M.K. & Dahn, J.R.: Characterization and PEMFC testing of Pt$_{1-x}$M$_x$ (M=Ru,Mo,Co,Ta,Au,Sn) anode electrocatalyst composition spreads. *J. Electrochem. Soc.* 154 (2007), pp. B566–B576.

Stottlemyer, A.L., Kelly, T.G., Meng, Q. & Chen, J.G.: Reactions of oxygen-containing molecules on transition metal carbides: Surface science insight into potential applications in catalysis and electrocatalysis. *Surf. Sci. Rep.* 67 (2012), pp. 201–232.

Strasser, P., Koh, S. & Greeley, J.: Voltammetric surface dealloying of Pt bimetallic nanoparticles: An experimental and DFT computational analysis. *Phys.* 10 (2008), pp. 3670–3683.

Strasser, P., Koh, S., Anniyev, T., Greeley, J., More, K., Yu, C.F., Liu, Z.C., Kaya, S., Nordlund, D., Ogasawara, H., Toney, M.F. & Nilsson, A.: Lattice-strain control of the activity in dealloyed core-shell fuel-cell catalysts. *Nature Chem.* 2 (2010), pp. 454–460.

Sun, Y., Mayers, B.T. & Xia, Y.: Template-engaged replacement reaction: A one-step approach to the large-scale synthesis of metal nanostructures with hollow interiors. *Nano Lett.* 2 (2002), pp. 481–485.

Sun, Y., Zhuang, L., Lu, J., Hong, X. & Liu, P.: Collapse in crystalline structure and decline in catalytic activity of Pt nanoparticles on reducing particle size to 1 nm. *J. Am. Chem. Soc.* 129 (2007), pp. 15,465–15,467.

Tans, S.J., Verschueren, A.R.M. & Dekker, C.: Room-temperature transistor based on a single carbon nanotube. *Nature* 393 (1998), pp. 49–52.

Templeton, A.C., Wuelfing, W.P. & Murray, R.W.: Monolayer-protected cluster molecules. *Acc. Chem. Res.* 33 (1999), pp. 27–36.

Teng, X., Maksimuk, S., Frommer, S. & Yang, H.: Three-dimensional PtRu nanostructures. *Chem. Mater.* 19 (2007), pp. 36–41.

Thompsett, D.: Catalysts for the proton exchange membrane fuel cell. In: G. Hoogers (ed): *Fuel cell technology handbook.* CRC Press LCC, 2003, Chapter 6.

Tian, N., Zhou, Z.Y., Sun, S.G., Ding, Y. & Wang, Z.L.: Synthesis of tetrahexahedral platinum nanocrystals with high-index facets and high electro-oxidation activity. *Science* 316 (2007), pp. 732–735.

Tian, N., Zhou, Z.Y. & Sun, S.G.: Platinum metal catalysts of high-index surfaces: From single-crystal planes to electrochemically shape-controlled nanoparticles. *J. Phys. Chem.* C 112 (2008), pp. 19,801–19,817.

Toshima, N. & Yonezawa, T.: Bimetallic nanoparticles—novel materials for chemical and physical applications. *New J. Chem.* 22 (1998), pp. 1179–1201.

Tripković, V., Abild-Pedersen, F., Studt, F., Cerri, I., Nagami, T., Bligaard, T. & Rossmeisl, J.: Metal oxide-supported platinum overlayers as proton-exchange membrane fuel-cell cathodes. *ChemCatChem* 4 (2012), pp. 228–235.

Tsiouvaras, N., Martínez-Huerta, M.V., Moliner, R., Lazaro, M.J., Rodriguez, J.L., Pastor, E., Peña, M.A. & Fierro, J.L.G.: CO tolerant PtRu-MoO$_x$ nanoparticles supported on carbon nanofibers for direct methanol fuel-cells. *J. Power Sources* 186 (2009), pp. 299–304.

Tsiouvaras, N., Martínez-Huerta, M.V., Paschos, O., Stimming, U., Fierro, J.L.G. & Peña, M.A.: PtRuMo/C catalysts for direct methanol fuel-cells: Effect of the pretreatment on the structural characteristics and methanol electrooxidation. *Int. J. Hydrogen Energy* 35 (2010a), pp. 11,478–11,488.

Tsiouvaras, N., Peña, M.A., Fierro, J.L.G., Pastor, E. & Martínez-Huerta, M.V.: The effect of the Mo precursor on the nanostructure and activity of PtRuMo electrocatalysts for proton exchange membrane fuel-cells. *Catal. Today* 158 (2010b), pp. 12–21.

Urian, R.C., Gullá, A.F. & Mukerjee, S.: Electrocatalysis of reformate tolerance in proton exchange membranes fuel-cells: Part I. *J. Electroanal. Chem.* 554–555 (2003), pp. 307–324.

Venkataraman, R., Kunz, H.R. & Fenton, J.M.: Development of new CO tolerant ternary anode catalysts for proton exchange membrane fuel-cells. *J. Electrochem. Soc.* 150 (2003), pp. A278–A284.

Vidakovic, T., Christov, M., Sundmacher, K., Nagabhushana, K.S., Fei, W., Kinge, S. & Bonnemann, H.: PtRu colloidal catalysts: Characterisation and determination of kinetics for methanol oxidation. *Electrochim. Acta* 52 (2007), pp. 2277–2284.

Vidal-Iglesias, F.J., Arán-Ais, R.M., Solla-Gullón, J., Herrero, E. & Feliu, J.M.: Electrochemical characterization of shape-controlled Pt nanoparticles in different supporting electrolytes. *ACS Catal.* 2 (2012), pp. 901–910.

Vielstich, W., Lamm, A. & Gasteiger, H.A.: *Handbook of fuel-cells-Fundamentals, technology and applications.* Wiley, NJ, 2003.

Viswanathan, V., Hansen, H.A., Rossmeisl, J. & Norskov, J.K.: Universality in oxygen reduction electrocatalysis on metal surfaces. *ACS Catal.* 2 (2012), pp. 1654–1660.

Wang, B., Tian, C., Wang, L., Wang, R. & Fu, H.: Chitosan: a green carbon source for the synthesis of graphitic nanocarbon, tungsten carbide and graphitic nanocarbon/tungsten carbide composites. *Nanotechnol.* 21 (2010), pp. 025606 (9 pp).

Wang, C., Daimon, H., Onodera, T., Koda, T. & Sun, S.: A general approach to the size- and shape-controlled synthesis of platinum nanoparticles and their catalytic reduction of oxygen. *Angew. Chem.* Int. Ed. 47 (2008), pp. 3588–3591.

Wang, C., Li, D., Chi, M., Pearson, J., Rankin, R.B., Greeley, J., Duan, Z., Wang, G., van der Vliet, D., More, K.L., Markovic, N.M. & Stamenkovic, V.R.: Rational development of ternary alloy electrocatalysts. *J. Phys. Chem. Lett.* 3 (2012a), pp. 1668–1673.

Wang, C., Markovic, N.M. & Stamenkovic, V.R.: Advanced platinum alloy electrocatalysts for the oxygen reduction reaction. *ACS Catal.* 2 (2012b), pp. 891–898.

Wang, C.H., Du, H.Y., Tsai, Y.T., Chen, C.P., Huang, C.J., Chen, L.C., Chen, K.H. & Shih, H.C.: High performance of low electrocatalysts loading on CNT directly grown on carbon cloth for DMFC. *J. Power Sources* 171 (2007), pp. 55–62.

Wang, H., Alden, L.R., DiSalvo, F.J. & Abruña, H.D.: Methanol electrooxidation on PtRu bulk alloys and carbon-supported PtRu nanoparticle catalysts: A quantitative DEMS study. *Langmuir* 25 (2009), pp. 7725–7735.

Wang, H., Yuan, X., Li, D. & Gu, X.: Dendritic PtCo alloy nanoparticles as high performance oxygen reduction catalysts. *J. Colloid Interf. Sci.* 384 (2012), pp. 105–109.

Wang, J.X., Inada, H., Wu, L.J., Zhu, Y.M., Choi, Y.M., Liu, P., Zhou, W.P. & Adzic, R.R.: Oxygen reduction on well-defined core-shell nanocatalysts: Particle size, facet, and Pt shell thickness effects. *J. Am. Chem. Soc.* 131 (2009), pp. 17,298–17,302.

Wang, S., Jiang, S.P. & Wang, X.: Microwave-assisted one-pot synthesis of metal/metal oxide nanoparticles on graphene and their electrochemical applications. *Electrochim. Acta* 56 (2011), pp. 3338–3344.

Wang, X., Li, W., Chen, Z., Waje, M. & Yan, Y.: Durability investigation of carbon nanotube as catalyst support for proton exchange membrane fuel-cell. *J. Power Sources* 158 (2006), pp. 154–159.

Wang, Y.J., Wilkinson, D.P. & Zhang, J.: Noncarbon support materials for polymer electrolyte membrane fuel-cell electrocatalysts. *Chem. Rev.* 111 (2011), pp. 7625–7651.

Wang, Z.B., Zuo, P.J. & Yin, G.P.: Investigations of compositions and performance of PtRuMo/C ternary catalysts for methanol electrooxidation. *Fuel-Cells* 2 (2009), pp. 106–113.

Waszczuk, P., Solla-Gullon, J., Kim, H.S., Tong, Y.Y., Montiel, V., Aldaz, A. & Wieckowski, A.: Methanol electrooxidation on platinum/ruthenium nanoparticle catalysts. *J. Catal.* 203 (2001), pp. 1–6.

Watanabe, M. & Motoo, S.: Electrocatalysis by ad-atoms: Part I. Enhancement of the oxidation of methanol on platinum and palladium by gold ad-atoms. *J. Electroanal. Chem.* 60 (1975a), pp. 259–266.

Watanabe, M. & Motoo, S.: Electrocatalysis by ad-atoms: Part II. Enhancement of the oxidation of methanol on platinum by ruthenium ad-atoms. *J. Electroanal. Chem.* 60 (1975b), pp. 267–273.

Watanabe, M. & Uchida, H.: Catalysts for the electro-oxidation of small molecules. In: *Handbook of fuel-cells-Fundamentals, technology and applications.* Vol. 5: *Advances in electrocatalysis, materials, diagnostics and durability.* Wiley, NJ, 2009, pp. 81–90.

Watanabe, M., Uchida, M. & Motoo, S.: Preparation of highly dispersed Pt + Ru clusters and the activity for the electrooxidation of methanol. *J. Electroanal. Chem.* 229 (1987), pp. 395–406.

Watanabe, M., Igarashi, H. & Fujino, T.: Design of CO tolerant anode catalysts for polymer electrolyte fuel-cell. *Electrochem.* 67 (1999), pp. 1194–1196.

Wesselmark, M., Wickman, B., Lagergren, C. & Lindbergh, G.: Electrochemical performance and stability of thin film electrodes with metal oxides in polymer electrolyte fuel-cells. *Electrochim. Acta* 55 (2010), pp. 7590–7596.

Wickman, B., Wesselmark, M., Lagergren, C. & Lindbergh, G.: Tungsten oxide in polymer electrolyte fuel-cell electrodes: A thin-film model electrode study. *Electrochim. Acta* 56 (2011), pp. 9496–9503.

Wilson, M.S., Garzon, F.H., Sickafus, K.E. & Gottesfeld, S.: Surface area loss of supported platinum in polymer electrolyte fuel-cells. *J. Electrochem. Soc.* 140 (1993), pp. 2872–2877.

Wu, G., Li, L. & Xu, B.Q.: Effect of electrochemical polarization of PtRu/C catalysts on methanol electrooxidation. *Electrochim. Acta* 50 (2004), pp. 1–10.

Wu, G., Nelson, M.A., Mack, N.H., Ma, S., Sekhar, P., Garzon, F.H. & Zelenay, P.: Titanium dioxide-supported non-precious metal oxygen reduction electrocatalyst. *Chem. Commun.* 46 (2010), pp. 7489–7491.

Wu, H., Wexler, D. & Liu, H.: Durability investigation of graphene-supported Pt nanocatalysts for PEM fuel-cells. *J. Solid State Electrochem.* 15 (2011), pp. 1057–1062.

Xia, B.Y., Wu, H.B., Wang, X. & Lou, X.W.: One-pot synthesis of cubic $PtCu_3$ nanocages with enhanced electrocatalytic activity for the methanol oxidation reaction. *J. Am. Chem. Soc.* 134 (2012), pp. 13,934–13,937.

Xia, D., Chen, G., Wang, Z., Zhang, J., Hui, S., Ghosh, D. & Wang, H.: Synthesis of ordered intermetallic $PtBi_2$ nanoparticles for methanol-tolerant catalyst in oxygen electroreduction. *Chem. Mater.* 18 (2006), pp. 5746–5749.

Xia, Y., Xiong, Y., Lim, B. & Skrabalak, S.: Shape-controlled synthesis of metal nanocrystals: Simple chemistry meets complex physics? *Angew. Chem.* Int. Ed. 48 (2009), pp. 60–103.

Xing, Y., Cai, Y., Vukmirovic, M.B., Zhou, W.P., Karan, H., Wang, J.X. & Adzic, R.R.: Enhancing oxygen reduction reaction activity via PdAu alloy sublayer mediation of Pt monolayer electrocatalysts. *J. Phys. Chem. Lett.* 1 (2010), pp. 3238–3242.

Xiong, L. & Manthiram, A.: Nanostructured Pt-M/C (M=Fe and Co) catalysts prepared by a microemulsion method for oxygen reduction in proton exchange membrane fuel-cells. *Electrochim. Acta* 50 (2005), pp. 2323–2329.

Xu, H., Zeiger, B.W. & Suslick, K.S.: Sonochemical synthesis of nanomaterials. *Chem. Soc. Rev.* 42 (2013), pp. 2555–2567.

Xu, Y., Bai, H., Lu, G., Li, C. & Shi, G.: Flexible graphene films via the filtration of water-soluble noncovalent functionalized graphene sheets. *J. Am. Chem. Soc.* 130 (2008), pp. 5856–5857.

Yang, H., Alonso-Vante, N., Leger, J.M. & Lamy, C.: Tailoring, structure, and activity of carbon-supported nanosized Pt-Cr alloy electrocatalysts for oxygen reduction in pure and methanol-containing electrolytes. *J. Phys. Chem.* B 108 (2004), pp. 1938–1947.

Ye, S.: CO-tolerant catalysts. Chapter 16 in: *PEM fuel-cell electrocatalysts and catalyst layers. Fundamentals and applications.* Springer, 2009, pp. 759–834.

Yin, M., Huang, Y., Liang, L., Liao, J., Liu, C. & Xing, W.: Inhibiting CO formation by adjusting surface composition in PtAu alloys for methanol electrooxidation. *Chem. Commun.* 47 (2011), pp. 8172–8174.

Zellner, M.B. & Chen, J.G.: Surface science and electrochemical studies of WC and $W_2C$ PVD films as potential electrocatalysts. *Catal.* 99 (2005), pp. 299–307.

Zhang, G., Shao, Z.G., Lua, W., Xie, F., Xiao, H., Qin, X. & Yi, B.: Core-shell Pt modified Pd/C as an active and durable electrocatalyst for the oxygen reduction reaction in PEMFCs. *Appl. Catal.* B *Environ.* 132–133 (2013), pp. 183–194.

Zhang, J., Sasaki, K., Sutter, E. & Adzic, R.R.: Stabilization of platinum oxygen-reduction electrocatalysts using gold clusters. *Science* 315 (2007), pp. 220–222.

Zhang, W., Ravi, S. & Silva, P.: Application of carbon nanotubes in polymer electrolyte based fuel-cells. *Rev. Adv. Mater. Sci.* 29 (2011), pp. 1–14.

Zhang, X. & Chan, K.Y.: Water-in-oil microemulsion synthesis of platinum-ruthenium nanoparticles, their characterization and electrocatalytic properties. *Chem. Mater.* 15 (2003), pp. 451–459.

Zhang, X., Zhang, F. & Chan, K.Y.: Synthesis of Pt-Ru-Mo ternary metal nanoparticles by microemulsions, their characterization and electrocatalytic properties. *J. Mater. Sci.* 39 (2004), pp. 5845–5848.

Zhao, X., Yin, M., Ma, L., Liang, L., Liu, C., Liao, J., Lu, T. & Xing, W.: Recent advances in catalysts for direct methanol fuel-cells. *Energy Environ. Sci.* 4 (2011), pp. 2736–2753.

Zhong, C.J., Luo, J., Njoki, P.N., Mott, D., Wanjala, B., Loukrakpam, R., Lim, S., Wang, L., Fang, B. & Xu, Z.: Fuel-cell technology: nano-engineered multimetallic catalysts. *Energy Environ. Sci.* 1 (2008), pp. 454–466.

Zhong, C.J., Luo, J., Fang, B., Wanjala, B.N., Njoki, P.N., Loukrakpam, R. & Yin, G.: Nanostructured catalysts in fuel-cells. *Nanotechnol.* 21 (2010), 062001 (20 pp).

Zhou, W.P., Lewera, A., Bagus, P.S. & Wieckowski, A.: Electrochemical and electronic properties of platinum deposits on Ru(0001): Combined XPS and cyclic voltammetric study. *J. Phys. Chem.* C 111 (2007), pp. 13,490–13,496.

Zhou, Y.G., Chen, J.J., Wang, F.B., Sheng, Z.H. & Xia, X.H.: A facile approach to the synthesis of highly electroactive Pt nanoparticles on graphene as an anode catalyst for direct methanol fuel-cells. *Chem. Commun.* 46 (2010), pp. 5951–5953.

Zhu, J., Cheng, F., Tao, Z. & Chen, J.: Electrocatalytic methanol oxidation of $Pt_{0.5}Ru_{0.5-x}Sn_x/C$ ($x = 0$–$0.5$). *J. Phys. Chem.* C 112 (2008), pp. 6337–6345.

# CHAPTER 5

## Ordered mesoporous carbon-supported nano-platinum catalysts: application in direct methanol fuel cells

Parasuraman Selvam & Balaiah Kuppan

### 5.1 INTRODUCTION

During the past century, chemists have primarily focused on making and breaking strong covalent bonds. With this approach, it is possible to combine atoms into molecules and extended structures with nearly arbitrary atomic scale configurations (Service, 2005). Molecules of increased size and complexity require more demanding synthetic methods and for many years, meso- or macro-scale designed configurations, generated by molecular assembly, had limited accessibility. However, chemists have not viewed hydrogen bonds, Van der Wall's forces, and medium to long range electrostatic forces, all of which are much weaker than covalent bonds, as chemical glue for assembling molecules into materials (Fan et al., 2008). This is in spite of the fact that nature is built on this approach; nearly all that surrounds us, from cells to trees, are knit together using weak interactions between molecules. The formation process for many of the meso-structured materials, whereby a collection of small molecules, electrolytes, polymers, and co-solvents spontaneously combine into larger, well-defined supra-molecular assemblies or aggregates due to the weak forces, has historically been termed by chemists and materials scientists as "self-assembly", "cooperative self-assembly", or "molecular assembly". Over the past two decades, researchers have made large advances in understanding the basic rules of molecular assembly, as well as in developing methods to simultaneously control intermolecular interactions and reaction kinetics to create material systems with hierarchical ordering and complexity (Davis, 2002). Methodologies, which make use of molecular assembly, have been recognized as the most promising approach for the fabrication of a wide variety of meso-structured materials.

At this juncture, it is important to note that platinum-supported carbon catalysts are generally used as electrode materials, e.g., methanol oxidation reaction at the anodes of direct methanol fuel cells (DMFCs) and oxygen reduction reaction at the cathodes of proton exchange membrane fuel cells (PEMFCs). These systems have received considerable attention as clean energy sources for various applications (Winter and Brodd, 2004). However, in order to achieve high dispersion, good stability, effective utilization and stupendous activity of platinum metal, porous carbons with ordered pore structure, high surface area, nano-scale morphology, tunable pore characteristics with varied surface functionality, and good electric conductivity are highly desirable for electro-catalysts (Dicks, 2006; Gasteiger and Marković, 2009). In this regard, the most commonly used electro-catalyst, both for cathode and anode, is platinum supported on carbon blacks (Escudero et al., 2002; Lizcano-Valbuena et al., 2003; Kim et al., 2003; Tian et al., 2004). However, it is necessary to obtain a more effective catalyst, both in catalytic performance and electronic conductivity. To achieve a higher efficiency of the electro-catalysts, platinum has to be well dispersed on the support. For this reason, it is desirable that the support material provides a suitable specific area and surface chemistry as well as good electrical conductivity. In this chapter, we address such issues in a greater detail by taking into account of our recent work (Kuppan, 2014; Kuppan and Selvam, 2012; Selvam and Kuppan, 2012) on nano-platinum-supported mesoporous carbons such as NCCR-41, CMK-3, NCCR-11 and CMK-1 for DMFCs.

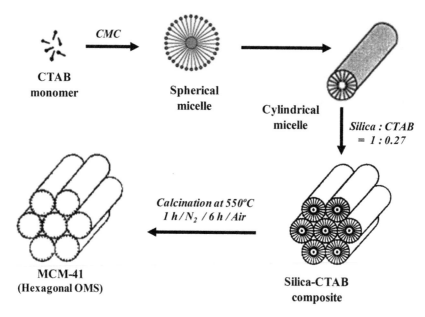

Figure 5.1.    Liquid-crystal templating mechanism for MCM-41 (after Beck *et al.*, 1992).

## 5.2    ORDERED MESOPOROUS SILICAS

In the early 1990s, a Japanese group (Yanagisawa *et al.*, 1990) and Mobil scientists (Kresge *et al.*, 1992) respectively reported the independent discovery of an FSM-type and an M41S-family of ordered mesoporous silicate (OMS) materials. These silicate/silica (the terms "silicate(s)" and "silica(s)" are interchangeably used in this chapter) framework structures are generated by the molecular assembly of many cationic surfactant molecules. Under basic conditions, the electrostatic interaction between solvated silicate anions and cationic surfactant assemblies (template molecules), combined with the hydrophobic interactions of the non-polar surfactant tails, drive the formation of mesoporous silica with a one-dimensional hexagonal (MCM-41; Fig. 5.1), three-dimensional cubic (MCM-48; Fig. 5.2), or lamellar (MCM-50) structures (Beck *et al.*, 1992; Kresge *et al.*, 1992). Among these, MCM-41 and MCM-48 have been identified as thermally stable while the lamellar MCM-50 phase is unstable. However, quite a large amount of research effort has been devoted to MCM-41 because of the ease in preparation while less attention has been paid to MCM-48 owing to the difficulty in the synthesis of good quality samples, as this phase has a very narrow homogeneity region as depicted in the phase diagram (Fig. 5.3).

On the other hand, Stucky and co-workers (Huo *et al.*, 1994a; 1994b) have found that ordered mesoporous silica structures could be obtained in acidic media at or below the isoelectric point of silica ($pH \sim 2$–3) as opposed to the classic basic $pH$ approach used for the synthesis of zeolites and mesoporous materials. They have reported a series of meso-structured silicas such as SBA-15 (Fig. 5.4), SBA-11 (Fig. 5.5), etc. (Selvam *et al.*, 2010a). Following this approach, several mesoporous metal oxides have also been prepared (He and Antonelli, 2002; Yang *et al.*, 1999). The molecular assembly of cationic silica species with cationic quaternary ammonium species requires that the assembly takes place through bridging anions, and the manner in which anions could be used to determine the assembly of the meso-structured phase was demonstrated (Huo *et al.*, 1994b). Utilizing silica precursors in acidic solutions with cationic surfactants, e.g., $C_{16}H_{33}(CH_3)_3N^+$, and tri-block copolymers consisting of polyethylene oxide units and polypropylene oxide units, e.g., $EO_{20}PO_{70}EO_{20}$ or P123, $EO_{106}PO_{70}EO_{106}$ or F127, etc. as the supra-molecular surfactant (soft) templates, a new series of (SBA-family) meso-structured

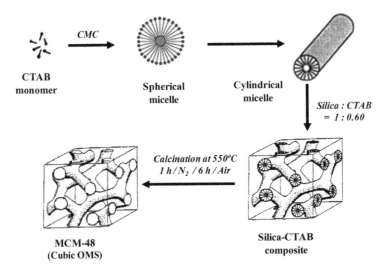

Figure 5.2.   Liquid crystal templating mechanism for MCM-48 (after Beck *et al.*, 1992).

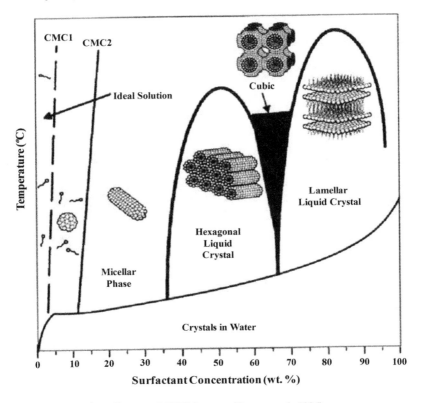

Figure 5.3.   Schematic phase diagram of CTAB in water (Raman *et al.*, 1996).

materials were obtained (Zhao *et al.*, 1998). Since the discovery of these classes of materials, excellent progress has been made with the focus in the following major directions (Davis, 2002): (i) the development of the basic principles of inorganic-organic molecular assembly processes; (ii) the application of these principles to synthesize new materials; and (iii) the design of new

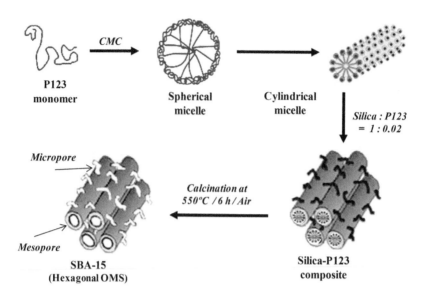

Figure 5.4.   Block copolymer assisted templating mechanism of SBA-15 (after Huo *et al.*, 1994a).

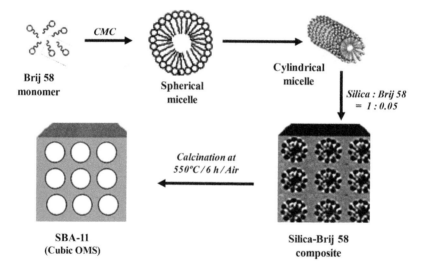

Figure 5.5.   Block copolymer assisted templating mechanism of SBA-11 (after Huo *et al.*, 1994a).

supra-molecular assembly systems and exploitation of such meso-structured materials for applications in different fields.

The mesoporous molecular sieve materials are new generation ordered nano-porous materials, analogous to micro-porous zeolites, having high surface area, large pore opening and a huge pore volume (Table 5.1). The unique flexibility in terms of synthetic conditions, pore size tuning, high surface area, large internal hydroxyl groups, framework substitution, etc. have created new avenues not only in catalysis but also in the areas of advanced materials, environmental pollution control strategies and separation processes (Corma, 1997; Selvam *et al.*, 2001; 2010a; Wan and Zhao, 2007).

A number of reviews have appeared on various aspects of different mesoporous materials (Corma, 1997; Fan *et al.*, 2008; Meynen *et al.*, 2009; Selvam *et al.*, 2001; 2010a; Wan and Zhao,

Table 5.1. Synthesis parameters and physical properties of OMS (after Selvam *et al.*, 2010a).

| | | | Crystallographic data | | Textural data | | |
|---|---|---|---|---|---|---|---|
| Material | Surfactant | pH | Dimensionality/ Crystal system | $a$ [Å]; $c$ [Å] | $V_p$ [nm] | $S_{BET}$ [m$^2$ g$^{-1}$] | D [nm] |
| MCM-41 | CTAB ($C_nH_{2n+1}(CH_3)_3N^+$)[1] | Basic | 2D Hexagonal | 40.4 | 3.70 | 1041 | 0.34 |
| MCM-48 | CTAB ($C_nH_{2n+1}(CH_3)_3N^+$)[1] | Basic | Cubic | 80.8 | 3.49 | 1010 | 0.86 |
| KIT-1 | $C_{16}H_{33}(C_2H_5)_3N^+$ | Basic | 3D Hexagonal | 48.0; | 3.40 | 1000 | 0.70 |
| HMM | $C_{18}H_{37}(CH_3)_3N^+$ | Basic | 2D Hexagonal | 57.0 | 3.10 | 750 | 2.60 |
| HMM | $C_{18}H_{37}(CH_3)_3N^+$ | Basic | 3D Hexagonal | 88.6; 5.54 | 2.70 | 1170 | 6.10 |
| FDU-2 | $C_{m-2-3-1}(m = 14–18)$[2] | Basic | Cubic | 120.0 | 3.02 | 964 | 8.98 |
| SBA-6 | $18B_{4-3-1}$[3] | Basic | Cubic | 146.0 | 2.00 | 686 | 6.15 |
| SBA-7 | Gemini $– C_{n-3-1}(n = 12,18)$[4] | Basic | 3D Hexagonal | 46.0; 7.50 | 2.30 | 555 | 1.50 |
| | Gemini $– C_{n-6-1}(n = 16, 18)$[4] | Basic | 3D Hexagonal | – | – | – | – |
| SBA-8 | Bolaformsurfactant $R_{12}$‡ | Basic | 2D Rectangular | 74.9; 4.98 | 2.91 | 1022 | 2.07 |
| SBA-1 | $C_{16}H_{33}(C_2H_5)_3N^+$ | Acidic | Cubic | 76.0 | 2.10 | 1355 | 2.14 |
| SBA-2 | $C_{n-3-1}(n = 12 – 18)$[4] | Acidic | 3D Hexagonal | 54.0; 87.0 | 2.22 | 990 | 3.18 2.9 |
| | $C_{n-6-1}(n = 16, 18)$[4] | Acidic | 3D Hexagonal | 75.8; 127 | 2.50 | 500–700 | |
| SBA-3 | $C_{16}H_{33}(CH_3)_3N^+$ | Acidic | 2D Hexagonal | 36.0 | 2.60 | 1430 | 1.00 |
| SBA-11 | Brij 56 ($C_{16}EO_{10}$) | Acidic | Cubic | 106.4 | 2.50 | 1070 | 3.44 |
| SBA-12 | Brij 76 ($C_{18}EO_{10}$) | Acidic | 3D Hexagonal | 54.0; 87.0 | 3.10 | 1150 | 2.30 |
| SBA-14 | Brij 30 ($C_{12}EO_4$) | Acidic | Cubic | 44.7 | 2.20 | 670 | 0.30 |
| SBA-15 | P123 ($EO_{20}PO_{70}EO_{20}$) | Acidic | 2D Hexagonal | 121.2 | 8.90 | 850 | 3.20 |
| SBA-16 | F127 ($EO_{106}PO_{70}EO_{106}$) | Acidic | Cubic | 176.0 | 5.40 | 740 | 9.84 |
| IITM-56 | Brij 56 ($C_{16}EO_{10}$) | Acidic | 2D Hexagonal | 59.4 | 3.80 | 772 | 2.10 |
| FDU-1 | B50-6600 ($EO_{39}BO_{47}EO_{39}$) | Acidic | Cubic | 207.0 | 9.00 | 650 | 8.90 |
| FDU-12 | F127 ($EO_{106}PO_{70}EO_{106}$) | Acidic | Cubic | 253.0 | 10.0 | 712 | 4.42 |
| KIT-5 | F127 ($EO_{106}PO_{70}EO_{106}$) | Acidic | Cubic | 190.0 | 9.30 | 715 | 1.53 |
| KIT-6 | P123 ($EO_{20}PO_{70}EO_{20}$) | Acidic | Cubic | 229.0 | 8.20 | 800 | 3.31 |
| HMS | $C_nH_{2n+1}NH_2(n = 8–22)$ | Neutral | 3D Hexagonal | 45.5 | 2.80 | 1070 | 2.70 |
| MSU-1 | Tergitol ($C_{11–15}(EO)_{12}$) | Neutral | 3D Hexagonal | 47.3 | 3.10 | 1005 | 1.00 |
| MSU-2 | TX-114 ($C_8Ph(EO)_8$) | Neutral | 3D Hexagonal | 70.4 | 2.00 | 780 | 4.10 |
| | TX-100 ($C_8Ph(EO)_{10}$) | Neutral | 3D Hexagonal | 71.6 | 3.50 | 715 | 2.70 |
| MSU-3 | P64L ($EO_{13}PO_{30}EO_{13}$) | Neutral | 3D Hexagonal | 70.4 | 5.80 | 1190 | 0.30 |
| MSU-4 | Tween-20,40,60,80[5] | Neutral | 3D Hexagonal | 60.1 | 3.40 | 773 | 1.80 |

[1]$(n = 12–18)$; $[(CH_3)_3N^+–(CH_2)_{12}–OC_6H_4C_6H_4 O–(CH_2)_{12}–N^+(CH_3)_3][2Br^-]$.
[2]Gemini surfactant $–C_mH_{2m+1}N^+(CH_3)_2CH_2CH_2N^+ (CH_3)_2CH_2CH_2CH_2N^+(CH_3)_3.[3Br^-]$.
[3]Gemini surfactant $–C_{18}H_{37}O–C_6H_4–(CH_2)_4N^+(CH_3)_2(CH_2)_3N^+ (CH_3)_3·[2Br^-]$.
[4]Divalent surfactant $–C_nH_{2n+1}N^+–(CH_3)_2(CH_2)_3–N^+(CH_3)_3$.
[5]Polysorbate $–C_{18}H_{34}O_6·EO_{20}$.

2007; Zhao *et al.*, 1998). It is interesting to note that different choices of the inorganic precursors, organic templates, reaction conditions, etc. have resulted in a variety of meso-structured materials. The physico-chemical properties of some of these silicates are tabulated in Table 5.1.

## 5.3  ORDERED MESOPOROUS CARBONS

Porous carbon materials with high surface areas and pore volumes prepared from porous inorganic templates are of current interest for energy storage, gas separation, heterogeneous catalysis, and many other applications including water purification, catalyst support as well as electrode material

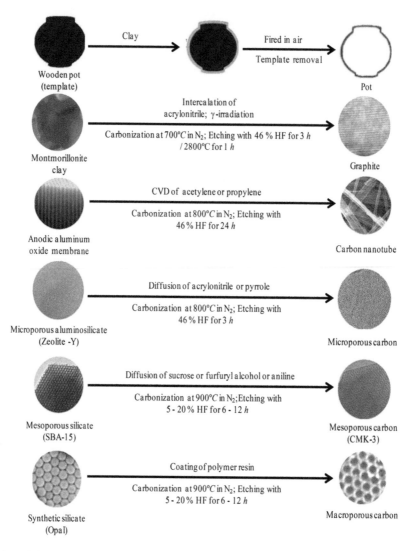

Figure 5.6.   Schematic representations of templated-synthesis of porous carbons (after Lee *et al.*, 2006).

for electrochemical double layer capacitors and fuel cells (Inagaki *et al.*, 2011; Jiang *et al.*, 2011; Lee *et al.*, 2006; Liu *et al.*, 2010; Sharma and Pollet, 2012; Yang *et al.*, 2011) Figure 5.6 depicts the overall concept of such procedures, which are essentially employed to make a pot but scaled down to the nanometer regime (Lee *et al.*, 2006). In this way, several attempts have been made for the template-synthesis of porous carbons (Gilbert *et al.*, 1982) such as graphite sheets, carbon nanotubes (CNTs), ordered mesoporous carbons (OMCs). Various inorganic materials including silica nano-particles (silica-sol), clays, anodic alumina membranes, zeolites, and mesoporous silica have been used as hard templates for the preparation of such ordered porous carbon materials. However, one of the most successful examples is the synthesis of OMCs (Inagaki *et al.*, 2011; Jun *et al.*, 2000; Ryoo *et al.*, 1999; 2001) using hard templates such as ordered mesoporous silicas (OMSs). Similarly, OMCs have also been prepared using polymeric soft-templates (Meng *et al.*, 2006).

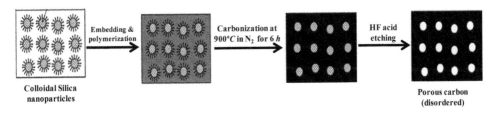

Figure 5.7. Formation of (disordered) carbon from colloidal silica via a hard-template route (Lee *et al.*, 2006).

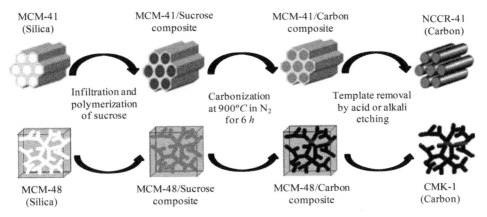

Figure 5.8. Formation of NCCR-41 and CMK-1, *via* a hard-template route (after Lee *et al.*, 2006; Selvam and Kuppan, 2012; Tian *et al.*, 2003).

### 5.3.1 *Hard-template approach*

The hard template approach can be classified into two distinct categories (Lee *et al.*, 2006; Ryoo *et al.*, 1999). In the first approach, an inorganic template such as colloidal silica is embedded in certain carbon precursor; carbonization followed by the removal of the template by acid etching generates disordered porous carbons (Fig. 5.7). In the second, the carbon precursor is infiltrated into the ordered meso-pores of OMSs followed by carbonization and the subsequent removal of the template by acid or alkali leaching generates OMCs (Fig. 5.8). Although the majority of studies have focused on MCM-41, there has been continued interest in developing new meso-phases. Many factors influence the final product of OMCs such as the quality of hard templates, the filling amount, the carbonization process, and more importantly the nature of the carbon source. For instance, with the same (SBA-15) template, tube-like OMCs are obtained from furfuryl alcohol whereas rod-like structure (Fig. 5.9) is obtained from sucrose (Ryoo *et al.*, 2001; Zhai *et al.*, 2008). A disordered cylindrical porous structure is obtained with MCM-41 and furfuryl alcohol (Selvam and Kuppan, 2012). Therefore, the pore size and morphologies of the final OMCs can be tuned by carefully choosing inorganic silica templates (OMSs) and various carbon precursors to get the desired final OMCs. Table 5.2 lists the various OMCs reported in literature.

Among this, we have adopted the second approach to synthesize OMCs using OMSs such as MCM-41 (Selvam and Kuppan, 2012) and SBA-15 (Kuppan and Selvam, 2012) as hard templates. Figure 5.10 presents XRD patterns of both MCM-41 and SBA-15. As expected, the silica samples exhibit typical reflections characteristics of MCM-41/SBA-15 structures indicating the formation of highly ordered mesoporous structures (Selvam *et al.*, 2001; 2010a). The nitrogen sorption isotherms of MCM-41 (Selvam and Kuppan, 2012) and SBA-15 (Kuppan and Selvam, 2012) exhibit type IV isotherms together with H1 hysteresis loop, and uniform meso-pores within the

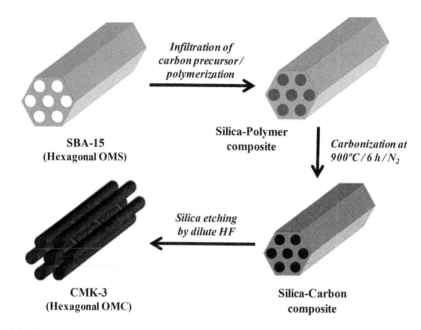

Figure 5.9.   Formation CMK-3, *via* nano-casting method (Roggenbuck and Tiemann, 2005).

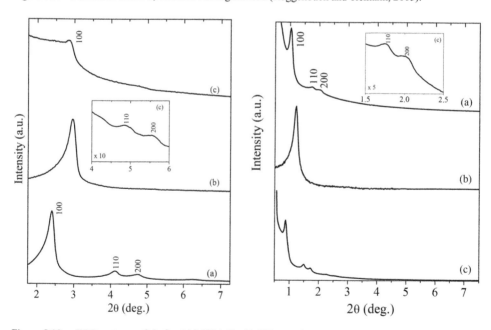

Figure 5.10.   XRD patterns of: Left – (a) MCM-41; (b) Silica-carbon composite; (c) NCCR-41. Right – (a) CMK-3; (b) Silica-carbon composite; (c) SBA-15.

matrix. Figure 5.11 displays HRTEM images of MCM-41 and SBA-15. Both the samples show cylindrical meso-pores with pore diameter 6.5 and 2.5 nm, respectively. Likewise Figure 5.12 shows HRTEM images of NCCR-41 and CMK-3. This is in good agreement with the literature (Jun *et al.*, 2000) as well as with XRD and BET data (see Table 5.3). For a comparison, the data for commercial activated carbon (AC) and E-TEK carbon are also included in this table.

Table 5.2. Synthesis parameters and textural properties of various OMCs.[†]

| OMC | Template | Carbon source | $a_0$ [nm] | $V_p$ [nm] | $S_{BET}$ [m$^2$g$^{-1}$] | $D$ [nm] | Ref. |
|---|---|---|---|---|---|---|---|
| *Hard-template route* | | | | | | | |
| NCCR-41 | MCM-41 | Sucrose | 3.7 | 0.83 | 1080 | 2.2 | Selvam and Kuppan (2012) |
| SNU-1 | MCM-48 | Phenol/ formaldehyde resin | 8.0 | – | 1257 | 2.3 | Lee *et al.* (1999) |
| CMK-1 | MCM-48 | Sucrose | 7.1 | 1.30 | 1380 | 3.0 | Ryoo *et al.* (1999) |
| CMK-4 | MCM-48 | Acetylene | 8.6 | 0.63 | 970 | 3.0 | Kaneda *et al.* (2002) |
| CMK-2 | SBA-1 | Sucrose/ Furfuryl alcohol | 8.3 | 1.30 | 1520 | 4.5 | Ryoo *et al.* (2001) |
| CMK-3 | SBA-15 | Sucrose | 9.7 | 1.04 | 1160 | 4.0 | Jun *et al.* (2000) Kuppan and Selvam (2012) |
| CMK-5 | SBA-15 | Furfuryl alcohol | 9.4 | 2.00 | 2000 | 5.0 | Joo *et al.* (2001) |
| MCN-1 | SBA-15 | Ethylenediamine | 9.5 | 0.55 | 505 | 4.0 | Vinu *et al.* (2005) |
| CMK-8 | KIT-6 | Sucrose/ Furfuryl alcohol | 18.0 | 1.26 | 1060 | 4.0 | Kim *et al.* (2005) |
| MCN-2 | SBA-16 | Ethylenediamine | 13.4 | 0.81 | 810 | 3.8 | Vinu *et al.* (2008) |
| C-MSU-H | MSU-H | Sucrose | 11.0 | 1.26 | 1228 | 3.9 | Kim and Pinnavaia (2001) |
| NCCR-56 | IITM-56 | sucrose | 5.1 | 2.40 | 1367 | 2.7 | Selvam *et al.* (2010b; 2010c) |
| CMK-3*Va* | SBA-15 | Acetonitrile | 6.8 | 0.70 | 678 | 4.0 | Xia *et al.* (2004) |
| *Soft-template route* | | | | | | | |
| FDU-14 | P123 | Phenol/ formaldehyde resin | 15.4 | 0.50 | 1000 | 2.8 | Zhang *et al.* (2006) |
| FDU-15 | P123/ Hexadecane | Phenol/ formaldehyde resin | 8.6 | 0.55 | 1040 | 4.0 | Zhang *et al.* (2006) |
| FDU-16 | F108/F127 | Phenol/ formaldehyde resin | 11.6 | 0.50 | 1030 | 3.2 | Zhang *et al.* (2006) |
| FDU-17 | PPO-PEO-PPO | Phenol/ formaldehyde resin | 32.6 | 0.47 | 870 | 3.5, 5.8 | Huang *et al.* (2007) |
| FDU-18 | PEO-b-PMMA | Phenol/ formaldehyde resin | 32.4 | 0.57 | 1050 | 10.5 | Deng *et al.* (2008) |
| COU-1 | F127/EOA | Phenol/ formaldehyde resin | 11.5 | 0.57 | 640 | 6.0 | Tanaka *et al.* (2005) |
| OMC-GL | F127/ Glutamic acid | Phenol/ formaldehyde resin | 11.3 | 0.55 | 714 | 5.0 | Lu *et al.* (2008) |

[†]Lattice constant, $a_0$; Surface area, $S_{BET}$; Pore volume, $V_p$; Pore size, $D$.

### 5.3.2 *Soft-template approach*

Supra-molecular self-assembly provide routes for the synthesis of ordered mesoporous carbon molecular sieves *via* a soft-template route as recently reported by Meng *et al.* (2006), Huang *et al.* (2007), Wan *et al.* (2007) and Xue *et al.* (2008). As an alternative to hard-template methods, researchers have focused on the synthesis of OMC by a single-step (soft-template) method, which can eliminate the pre-formed silica (hard) template and tedious infiltration steps. In this regard, commercially available amphiphilic triblock copolymers such as pluronics or synperonics are used for the preparation of OMCs. The synthesis procedure involves the preparation

Figure 5.11.   TEM images of: (a) MCM-41 and (b) SBA-15.

Table 5.3.   Structural and textural properties of various silica and carbon.

| Material | Structural parameters | | | | |
| | $a_0^{1)}$ [nm] | $t^{2)}$ [nm] | $S_{BET}$ [$m^2g^{-1}$] | $V_p$ [$cm^3\,g^{-1}$] | $D$ [nm] |
| --- | --- | --- | --- | --- | --- |
| MCM-41 | 4.2 | 1.7 | 966 | 0.86 | 2.5 |
| SBA-15 | 10.4 | 3.9 | 720 | 1.10 | 6.5 |
| NCCR-41 | 3.7 | 1.5 | 1080 | 0.83 | 2.2 |
| CMK-3 | 9.6 | 5.6 | 997 | 1.30 | 4.0 |
| AC$^{3)}$ | – | – | 1080 | 0.79 | 1.3 |
| E-TEK$^{4)}$ | – | – | 193 | 0.71 | – |

$^{1)}$For the hexagonal system: $a_0 = 2d_{100}/\sqrt{3}$; For the cubic system: $a_0 = d \times (h^2 + l^2 + k^2)^{1/2}$.
$^{2)}$Wall thickness. For the hexagonal system: $t = a_0 - D_{BJH}$; For the cubic system: $t = (0.23a_0 - 0.5D_{BJH})$.
$^{3)}$Activated carbon;
$^{4)}$VULCAN XC-72 Carbon.

of resol precursors and the formation of ordered hybrid meso-phases by organic-organic self-assembly with triblock-copolymer, as structure directing agent. The preferential evaporation of ethanol progressively enriches the concentration of the copolymer and drives the organization of resol-copolymer composites into an ordered liquid-crystalline meso-phase. The high curvature of resol-copolymer composites can be introduced by increasing the ratio of phenol-formaldehyde resol/triblock-copolymer. Furthermore, the ordered meso-phase is solidified by the cross-linking of resols, which can be easily induced by thermo-polymerization (Figs. 5.13 and 5.14).

Figure 5.15 illustrates HRTEM images of FDU-17 viewed along different directions, which confirm highly ordered meso-structure. Although, these classes of materials show disordered pores with wider pore size distribution, their synthesis is simple and mesoporous structures can be controlled by varying the molar ratio of the respective precursors. The resulting samples has been reported with a specific surface area of 510–870 $m^2\,g^{-1}$, pore diameter of 3.9–6.9 nm, and pore volume of 0.33–0.54 $cm^3\,g^{-1}$ which was achieved by the co-assembly of a tri-block copolymer, *viz.*, pluronic-F127 as a structure directing agent, and a mixture of phloroglucinol and formaldehyde as carbon precursor.

## 5.4   DIRECT METHANOL FUEL CELL

The subject of energy has great significance in recent times not only due to the demand factor, coupled with dwindling oil or conventional fuel resources, but also due to an increasing awareness

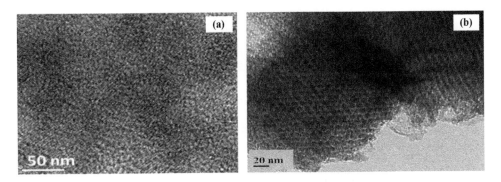

Figure 5.12.   TEM images of: (a) NCCR-41 and (b) CMK-3.

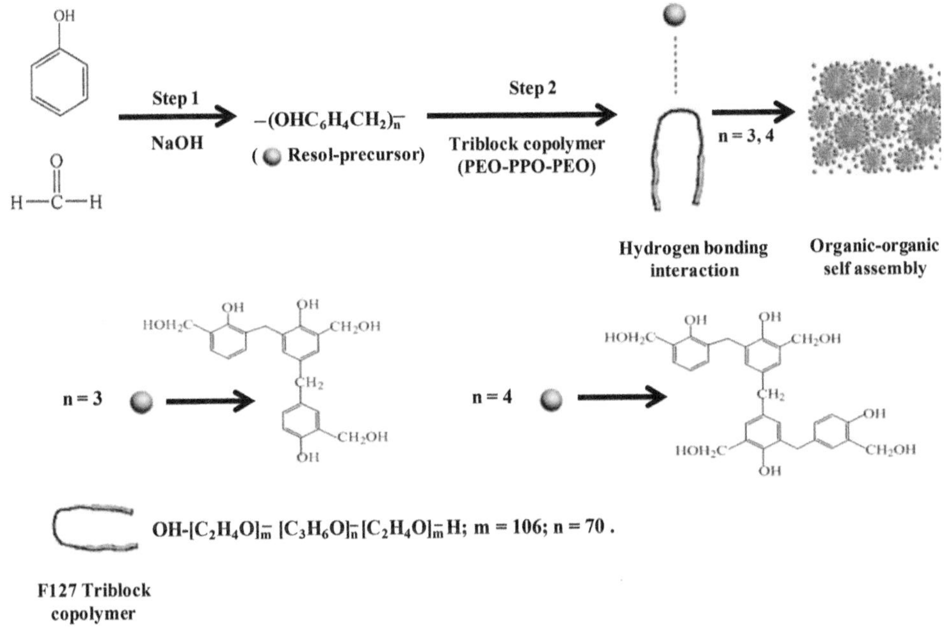

Figure 5.13.   The formation of organic-organic self-assembly via soft-template route (after Meng *et al.*, 2006).

and consciousness of the dangers of an unlimited use of conventional fuels which result in the formation of greenhouse effects and environmental pollution-related degradation (McNicol *et al.*, 1999; Yang *et al.*, 2011). A fuel cell is defined as an electrochemical device that can continuously convert chemical energy into electrical energy. In such a device, hydrogen is a commonly used fuel consumed at the anode and oxygen is the oxidant at the cathode. In acidic electrolyte, the cell reaction is as follows:

$$\text{Anode:} \quad H_2 \rightarrow 2H^+ + 2e^- \tag{5.1}$$

$$\text{Cathode:} \quad O_2 + 4H^+ + 4e^- \rightarrow 2H_2O \tag{5.2}$$

The main advantage of fuel cells over heat engines is that the former is theoretically more efficient and produces no toxic emissions (Jewulski and Rak, 2006). Further, fuel cells operate very quietly thereby reducing the undesired noise (Dyer, 2002). Owing to these advantages, they

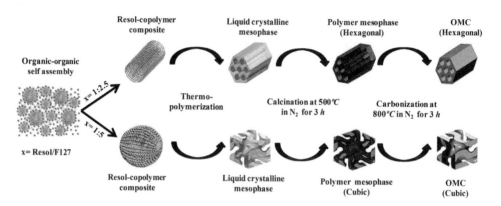

Organic-organic
self assembly

x= Resol/F127

Resol-copolymer
composite

Liquid crystalline
mesophase

Polymer mesophase
(Hexagonal)

OMC
(Hexagonal)

Thermo-
polymerization

Calcination at 500 °C
in N₂ for 3 h

Carbonization at
800 °C in N₂ for 3 h

Resol-copolymer
composite

Liquid crystalline
mesophase

Polymer mesophase
(Cubic)

OMC
(Cubic)

Figure 5.14.   The formation of OMC via soft-template route (after Meng *et al.*, 2006).

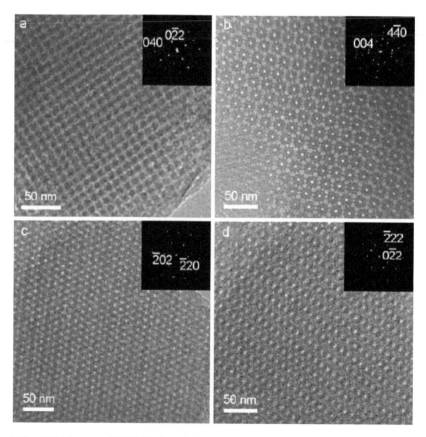

Figure 5.15.   TEM images of FDU-17 viewed along different direction; (a) [100]; (b) [110]; (c) [111]; (d) [211].

are being developed for numerous applications, e.g., automobiles, portable electronic devices, and mobile and stationary power generation. Among the various fuels, it can be seen from Table 5.4, that hydrogen is the most promising because of its high energy density and the maximum voltage that can be derived closer to the theoretical value. However, the use of hydrogen as fuel is always associated with complexity in storage and handling. In this regard, methanol is considered as a

Table 5.4. Chemical and electrochemical data of various fuels (Viswanathan and Aulice, 2008).

| Fuel | $\Delta G°$ [kcal mol$^{-1}$] | $E_{0\,(theo)}$ [V] | $E_{0\,(max)}$ [V] | Energy density [kWh kg$^{-1}$] |
|---|---|---|---|---|
| Hydrogen | −56.69 | 1.23 | 1.15 | 32.67 |
| Methanol | −166.80 | 1.21 | 0.98 | 6.13 |
| Ethanol | −65.77 | 1.14 | 0.25 | 6.60 |
| Ammonia | −80.80 | 1.17 | 0.62 | 5.52 |
| Hydrazine | −143.90 | 1.56 | 1.28 | 5.22 |
| Formaldehyde | −124.70 | 1.35 | 1.15 | 4.82 |
| Carbon monoxide | −61.60 | 1.33 | 1.22 | 2.04 |
| Formic acid | −68.20 | 1.48 | 1.14 | 1.72 |
| Methane | −195.50 | 1.06 | 0.58 | – |
| Propane | −503.2 | 1.08 | 0.65 | – |

favorable fuel from the point of view of cost, efficiency, availability, existence in a liquid state, stability, oxidizing ability, and electrical yield in terms of its weight.. Although methanol has a much lower energy density than hydrogen, the direct methanol fuel cell (DMFC) is attractive over other fuels, which poses numerous problems during operation. Fuel cells are generally classified according to the type of electrolyte and the operating temperature (Carrette *et al.*, 2001).

The direct methanol fuel cell is a special form of low temperature fuel cell based on Proton Exchange Membrane (PEM) technology. Among many types of fuel cells, the DMFC deserves special attention due to relatively cheap, abundant, easy for handling and storage feeds (from one side methanol and other side oxygen/air). The typical design of a DMFC consists of an anode at which methanol is electro-oxidized to $CO_2$ through the reaction:

$$CH_3OH + H_2O \rightarrow CO_2 + 6H^+ + 6e^- \qquad E° = 0.016\,V/SHE \qquad (5.3)$$

and the cathode at which oxygen (usually air) is reduced to form water.

$$3/2\,O_2 + 6H^+ + 6e^- \rightarrow 3H_2O \qquad E° = 1.229\,V/SHE \qquad (5.4)$$

There are many advantages of employing methanol as a fuel as compared with hydrogen gas. Methanol is liquid; hence, its storage and transportation are less complicated and also methanol could be supplied through the existing gasoline infrastructure. Hydrogen can be generated in-situ from methanol using a reforming process; the reformer unit further complicates the fuel cell system as the reformed hydrogen contains significant amount of CO that can poison the Pt catalyst (Arico *et al.*, 2001).

Methanol is electrochemically oxidized at the anode resulting in the production of $CO_2$ and the delivery of $6H^+$ ions. On the cathode side, oxygen reduction and water production takes place. The theoretical overall cell potential amounts to 1.18 V, which is slightly lower than that of $H_2$-PEFC. On the other hand, a slightly higher operating temperature of 100–120°C is realized in DMFC due to the higher water content at the anode (feeding the anode with a liquid methanol/water mixture), which keeps the membrane in a well humidified state. There is an option to use the liquid water methanol feed stream to cool the stack, thereby making a separate cooling cycle needless. However, despite the similar standard cell potential (1.18 V *vs.* 1.23 V), a slightly increased temperature results in tremendous polarization losses ($\eta \sim 0.3\,V$ *vs.* $\eta \sim 0.05\,V$ for the $H_2$ oxidation) leading to a 0.2 to 0.4 V reduced operation voltage of a DMFC. In addition, methanol shows a high diffusion rate through the polymer membranes (methanol crossover). Further, the catalysts are easily poisoned by impurities and more seriously by the products of the anodic reaction itself.

In spite of many attempts in the last two decades by researchers and fuel cell developers to create non-Pt catalyst, Pt remains the catalyst of choice since apart from its passivity in an acidic electrolyte, it has demonstrated appreciable activity for the methanol electro-oxidation reaction. However, there are several advantages that are associated with fuel cell technology. First, the

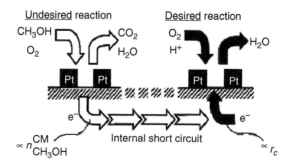

Figure 5.16.    Cathodic oxygen reduction and undesired methanol oxidation with internal short circuit created by crossover (Zhou *et al.*, 2001).

Pt is more than 95% recoverable at low cost (Zhao *et al.*, 2007). Second, there is substantial scope for improving the performance of the DMFC catalyst. On the other hand, various surface intermediates are formed during methanol electro-oxidation. Methanol is mainly decomposed to CO, which is then further oxidized to $CO_2$; other CO-like species are also formed: $COH_{ads}$, $HCO_{ads}$, $HCOO_{ads}$ (Sundmacher *et al.*, 2001). Principle by-products are formaldehyde and formic acid; some of the intermediates are not readily oxidizable and remain strongly adsorbed on the catalyst's surface. Consequently, they prevent fresh methanol molecules from adsorbing and undergoing further reaction. Thus, electro-oxidation of intermediates is the rate determine step.

In DMFC, the fuel diffuses through the state-of-the-art Nafion membrane. Owing to the presence of an hydroxyl group and its hydrophilic properties, methanol interacts with the ion exchange sites and is dragged by hydronium ions in addition to a diffusion resulting from a concentration gradient between anode and cathode. Methanol that crosses over reacts directly with oxygen at the cathode. Electrons are brought directly from the anode to the cathode along with methanol resulting in an internal short-circuiting and consequently a loss of current (Fig. 5.16). In-addition to that, the cathode catalyst, which is pure Pt, is contaminated by methanol oxidation intermediates similar to the anode (Zhou *et al.*, 2001). Furthermore, the performance of a catalyst vary with the nature of the electrolyte due to difference in the ionic conductivity, the degree of adsorption of acid radicals on the catalyst surface and the influence of corrosion on stability. In general, high acid concentrations tend to reduce the activity of Pt catalysts, especially above 5 M. Phosphoric acid at high concentrations ($>5$ M) has been shown to facilitate better activity to methanol oxidation than sulfuric acid at the same concentration (Andrew *et al.*, 1977), however the best electrolyte at low temperatures has been shown to be 3 M sulfuric acid ($H_2SO_4$). The oxidation of methanol at higher temperatures (e.g. 200°C) in phosphoric acid (96%) on Pt and Pt alloy catalysts (with Sn, Ru, and Ti) has been shown to be reasonably fast (Lee *et al.*, 1999).

## 5.5    ELECTROCATALYSTS FOR DMFC

### 5.5.1    *Bulk platinum catalyst*

The reported studies on bulk Pt lead to the following essential features: (i) electro-sorption of methanol is low on bulk Pt at lower potentials with an activation energy of 35 kJ mol$^{-1}$; (ii) sequential proton stripping from methanol takes place, giving a series of multiply-bonded intermediates that eventually convert to linearly bonded CO (Wasmus and Kuver, 1999); (iii) at potentials above $-0.5$ V/NHE, there is a steady loss of adsorbed CO from the surface of Pt and a gain of $CO_2$; beyond 0.7 V, the surface is almost free from $CO_{ads}$ (Sriramulu *et al.*, 1999); (iv) morphology appears to play an important role in determining the activity of Pt, with roughened Pt showing higher activity (McNicol *et al.*, 1999); (v) currents are always higher on roughened Pt than on equilibrated Pt, at least for potential below about 0.55–0.60 V/NHE. Above this potential, $CO_{ads}$ is oxidized rapidly from the Pt surface; (vi) studies on methanol adsorption and oxygen on single crystal surfaces of Pt show considerable sensitivity to the oriented planes of Pt surface (Parsons and Vander Noot, 1988); (vii) kinetics analyses suggest two extreme types of mechanism for methanol oxidation on Pt; one involves attack of water on the outside of the chemisorbed CO from the edges of the islands, and the second involves rate-limiting migration of CO from the edges of the islands to active sites; (viii) the posioning of CO is much lower on the Pt(111) surface (Xia *et al.*, 1996).

### 5.5.2   *Platinum alloy catalyst*

The most successful binary catalyst for methanol oxidation is Pt-Ru alloy (Lin *et al.*, 2009; Liu *et al.*, 2008). On the other hand, investigations on a highly dispersed Pt-Ru alloy on a carbon-supported catalyst show that surface metal alloy domains of Ru:Pt catalysts are the key for higher activity in DMFCs applications (Antolini and Cardellini, 2001; Sharma and Pollet, 2012). A proposed role for Ru in conjunction with Pt for methanol oxidation will be its function in enabling desorption of catalyst poisons from the Pt surface. For an optimum activity, Ru should be in solid solution with Pt (Burstein *et al.*, 1997).

### 5.5.3   *Nano-platinum catalyst*

In the context of improving the efficiency of DMFCs, the concept of nano-platinum electrodes gained significant importance. The essential findings are: (i) controversies exist regarding the "size-effects" towards methanol oxidation (Goodenough *et al.*, 1988; Wantanabe, 1987; Wasmus and Kuver, 1999); (ii) spectro-electrochemical studies of Pt nano-particles revealed a decrease in linearly bonded CO with an increase in the bridged form at higher coverages (Beden *et al.*, 1991); (iii) the rate of chemisorption of methanol on Pt particles is faster than that on bulk Pt at lower potentials, suggesting that methanol chemisorption is favored on steps or defective crystallographic sites (Christensen *et al.*, 1993); (iv) the specific activity was found to decrease with decrease in the particle size in the range 4.5 to 1.2 nm (Frelink *et al.*, 1995).

### 5.5.4   *Catalyst promoters*

The promotion of Pt by a range of metals and metal oxides to enhance methanol oxidation is well known. The species have positive effect on methanol oxidation, and the results are summarized in Table 5.5. Though elements like Bi, Pb, Mo, Sb, In, and their oxides or alloys enhance the electro-catalytic activity of Pt for methanol oxidation, they dissolve in an acid medium during the course of the reaction. The role of hydrous $RuO_2$ as a catalyst promoter has been highlighted in terms of its mixed electronic and proton conductivity (Rolison and White, 1999). Commercial Pt-Ru black catalysts were shown to contain a higher portion of $RuO_2$, which was essential for their high electro-catalytic activity.

## 5.6   OMC-SUPPORTED PLATINUM CATALYST

The support for the metal nano-particles turns out to be as important as the nano-particles for providing their dispersion and stability. Studies on electron transfer are particularly important for carbon-based materials because those materials, such as glassy carbon, graphite, fullerene, and diamond with different electronic and structural properties, have been proved to possess distinctly different electrochemical properties from each other (Ramesh and Sampath, 2003). Carbon supports provide high electronic conductivity, uniform catalyst dispersion, corrosion resistance, and sufficient access of gas reactants to the catalyst (Ismagilov *et al.*, 2005). In addition to electrical conductivity and surface area, hydrophobicity, morphology, porosity, and

Table 5.5.    Effect of catalyst promoters on methanol oxidation (Viswanathan and Aulice, 2008).

| Materials/treatment | Catalyst promoters | Improvements |
|---|---|---|
| Alloying and dissolution to produce highly reticulated surfaces | Cr, Fe, Sn | Typically less than 100 mV lower potential than Pt |
| Surface atoms | Sn, Bi | Enhance the catalytic activity |
| Alloys | Ru, Sn, Mo, Os, Ir, Ti, Re | Ru serves as better alloying metal. Sn, Mo, Os, and Re are substantial promoters |
| Metal oxides | Ru | Hydrous Ru oxides are the most active catalyst |
| Base metal oxides | W, Ti, Nb, Zr, Ta | Tungsten oxide is a notable promoter in DMFC. Other metal oxides show relatively smaller effect |

corrosion resistance are also important factors in choosing a good catalyst support. In this regard, carbon is one of the best catalyst support materials for low temperature fuel cells. Carbon black and activated carbons have been extensively used as catalyst supports, with Vulcan XC-72 being the most common carbon material used. The most common supported catalyst is Pt -supported on high surface area carbon, and is used in both the cathode and anode (Dicks, 2006). In the last decade, a number of new catalyst supports with various meso-structures and nano-structures has been reported (Jiang *et al.*, 2011; Wu *et al.*, 2012; You *et al.*, 2012). The use of carbon as a catalyst support is continuously growing and there are a number of reports available (Auer *et al.*, 1998) to relate metal dispersion with the properties of the support. The main reasons are: (i) Carbon has sufficient electronic conductivity and chemical stability under fuel cell operating conditions; (ii) the surface area of noble metal catalysts is greatly increased by using carbon support; (iii) the primary function of the support is to improve the stability of electro-catalysts; (iv) the surface functional groups of the carbon support have a pronounced effect on the dispersion of the catalysts; (v) it is relatively easy to obtain uniform and highly dispersed catalysts, even when catalyst loading is more than 30 wt% (to obtain better PEM fuel cell reactions, higher loading catalysts are expected); (vi) the activity of electro-catalysts has been improved through the interaction between metal and support; (vii) carbon-supported catalysts are more stable than non-supported catalysts in relation to the catalysts' agglomerization under fuel cell operating conditions.

It is important to optimize carbon support and metal loading in order to obtain sufficient reaction in the catalyst layer. The structure of the carbon support in the catalyst layer can directly affect the Pt utilization, as well as degradation mechanisms that lead to decrease in performance. It is, however, difficult to maintain metal nano-particles with high metal dispersion combined with stability inside the pores. On the other hand, there are well known methods for producing highly dispersed metal nano-particles on disordered carbons with non-uniform pores, *viz.*, amorphous carbons, carbon nanotubes, etc., but the resulting materials cannot be used as catalysts as they tend to sinter through Ostwald ripening process (see Fig. 5.17). For example, Pt nano-particles with 20 wt% loading was supported on activated carbon (AC) and commercial carbon (E-TEK; Vulcan XC-72 carbon) by wet chemical reduction method (Li and Xing, 2007).

Dispersed nano-particles of noble metals supported on high surface area carbon materials are of considerable interest to catalysis owing to their unique physico-chemical properties (Shukla *et al.*, 1994; Watanabe *et al.*, 1987). The primary function of the support is to separate the individual particles physically in order to diminish the rate of agglomerization. Support plays a major role in determining the mechanical and thermal stability of particles while helping them in highly dispersed state (Attwood *et al.*, 1980). The choice of suitable support is a factor, which affects the performance of supported catalysts. Interaction between the catalysts and the support has

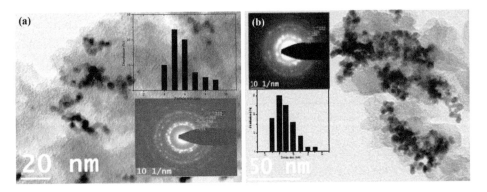

Figure 5.17.   TEM images, SAED, and particle size distribution of: (a) Pt/AC: (b) Pt/E-TEK.

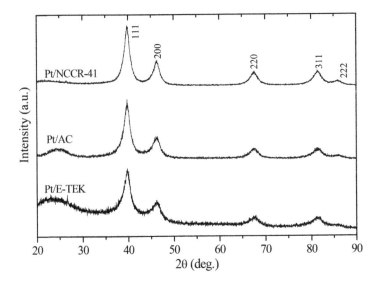

Figure 5.18.   XRD patterns of prepared (Pt/AC and Pt/NCCR) and commercial (Pt/E-TEK) catalysts.

been identified to modify the intrinsic catalytic activity. In studies of methanol oxidation, supports so far employed for dispersing active metal particles are carbon and heteroatom containing carbons (Lei *et al.*, 2009; 2011; Liu *et al.*, 2010). For instance, in the case of carbon supports, nature of functional groups and carbon with sulfur and nitrogen based functionalities have shown enhanced catalytic activity. Therefore, we have recently adopted different methodologies for a uniform dispersion of small-size Pt over OMCs. One of the approaches (Wu *et al.*, 2007), *viz.*, paraformaldehyde reduction method, give smaller particles size and highly dispersed Pt on OMCs, designated as Pt/OMCs.

### 5.6.1   *Pt/NCCR-41*

The mesoporous carbon, NCCR-41 was used as a support for the preparation of uniform and well-dispersed nano-sized Pt nano-particles, designated as Pt/NCCR-41. The supported catalysts with narrow size distribution of the active Pt, referred as Pt/NCCR-41, show excellent electro-catalytic activity for methanol oxidation reaction in comparison with Pt-supported activated carbon (Pt/AC) as well as commercial carbon (Pt/E-TEK) catalysts. Figure 5.18 shows XRD patterns of

Table 5.6.   Physico-chemical and electrochemical data of Pt/NCCR-41.

| Catalyst system (20 wt% Pt/Carbon) | Pt Crystallite size [nm] | | Pt dispersion [%] | EAS [m² g⁻¹) | Onset potential [V] | Current, I at 0.7 V [mA (mg Pt)⁻¹] | Activity loss [%] | $I_f/I_b$ |
|---|---|---|---|---|---|---|---|---|
| | XRD | TEM | | | | | | |
| Pt/NCCR-41[1] | 4.7±0.2 | 4.5±1.3 | 13.7 | 62 | 0.16 | 158 | 63 | 1.58 |
| Pt/AC | 5.0±0.3 | 5.5±1.0 | 11.7 | 60 | 0.12 | 139 | 70 | 1.29 |
| Pt/E-TEK[2] | 5.7±0.5 | 4.9±2.0 | 11.8 | 64 | 0.12 | 128 | 93 | 0.72 |

[1] $d_{111}$ (XRD) = 0.226 nm; $d_{111}$(TEM) = 0.230 nm.
[2] Commercial catalyst.

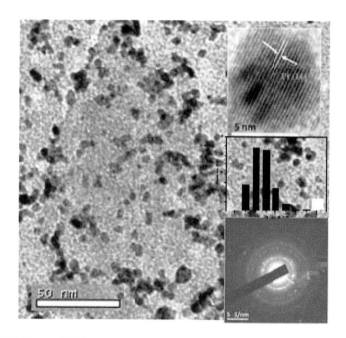

Figure 5.19.   TEM image, SAED pattern, and particle size distribution of 20%Pt/NCCR-41.

Pt-loaded on various carbon supports, viz., NCCR-41, AC, and E-TEK, which exhibit characteristic reflections consistent with face-centered cubic lattice of Pt crystallites (Table 5.6). Further, the average particle size of the deposited Pt nano-clusters was calculated, and the computed values along with the textural properties are summarized in Table 5.6.

Figure 5.19 illustrates the high resolution transmission electron microscopy (HRTEM) images, selected area electron diffraction (SAED) patterns, and particle size distribution of Pt/NCCR-41 TEM images, which show dispersed smaller size (2–4 nm) Pt particles. At this juncture, it is to be noted here that the cluster size increases with increase in metal loading, however, the increase is marginal as compared to the commercial (amorphous/disordered) carbons. It can be seen from these figures that the Pt nano-particles are, although, uniform with a size distribution of about 5 nm (see Table 5.6), in agreement with XRD data, but the distribution of the nano-Pt clusters are much more uniform than the corresponding Pt/AC and Pt/E-TEK catalysts (see also Fig. 5.17). The Pt nano-particles supported on AC (Fig. 5.17a) and E-TEK (Fig. 5.17b) show predominantly clustering as evidenced from TEM.

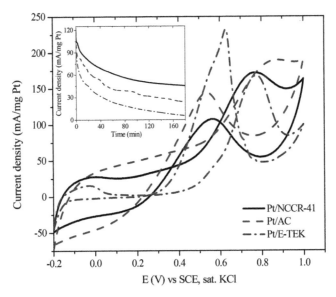

Figure 5.20.   CVs recorded in mixture of 1 M $H_2SO_4$ and 1 M $CH_3OH$ with a scan rate of 25 mV s$^{-1}$ at 298 K: (a) 20 wt% Pt/NCCR-41, (b) 20 wt% Pt/AC and (c) commercial E-TEK catalyst consisting of 20 wt% Pt. Inset: Corresponding chronoamperometric data.

Figure 5.20 presents the electro-catalytic behavior of various Pt-supported carbon catalysts under study for methanol oxidation reaction at room temperature. All the samples indicate the irreversible nature of the methanol electro-oxidation, and the CVs were reproducible. The electrochemical activities of all the catalysts are given in Table 5.6. The potential of 0.70 V was chosen for comparing the electro-catalytic activities for methanol oxidation because the capacitive current at this potential is negligible compared to that of methanol oxidation current. It is clear from both Figure 5.20 and Table 5.6 that the Pt-supported mesoporous carbon exhibited higher current density than the commercial (E-TEK) catalyst and the catalyst prepared using activated carbon. These results suggest that in the case of Pt/NCCR-41 a better dispersion and utilization of active Pt catalyst may possibly play a crucial role in determining the activity. The better dispersion of Pt nano-particles may be attributed to higher surface area of NCCR-41 as well as the very different surface functionalities of the mesoporous carbon. Further, as can be seen from Figure 5.20, the ratio of the anodic peak current densities in the forward ($I_f$) and reverse ($I_b$) scans give another measure of the catalytic activity behavior. That is, a higher $I_f/I_b$ ratio indicates a better oxidation performance of methanol oxidation activities during the anodic scan and low accumulation of carbonaceous species on the electro-catalyst surface, suggesting a better carbon monoxide (CO) tolerance. Further, the higher $I_f/I_b$ ratio also suggests that there has been a more effective removal of the poisoning species, *viz.*, CO, on the electro-catalysts surface. It can be seen from Table 5.6 that the $I_f/I_b$ ratio for Pt/NCCR-41 is 1.58, which is much higher than both Pt/AC (1.29) and Pt/E-TEK commercial catalyst (0.72) and therefore the better catalyst tolerance of Pt/NCCR-41. Since the mesoporous structure has a high surface area, the large pore size will facilitate the transport of substrate and reaction products more easily. This may be attributed to the fine dispersion of metal nano-particles and the elimination of diffusion problems in mesoporous carbon materials. Unlike Pt/AC and Pt/E-TEK catalysts wherein Pt forms mostly as clusters as compared to fine dispersion of Pt in Pt/NCCR-41. Further, the metal dispersion values (Table 5.6) suggest that Pt/NCCR-41 catalyst is better than the other two catalysts, indicating that Pt/NCCR-41 is more effective against CO poisoning.

Thus, Pt/NCCR-41, exhibited higher activity and better stability than that of Pt/E-TEK commercial catalyst. The enhanced activity and the superior performance of Pt/NCCR-41 is due to

Table 5.7.   Micro-structural and electrochemical data of various Pt/CMK-3 electro-catalysts.

| Catalyst[1] | Reducing agent | Crystallite size, Pt | | EAS $[m^2 g^{-1}]$ | Onset potential [V] | Current, $I$ at 0.7 V $[mA (mg\ Pt)^{-1}]$ | Activity loss [%] | $I_f/I_b$ |
|---|---|---|---|---|---|---|---|---|
| | | XRD [nm] | TEM [nm] | | | | | |
| Pt/CMK-3 | Para formaldehyde | 4.3 | 4.5 | 84 | 0.14 | 185 | 65 | 1.33 |
| Pt/CMK-3 | Sodium borohydride | 6.0 | 6.0 | 68 | 0.14 | 157 | 58 | 1.57 |
| Pt/CMK-3 | Ethylene glycol | 5.5 | 6.0 | 72 | 0.16 | 165 | 95 | 0.76 |
| Pt/CMK-3 | Hydrogen | 18.0 | 20.0 | 72 | 0.11 | 105 | 92 | 1.55 |

[1]20 wt% Pt-supported mesoporous carbon.

improved dispersion of uniform Pt nano-particles as well as better utilization of the catalyst, which may possibly originate from a higher surface area, large pore volume, and functional groups present in the newly synthesized carbon. The study also demonstrates that optimized carbon supports can offer significant improvement by way of lowering the noble metal loading.

### 5.6.2   Pt/CMK-3

Pt-supported ordered mesoporous carbon catalysts, designated as Pt/CMK-3, were prepared employing colloidal Pt, reduced by four different reducing agents, viz., paraformaldehyde, sodium borohydride, ethylene glycol and hydrogen, and deposited over CMK-3. The resulting Pt nano-particles supported mesoporous carbon, designated as Pt/CMK-3; catalysts were tested for the electro-catalytic oxidation of methanol. The effect of the various reduction methods on the influence of particle size vis-à-vis on the electro-catalytic effect is investigated. Table 5.7 summarizes the results of the investigation. Their electro-catalytic performance studies indicate that the Pt/CMK-3 catalysts are superior to that prepared with activated carbon (Pt/AC) as well as with that of the commercial Pt-supported carbon catalyst (Pt/E-TEK). In particular, the catalyst, Pt/CMK-3, prepared using paraformaldehyde reduced Pt, showed much higher activity and long-term stability as compared with the other reducing methods.

Figure 5.21 shows that the XRD patterns of Pt nano-particles supported mesoporous carbon by different deposition methods. The diffraction patterns represent all the reflections corresponding to the face centered cubic lattice of Pt, supported on mesoporous carbon. The computed values of the crystallite size of the Pt nano-particles are listed in Table 5.7. It is clear from Figure 5.21 and Table 5.7 that the paraformaldehyde reduction method results in a relatively smaller particle size while the hydrogen reduction method yields much larger Pt particles, which is in good agreement with the literature. Figure 5.22 display HRTEM image, SAED pattern, and particle size distribution of Pt/CMK-3, electro-catalyst. It can be seen from the figure that the Pt nano-particles are uniformly distributed with a size of about 3–4 nm which is in agreement with XRD data (see Table 5.7). Furthermore, the distribution of the nano-Pt clusters is uniform on the NCCR-41, AC and E-TEK carbons.

It is clear from these images that the paraformaldehyde reduced Pt depicts a fine dispersion on CMK-3 with a narrow size distribution in the range, 4–5 nm (Fig. 5.22). Whereas the sodium borohydride (not shown here) and ethylene glycol (not reproduced here) reduction methods give ~6–7 nm particle sizes with a similar size distribution. On the other hand, as expected, the hydrogen reduction method gives much larger particle size (~18–20 nm; not shown here). Thus, it is evident that a narrow particle size distribution of ~4–5 nm is achieved in the paraformaldehyde reduction method, and therefore it forms one of the most preferable methods over other reduction methods for the preparation of uniform and smaller Pt clusters.

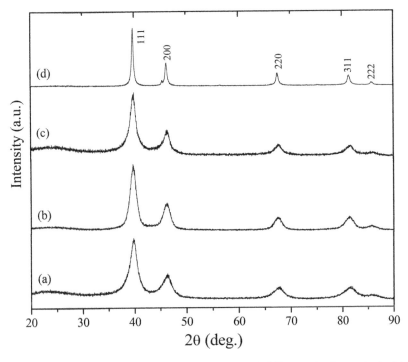

Figure 5.21.   XRD patterns of 20 wt% Pt/CMK-3 prepared with different reduction methods: (a) Sodium borohydride; (b) Paraformaldehyde; (c) Ethylene glycol; (d) Hydrogen.

Figure 5.22.   TEM image, SAED pattern, and particle size distribution of 20%Pt /CMK-3.

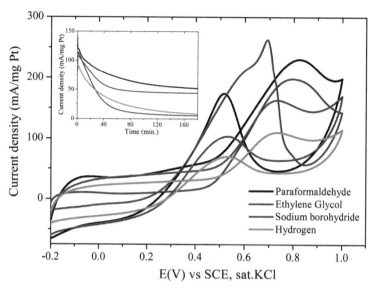

Figure 5.23.   CVs recorded in mixture of 1 M $H_2SO_4$ and 1 M $CH_3OH$ with a scan rate of 25 mV s$^{-1}$ at 298 K for 20 wt% Pt/CMK-3 catalyst prepared employing various reducing agents.

The electro-catalytic properties of the various Pt/CMK-3 catalysts towards the methanol oxidation reactions were tested. To investigate the electro-catalytic activity of the prepared Pt/CMK-3 electro-catalysts, room temperature CV measurements of the methanol oxidation reaction were carried out in 1M $H_2SO_4$ and 1 M $CH_3OH$ aqueous solutions at a potential scan rate of 25 mV s$^{-1}$ (Fig. 5.23). The forward anodic peak, $I_f$ at around 0.7–0.8 V *versus* SCE, is due to the oxidation of methanol. In the backward scan, $I_b$ the oxidation peak at around 0.5–0.6 V *versus* SCE is attributed to the oxidation of $CO_{ads}$ like species, generated *via* incomplete oxidation of methanol during the forward scan. The $I_f$ value relates directly to the methanol oxidation activity of the electro-catalysts. Among the prepared electro-catalysts, the paraformaldehyde reduction method gives a better activity over the other methods. The ratio of the anodic peak current densities in the forward and reverse scans could give another measure of its catalytic performances. That is, a higher $I_f/I_b$ ratio indicates superior oxidation activity of the methanol during the anodic scan and less accumulation of carbonaceous species on the nano-catalyst surface, and thus shows better CO tolerance. On the other hand, the entire Pt/CMK-3 catalysts show higher $I_f/I_b$ ratio than commercial Pt-supported carbon catalysts, for example, the commercial Pt/E-TEK catalyst shows a much lower ratio (0.72). The higher electro-catalytic activity of the former is due to a high dispersion of uniform sized Pt nano-particles on the mesoporous carbon support.

Furthermore, it is also important to address the long time stability of the electro-catalysts for useful application. In this regard, the stability of all the Pt/CMK-3 catalysts were tested by chronoamperometric measurement studies at the corresponding forward peak potential (0.7 V) for 3 h in 1 M $H_2SO_4$ and 1 M $CH_3OH$. The results are presented as an inset in Figure 5.23. It can be seen from this figure that all the electrodes display an initial fast current decay followed by slower attenuation over a long period of operation, reaching a quasi-equilibrium steady state. This initial fast decay is attributed to the formation of intermediate species such as $CO_{ads}$, $CH_3OH_{ads}$, and $CHO_{ads}$ during the electrochemical methanol oxidation reaction. The slow attenuation over longer periods may be due to the adsorption of sulfate anions on the electro-catalyst surface. The lower stability of some of the catalysts can be attributed partly due to larger and/or agglomerated Pt nano-particles. However, the long-term stability of the Pt/CMK-3, in general, is consistent with the view that the high dispersion of Pt nano-particles on the high surface area mesoporous carbons provide much better dispersion of metal nanoparticles and better utilization and therefore accelerates the rate of $CO_{ads}$ oxidation.

Thus, having Pt nano-particles deposited over mesoporous CMK-3 carbon by a paraformalde-hyde reduction method is superior, compared to the other reduction methods employed in this study, *viz.*, sodium borohydride, ethylene glycol and hydrogen reduction methods. Furthermore, the paraformaldyde method is more suitable for the preparation of the highly dispersed uniform sized Pt nano-particles on the mesoporous support with an average particle size of ∼4 nm. Thus, the dispersion and particle size distribution of platinum nano-particles on CMK-3 is strongly dependent on both the deposition method and the carbon support properties. On the other hand, the performance of the Pt/CMK-3 catalysts exhibit higher activity and better stability than commercial Pt-supported carbon catalysts. The enhanced activity of the Pt-supported mesoporous carbon is attributed to a better dispersion and utilization of the Pt, which essentially originated from a higher surface area, large pore volume, and narrow pore size distribution of the meso-porous carbon materials. Thus, the present study demonstrates that optimized reduction methods and appropriate choice of carbon supports can offer significant cost savings by lowering the catalyst loading.

## 5.7  SUMMARY AND CONCLUSION

In this chapter, we have addressed the following issues:

- Ordered mesoporous carbons, *viz.*, NCCR-41 and CMK-3, with high surface areas, large pores, and pore volumes were successfully synthesized employing mesoporous silicas, *viz.*, MCM-41 and SBA-15, respectively, as hard templates, and sucrose as carbon source.
- All the samples, including mesoporous silica, were systematically characterized by various analytical, spectroscopic, and imaging techniques. The results indicate that the samples under study were high quality in terms of structural and textural characteristics.
- Among different reduction techniques, employed for the preparation of Pt- supported carbons, the one that was based on a paraformaldehyde reduction method resulted in very fine dispersion and smaller particle size of the metal inside the meso-pore.
- The finely dispersed Pt with particle sizes ranging from 2–5 nm supported over mesoporous carbon materials showed excellent electro-catalytic properties. Further, the activities were much higher than that obtained with Pt-supported activated carbon (Pt/AC), commercial Pt-supported Vulcan (XC-72) carbon, Pt/E-TEK catalysts.
- Lastly, the study also demonstrates that optimized carbon supports can offer significant cost savings by lowering the catalyst loading.

## ACKNOWLEDGEMENT

This work is partially supported by a grant-in-aid for scientific research by the Ministry of New and Renewable Energy under grant No.103/140/2008-NT. Department of Science and Technology, New Delhi is gratefully acknowledged for funding NCCR, IIT-Madras. We are also thankful to Professor B. Viswanathan for encouragement and support.

## REFERENCES

Andrew, M.R., McNicol, B.D., Short, R.T. & Drury, J.S.: Electrolytes for methanol-air fuel-cells. I. The performance of methanol electro-oxidation catalysts in sulfuric acid and phosphoric acid electrolytes. *J. Appl. Electrochem.* 7 (1977), pp. 153–160.
Antolini, E. & Cardellini, F.: Formation of carbon supported Pt-Ru alloys: an XRD analysis. *J. Alloys Compd.* 315 (2001), pp. 118–122.

Arico, A.S., Srinivasan, S. & Antonucci, V.: DMFCs: from fundamental aspects to technology development. *Fuel-cells* 1 (2001), pp. 133–161.

Attwood, P.A., McNicol, B.D., Short, R.T. & Van, A.J.A.: Platinum on carbon fiber paper catalysts for methanol electrooxidation. Part 1. Influence of activation conditions on catalytic activity. *J. Chem. Soc., Faraday Trans.* 1:76 (1980), pp. 2310–2321.

Auer, E., Freund, A., Pietsch, J. & Tacke, T.: Carbons as supports for industrial precious metal catalysts. *Appl. Catal.* A 173 (1998), pp. 259–271.

Beck, J.S., Vartuli, J.C., Roth, W.J., Leonowicz, M.E., Kresge, C.T., Schmitt, K.D., Chu, C.T.W., Olson, D.H., Sheppard, E.W., McCullen, S.B., Higgins, J.B. & Schlenker, J.L.: A new family of mesoporous molecular sieves prepared with liquid crystal templates. *J. Am. Chem. Soc.* 114 (1992), pp. 10,834–10,843.

Beden, B., Hahn, F., Leger, J.M., Lamy, C., Perdriel, C.L., De, T.N.R., Lezna, R.O. & Arvia, A.J.: Electro-modulated infrared spectroscopy of methanol electrooxidation on electrodispersed platinum electrodes. Enhancement of reactive intermediates. *J. Electroanal. Chem. Interfacial Electrochem.* 301 (1991), pp. 129–138.

Burstein, G.T., Barnett, C.J., Kucernak, A.R. & Williams, K.R.: Aspects of the anodic oxidation of methanol. *Catal. Today* 38 (1997), pp. 425–437.

Carrette, L., Friedrich, K.A. & Stimming, U.: Fuel-cells – fundamentals and applications. *Fuel-cells* 1 (2001), pp. 5–39.

Christensen, P.A., Hamnett, A. & Troughton, G.L.: The role of morphology in the methanol electro-oxidation reaction. *J. Electroanal. Chem.* 362 (1993), pp. 207–218.

Corma, A.: From microporous to mesoporous molecular sieve materials and their use in catalysis. *Chem. Rev.* 97 (1997), pp. 2373–2419.

Davis, M.E.: Ordered porous materials for emerging applications. *Nature* 417 (2002), pp. 813–821.

Deng, Y., Liu, C., Gu, D., Yu, T., Tu, B. & Zhao, D.: Thick wall mesoporous carbons with a large pore structure templated from a weakly hydrophobic PEO–PMMA diblock copolymer. *J. Mater. Chem.* 18 (2008), pp. 91–97.

Dicks, A.L.: The role of carbon in fuel-cells. *J. Power Sources* 156 (2006), pp. 128–141.

Dyer, C.K.: Fuel-cells for portable applications. *J. Power Sources* 106 (2002), pp. 31–34.

Escudero, M.J., Hontanon, E., Schwartz, S., Boutonnet, M. & Daza, L.: Development and performance characterization of new electrocatalysts for PEMFC. *J. Power Sources* 106 (2002), pp. 206–214.

Fan, J., Boettcher, S.W., Tsung, C.-K., Shi, Q., Schierhorn, M. & Stucky, G.D.: Field-directed and confined molecular assembly of mesostructured materials: Basic principles and new opportunities. *Chem. Mater.* 20 (2008), pp. 909–921.

Frelink, T., Visscher, W. & van Veen, J.A.R.: Particle size effect of carbon-supported platinum catalysts for the electrooxidation of methanol. *J. Electroanal. Chem.* 382 (1995), pp. 65–72.

Gasteiger H.A. & Marković, N.M.: Chemistry. Just a dream or future reality? *Science* 324 (2009), pp. 48–49.

Gilbert, M.T., Knox, J.H. & Kaur, B.: Porous glassy carbon, a new columns packing material for gas chromatography and high-performance liquid chromatography. *Chromatographia* 16 (1982), pp. 138–146.

Goodenough, J.B., Hamnett, A., Kennedy, B.J., Manoharan, R. & Weeks, S.A.: Methanol oxidation on unsupported and carbon supported platinum + ruthenium anodes. *J. Electroanal. Chem. Interfacial Electrochem.* 240 (1988), pp. 133–145.

He, X. & Antonelli, D.M.: Recent advances in transition metal containing mesoporous molecular sieves. *Angew. Chem., Int. Ed.* 41 (2002), pp. 214–229.

Huang, Y., Cai, H., Yu, T., Zhang, F., Zhang, F., Meng, Y., Dong, G., Wan, Y., Sun, X., Tu, B. & Zhap, D.Y.: Formation of mesoporous carbon with a face-centered-cubic Fd¯3m structure and bimodal architectural pores from the reverse amphiphilic triblock copolymer PPO-PEO-PPO. *Angew. Chem., Int. Ed.* 46 (2007), pp. 1089–1093.

Huo, Q., Margolese, D.I., Ciesla, U., Demuth, D.G., Feng, P., Gier, T.E., Sieger, P., Firouzi, A., Chmelka, B.F. & Stucky, G.D.: Organization of organic molecules with inorganic molecular species into nanocomposite biphase arrays. *Chem. Mater.* 6 (1994a), pp. 1176–1191.

Huo, Q., Margolese, D.I., Ciesla, U., Feng, P., Gier, T.E., Sieger, P., Leon, R., Petroff, P.M., Schueth, F. & Stucky, G.D.: Generalized synthesis of periodic surfactant/inorganic composite materials. *Nature* 368 (1994b), pp. 317–321.

Inagaki, M., Orikasa, H. & Morishita, T.: Morphology and pore control in carbon materials via templating. *RSC Adv.* 1 (2011), pp. 1620–1640.

Ismagilov, Z.R., Kerzhentsev, M.A., Shikina, N.V., Lisitsyn, A.S., Okhlopkova, L.B., Barnakov, C.N., Sakashita, M., Iijima, T. & Tadokoro, K.: Development of active catalysts for low Pt loading cathodes of PEMFC by surface tailoring of nanocarbon materials. *Catal. Today* 102–103 (2005), pp. 58–66.

Jewulski, J.R. & Rak, Z.S.: Fuel-cells – the opportunity for environmental protection. *Environ. Prot. Eng.* 32 (2006), pp. 189–194.

Jiang, H., Zhao, T., Li, C. & Ma, J.: Functional mesoporous carbon nanotubes and their integration in situ with metal nanocrystals for enhanced electrochemical performances. *Chem. Commun.* 47 (2011), pp. 8590–8592.

Joo, S.H., Choi, S.J., Oh, I., Kwak, J., Liu, Z., Terasaki, O. & Ryoo, R.: Ordered nanoporous arrays of carbon supporting high dispersions of platinum nanoparticles. *Nature* 412 (2001), pp. 169–172.

Jun, S., Joo, S.H., Ryoo, R., Kruk, M., Jaroniec, M., Liu, Z., Ohsuna, T. & Terasaki, O.: Synthesis of new, nanoporous carbon with hexagonally ordered mesostructure. *J. Am. Chem. Soc.* 122 (2000), pp. 10,712–10,713.

Kaneda, M., Tsubakiyama, T., Carlsson, A., Sakamoto, Y., Ohsuna, T., Terasaki, O., Joo, S.H. & Ryoo, R.: Structural study of mesoporous MCM-48 and carbon networks synthesized in the spaces of MCM-48 by electron crystallography. *J. Phys. Chem.* B 106 (2002), pp. 1256–1266.

Kim, D.J., Lee, H.I., Yie, J.E., Kim, S.-J. & Kim, J.M.: Ordered mesoporous carbons: Implication of surface chemistry, pore structure and adsorption of methyl mercaptan. *Carbon* 43 (2005), pp. 1868–1873.

Kim, H., Park, J.-N. & Lee, W.-H.: Preparation of platinum-based electrode catalysts for low temperature fuel-cell. *Catal. Today* 87 (2003), pp. 237–245.

Kim, S.-S. & Pinnavaia, T.J.: A low cost route to hexagonal mesostructured carbon molecular sieves. *Chem. Commun.* (2001), pp. 2418–2419.

Kresge, C.T., Leonowicz, M.E., Roth, W.J., Vartuli, J.C. & Beck, J.S.: Ordered mesoporous molecular sieves synthesized by a liquid-crystal template mechanism. *Nature* 359 (1992), pp. 710–712.

Kuppan, B.: *Ordered mesoporous carbon-supported platinum electrocatalysts for direct methanol fuel-cell application.* PhD Thesis, IIT-Madras, India, 2014.

Kuppan, B. & Selvam, P.: Platinum-supported mesoporous carbon (Pt/CMK-3) as anodic catalyst for direct methanol fuel-cell applications: The effect of preparation and deposition methods. *Prog. Nat. Sci. Mater. Int.* 22 (2012), pp. 616–624.

Lee, J., Yoon, S., Hyeon, T., Oh, S.M. & Kim, K.B.: Synthesis of a new mesoporous carbon and its application to electrochemical double-layer capacitors. *Chem. Commun.* (1999), pp. 2177–2178.

Lee, J., Kim, J. & Hyeon, T.: Recent progress in the synthesis of porous carbon materials. *Adv. Mater.* 18 (2006), pp. 2073–2094.

Lei, Z., An, L., Dang, L., Zhao, M., Shi, J., Bai, S. & Cao, Y.: Highly dispersed platinum supported on nitrogen-containing ordered mesoporous carbon for methanol electrochemical oxidation. *Micropor. Mesopor. Mater.* 119 (2009), pp. 30–38.

Lei, Z., Bai, D., & Zhao, X.S.: Improving the electrocapacitive properties of mesoporous CMK-5 carbon with carbon nanotubes and nitrogen doping. *Micropor. Mesopor. Mater.* 147(2011), pp. 86–93.

Li, L. & Xing, Y.: Pt-Ru nanoparticles supported on carbon nanotubes as methanol fuel-cell catalysts. *J. Phys. Chem.* C 111 (2007), pp. 2803–2808.

Liu, H.-J., Wang, X.-M., Cui, W.-J., Dou, Y.-Q., Zhao, D.-Y. & Xia, Y.-Y.: Highly ordered mesoporous carbon nanofiber arrays from a crab shell biological template and its application in supercapacitors and fuel-cells. *J. Mater. Chem.* 20 (2010), pp. 4223–4230.

Liu, S.-H., Yu, W.-Y., Chen, C.-H., Lo, A.-Y., Hwang, B.-J., Chien, S.-H. & Liu, S.-B.: Fabrication and characterization of well-dispersed and highly stable PtRu nanoparticles on carbon mesoporous material for applications in direct methanol fuel-cell. *Chem. Mater.* 20 (2008), pp. 1622–1628.

Lizcano-Valbuena, W.H., Paganin, V.A., Leite, C.A.P., Galembeck, F. & Gonzalez, E.R.: Catalysts for DMFC: Relation between morphology and electrochemical performance. *Electrochim. Acta* 48 (2003), pp. 3869–3878.

Lu, A., Spliethoff B. & Schüth F.: Aqueous synthesis of ordered mesoporous carbon via self-assembly catalyzed by amino acid. *Chem. Mater.* 20 (2008), pp. 5314–5319.

McNicol, B.D., Rand, D.A.J. & Williams, K.R.: Direct methanol-air fuel-cells for road transportation. *J. Power Sources* 83 (1999), pp. 15–31.

Meng, Y., Gu, D., Zhang, F., Shi, Y., Cheng, L., Feng, D., Wu, Z., Chen, Z., Wan, Y., Stein, A. & Zhao, D.: A family of highly ordered mesoporous polymer resin and carbon structures from organic-organic self-assembly. *Chem. Mater.* 18 (2006), pp. 4447–4464.

Meynen, V., Cool, P. & Vansant, E.F.: Verified syntheses of mesoporous materials. *Micropor. Mesopor. Mater.* 125 (2009), pp. 170–223.

Parsons, R. & VanderNoot, T.: The oxidation of small organic molecules. A survey of recent fuel-cell related research. J. Electroanal. *Chem. Interfacial Electrochem.* 257 (1988), pp. 9–45.

Raman, N.K., Anderson, M.T. & Brinker, C.J.: Template Based Approaches to the Preparation of Amorphous, Nanoporous Silicas. *Chem. Mater.* 8 (1996), pp. 1682–1701.

Ramesh, P. & Sampath, S.: Electrochemical characterization of binderless, recompressed exfoliated graphite electrodes: Electron-transfer kinetics and diffusion characteristics. *Anal. Chem.* 75 (2003), pp. 6949–6957.

Roggenbuck, J. & Tiemann, M.: Ordered mesoporous magnesium oxide with high thermal stability synthesized by exotemplating using CMK-3 carbon. *J. Am. Chem. Soc.* 127 (2005), pp. 1096–1097.

Rolison, D.R. & White, H.S.: Electrochemistry at nanostructured materials. *Langmuir* 15 (1999), pp. 649.

Ryoo, R., Joo, S.H. & Jun, S.: Synthesis of highly ordered carbon molecular sieves via template-mediated structural transformation. *J. Phys. Chem.* B 103 (1999), pp. 7743–7746.

Ryoo, R., Joo, S.H., Kruk, M. & Jaroniec, M.: Ordered mesoporous carbons. *Adv. Mater.* 13 (2001), pp. 677–681.

Selvam, P. & Kuppan, B.: Synthesis, characterization and electrocatalytic properties of nano-platinum-supported mesoporous carbon molecular sieves, Pt/NCCR-41. *Catal. Today* 198 (2012), pp. 85–91.

Selvam, P., Bhatia, S.K. & Sonwane, C.G.: Recent advances in processing and characterization of periodic mesoporous mcm-41 silicate molecular sieves. *Ind. Eng. Chem. Res.* 40 (2001), pp. 3237–3261.

Selvam, P., Krishna, N.V. & Viswanathan, B.: Architecting mesoporous AISBA-15: An overview on the synthetic strategy. *J. Indian Inst. Sci.* 90 (2010a), pp. 271–285.

Selvam, P., Murthy, P.R., Krishna, N.V. & Viswanathan, B.: Ordered nanoporous carbons (NCCR-56): Synthesis, characterization and applications. *3rd Int. Conf. on Adv. Mater. Sys. (ICAMS-2010)*, Bucharest, Romania, 2010b, pp. 125–130.

Selvam, P., Krishna, N.V. & Viswanathan, B.: Ordered nanoporous silicates (IITM-56): synthesis, characterization and applications. *3rd Int. Conf. on Adv. Mater. Sys. (ICAMS-2010)*, Bucharest, Romania, 2010c, pp. 119–124.

Service, R.F.: How far can we push chemical self-assembly? *Science* 309 (2005), p. 95.

Sharma, S. & Pollet, B.G.: Support materials for PEMFC and DMFC electrocatalysts-A review. *J. Power Sources* 208 (2012), pp. 96–119.

Shukla, A.K., Ravikumar, M.K., Roy, A., Barman, S.R., Sarma, D.D., Arico, A.S., Antonucci, V., Pino, L. & Giordano, N.: Electrooxidation of methanol in sulfuric acid electrolyte on platinized-carbon electrodes with several functional-group characteristics. *J. Electrochem. Soc.* 141 (1994), pp. 517–522.

Sriramulu, S., Jarvi, T.D. & Stuve, E.M.: Reaction mechanism and dynamics of methanol electrooxidation on platinum (111). *J. Electroanal. Chem.* 467 (1999), pp. 132–142.

Sundmacher, K., Schultz, T., Zhou, S., Scott, K., Ginkel, M. & Gilles, E.D.: Dynamics of the direct methanol fuel-cell (DMFC): Experiments and model-based analysis. *Chem. Eng. Sci.* 56 (2001), pp. 333–341.

Tanaka, S., Nishiyama, N., Egashira, Y. & Ueyama, K.: Synthesis of ordered mesoporous carbons with channel structure from an organic–organic nanocomposite. *Chem. Commun.* (2005), pp. 2125–2127.

Tian, B., Che, S., Liu, Z., Liu, X., Fan, W., Tatsumi, T., Terasaki, O. & Zhao, D.: Novel approaches to synthesize self-supported ultrathin carbon nanowire arrays templated by MCM-41. *Chem. Commun.* (2003), pp. 2726–2727.

Tian, J.M., Wang, F.B., Shan, Z.H.Q., Wang, R.J. & Zhang, J.Y.: Effect of preparation conditions of Pt/C catalysts on oxygen electrode performance in proton exchange membrane fuel-cells. *J. Appl. Electrochem.* 34 (2004), pp. 461–467.

Vinu, A.: Two-dimensional hexagonally-ordered mesoporous carbon nitrides with tunable pore diameter, surface area and nitrogen content. *Adv. Funct. Mater.* 18 (2008), pp. 816–827.

Vinu, A., Ariga, K., Mori, T., Nakanishi, T., Hishita, S., Golberg, D. & Bando, Y.: Preparation and characterization of well-ordered hexagonal mesoporous carbon nitride. *Adv. Mater.* 17 (2005), pp. 1648–1652.

Viswanathan, B & Aulice, M.: *Fuel-cells: Principles and applications.* Taylor & Francis Ltd., London, UK, 2008.

Wan, Y. & Zhao, D.Y.: On the controllable soft-templating approach to mesoporous silicates. *Chem. Rev.* 107 (2007), pp. 2821–2860.

Wasmus, S. & Kuver, A.: Methanol oxidation and direct methanol fuel-cells: a selective review. *J. Electroanal. Chem.* 461 (1999), pp. 14–31.

Watanabe, M., Uchida, M. & Motoo, S.: Preparation of highly dispersed platinum + ruthenium alloy clusters and the activity for the electrooxidation of methanol. *J. Electroanal. Chem. Interfacial Electrochem.* 229 (1987), pp. 395–406.

Winter, M. & Brodd, R.J.: What are batteries, fuel-cells, and supercapacitors? *Chem. Rev.* 104 (2004), pp. 4245–4269.

Wu, W., Cao, J., Chen, Y. & Lu, T.: Preparation of Pt/CMK-3 anode catalyst for methanol fuel-cells using paraformaldehyde as reducing agent. *Chinese J. Catal.* 28 (2007), pp. 17–21.

Wu, Z., Lv, Y., Xia, Y., Webley, P.A. & Zhao, D.: Ordered mesoporous platinum@graphitic carbon embedded nanophase as a highly active, stable, and methanol-tolerant oxygen reduction electrocatalyst. *J. Am. Chem. Soc.* 134 (2012), pp. 2236–2245.

Xia, X.H., Iwasita, T., Ge, F. & Vielstich, W.: Structural effects and reactivity in methanol oxidation on polycrystalline and single crystal platinum. *Electrochim. Acta* 41 (1996), pp. 711–718.

Xia, Y.D. & Mokaya, R.: Synthesis of ordered mesoporous carbon and nitrogen-doped carbon materials with graphitic pore walls via a simple chemical vapor deposition method. *Adv. Mater.* 16 (2004), pp. 1553–1558.

Xue, C., Tu, B. & Zhao, D.: Evaporation-induced coating and self-assembly of ordered mesoporous carbon-silica composite monoliths with macroporous architecture on polyurethane foams. *Adv. Funct. Mater.* 18 (2008), pp. 3914–3921.

Yanagisawa, T., Shimizu, T., Kuroda, K. & Kato, C.: The preparation of alkyltrimethylammonium-kanemite complexes and their conversion to microporous materials. *Bull. Chem. Soc. Jpn.* 63 (1990), pp. 988–992.

Yang, P., Zhao, D., Margolese, D.I., Chmelka, B.F. & Stucky, G.D.: Block copolymer templating syntheses of mesoporous metal oxides with large ordering lengths and semicrystalline framework. *Chem. Mater.* 11 (1999), pp. 2813–2826.

Yang, Y., Chiang, K. & Burke, N.: Porous carbon-supported catalysts for energy and environmental applications: A short review. *Catal. Today* 178 (2011), pp. 197–205.

You, D.J., Kwon, K., Joo, S.H., Kim, J.H., Kim, J.M., Pak, C. & Chang, H.: Carbon-supported ultra-high loading Pt nanoparticle catalyst by controlled overgrowth of Pt: Improvement of Pt utilization leads to enhanced direct methanol fuel-cell performance. *Int. J. Hydrogen Energy* 37 (2012), pp. 6880–6885.

Zhai, Y., Wan, Y., Cheng, Y., Shi, Y., Zhang, F., Tu, B. & Zhao, D.: The influence of carbon source on the wall structure of ordered mesoporous carbons. *J. Porous Mater.* 15 (2008), pp. 601–611.

Zhao, D., Feng, J., Huo, Q., Melosh, N., Frederickson, G.H., Chmelka, B.F. & Stucky, G.D.: Triblock copolymer syntheses of mesoporous silica with periodic 50 to 300 angstrom pores. *Science* 279 (1998), pp. 548–552.

Zhao, J., He, X., Tian, J., Wan, C. & Jiang, C.: Reclaim/recycle of Pt/C catalysts for PEMFC. *Energy Convers. Manage.* 48 (2007), pp. 450–453.

Zhou, S., Schultz, T., Peglow, M. & Sundmacher, K.: Analysis of the nonlinear dynamics of a direct methanol fuel-cell. Phys. Chem. Chem. Phys. 3 (2001), pp. 347–355.

# CHAPTER 6

## Modeling the coupled transport and reaction processes in a micro-solid-oxide fuel cell

Meng Ni

### 6.1 INTRODUCTION

A typical solid oxide fuel cell (SOFC) includes current collectors (interconnector), a porous anode, dense electrolyte, and a porous cathode (Singhal and Kendall, 2001), as shown in Figure 6.1. The cathode-electrolyte-anode is also called membrane-electrode-assembly (MEA), which is the key component of SOFCs. The commonly used materials for the electrolyte, anode and cathode, are yttria-stabilized zirconia (YSZ), YSZ-Ni composite, and a lanthanum strontium manganate (LSM)-YSZ composite. In operation, hydrogen and carbon monoxide can react with oxygen ions to produce electrons, steam, and carbon dioxide. The produced electrons flow through the external circuit to produce power and then flow to the cathode side. In the porous cathode, oxygen molecules react with electrons to produce oxygen ions, which are subsequently transported through the dense electrolyte to the porous anode to complete the cycle.

   The working mechanisms of SOFCs are similar to those of batteries, since both devices produce electric energy via electrochemical reactions. Batteries are essentially energy storage devices, as the reactants are stored in batteries (Yang *et al.*, 2011). Thus, the capacity and lifetime of the batteries depend on their size. Although rechargeable batteries can be recharged and used repeatedly, the charging process is usually time-consuming. For comparison, SOFCs are energy conversion devices as the fuels and oxidants are stored outside of the cell. SOFCs can continuously generate electric energy as long as fuel and an oxidant are supplied.

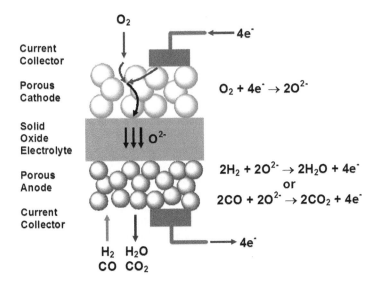

Figure 6.1.   Working principle of an SOFC.

Conventional heat engines are also energy conversion devices that can convert the chemical energy of a fuel into electric energy. In a typical thermal power plant, the chemical energy of a fuel is first converted into thermal energy by combustion, then converted to mechanical energy (rotation of the turbine blade), followed by conversion to electric energy using a generator. Due to the inefficiency of the combustion process and the loss of energy involved in each process, the overall efficiency of a thermal power plant is typically about 30%. In comparison, SOFCs can directly convert the chemical energy of a fuel into electric energy via electrochemical energy. This straightforward energy conversion process is efficient and can achieve a fuel-to-electricity efficiency as high as 50%.

Since SOFCs work at high temperatures (i.e. 400–1000°C), internal reforming of hydrocarbon fuels or thermal cracking of ammonia can occur in the porous anode of an SOFC to produce hydrogen-rich syngas. Thus, an SOFC can utilize a wide range of fuels, including hydrogen, natural gas, bio-ethanol, ammonia and syngas produced from coal or biomass (Achenbach *et al.*, 1994; Kawano *et al.*, 2008; Nahar and Kendall, 2011; Ni, 2011). This makes SOFCs advantageous, as low temperature fuel cells, like proton exchange membrane fuel cells (PEMFC), require very pure hydrogen fuel. It has been reported that the PEMFC catalyst can be poisoned by CO, even at ppm level (Shi *et al.*, 2007). In comparison, CO can be used as a fuel for power generation in SOFCs. In addition, a high operating temperature facilitates electrochemical reactions. Thus, SOFCs can use a low cost catalyst, such as Ni as anode, while low temperature fuel cells normally require Pt as a catalyst (Zhao *et al.*, 2007). Last but not least, the waste heat from a high temperature SOFC stack is of high quality and can be recovered by integrating SOFC stacks with additional thermal cycles for combined heat and power cogeneration to achieve high overall energy efficiency. It has been reported from some thermodynamic analyses that the overall efficiency of an SOFC cogeneration system can be as high as 70 to 80% (Veyo *et al.*, 2002).

Owing to the great potential of SOFCs for stationary power generations, extensive efforts have been made in the past years to solve the challenges of SOFCs, such as long term stability, system integration, scale-up etc. To improve the cell performance and durability, a fundamental understanding of the coupled transport and reaction phenomena in SOFCs is essential. As fuel flexibility is a major attraction of SOFCs, this chapter provides a numerical investigation of the performance of a micro-SOFC, fueled with a pre-reformed methane fuel mixture. The detailed distributions of gas species, local current density, reaction rates, temperature, etc. are presented and discussed.

## 6.2   MODEL DEVELOPMENT

In this chapter, the transport and reaction phenomena in a planar micro-SOFC, running on pre-reformed methane fuel mixture, are studied. Without considering the 3D effect, the physical-chemical processes and the computational domain can be shown in Figure 6.2. To capture the processes involved, a 2D numerical model is developed. The model consists of 3 sub-models: (i) computational fluid dynamics (CFD) model simulating the heat and mass transfer in SOFCs; (ii) electrochemical model calculating the local current density; and (iii) chemical model calculating the rates of methane steam reforming (MSR) reaction and water gas shift reaction (WGSR).

In order to avoid carbon deposition, the steam-to-carbon ratio is normally higher than 2 (Ni, 2012). However, a high steam concentration could decrease the SOFC's performance. A common practice for methane fueled SOFCs is to pre-reform methane fuel to obtain a gas mixture with sufficient hydrogen concentration. In this chapter, a commonly used 30% pre-reformed methane fuel mixture is employed (Ni and Leung, 2008). In operation, the fuel mixture is fed to the anode micro-channel while air is supplied to the cathode micro-channel. MSR and WGSR occur in the porous anode, as shown in Equations (6.1) and (6.2), respectively:

$$CH_4 + H_2O \leftrightarrow CO + 3H_2 \tag{6.1}$$

$$CO + H_2O \leftrightarrow CO_2 + H_2 \tag{6.2}$$

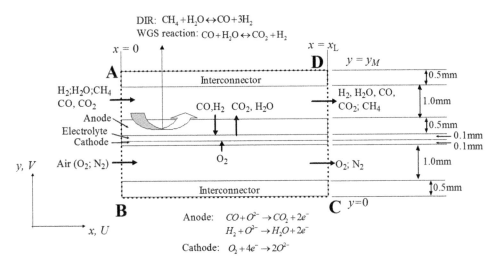

Figure 6.2.   Computational domain and working principle of a CH$_4$ fueled planar micro-SOFC.

H$_2$ and CO molecules are then transported to the triple-phase-boundary (TPB) at the anode-electrolyte interface, where they react with oxygen ions to produce electrons, H$_2$O and CO$_2$.

### 6.2.1   Computational fluid dynamic (CFD) model

The heat and mass transfer and fluid flow phenomena in the planar micro-SOFC is described by the CFD model. Due to the low gas velocity and small size of the SOFC, the Reynolds number in the micro-channel is usually much lower than 100 (Yuan *et al.*, 2003). Thus, the gas flow in an SOFC is typically laminar. From a heat transfer analysis, it is found that the local thermal equilibrium assumption is valid for the porous electrodes of an SOFC (Zheng *et al.*, 2013). The governing equations for the CFD model include mass conservation, momentum conservation, energy conservation, and species conservation (Wang, 2004):

$$\frac{\partial(\rho U)}{\partial x} + \frac{\partial(\rho V)}{\partial y} = S_m \tag{6.3}$$

$$\frac{\partial(\rho UU)}{\partial x} + \frac{\partial(\rho VU)}{\partial y} = -\frac{\partial P}{\partial x} + \frac{\partial}{\partial x}\left(\mu\frac{\partial U}{\partial x}\right) + \frac{\partial}{\partial y}\left(\mu\frac{\partial U}{\partial y}\right) + S_x \tag{6.4}$$

$$\frac{\partial(\rho UV)}{\partial x} + \frac{\partial(\rho VV)}{\partial y} = -\frac{\partial P}{\partial y} + \frac{\partial}{\partial x}\left(\mu\frac{\partial V}{\partial x}\right) + \frac{\partial}{\partial y}\left(\mu\frac{\partial V}{\partial y}\right) + S_y \tag{6.5}$$

$$\frac{\partial(\rho c_P UT)}{\partial x} + \frac{\partial(\rho c_P VT)}{\partial y} = \frac{\partial}{\partial x}\left(k\frac{\partial T}{\partial x}\right) + \frac{\partial}{\partial y}\left(k\frac{\partial T}{\partial y}\right) + S_T \tag{6.6}$$

$$\frac{\partial(\rho UY_i)}{\partial x} + \frac{\partial(\rho VY_i)}{\partial y} = \frac{\partial}{\partial x}\left(\rho D_{i,m}^{eff}\frac{\partial Y_i}{\partial x}\right) + \frac{\partial}{\partial y}\left(\rho D_{i,m}^{eff}\frac{\partial Y_i}{\partial y}\right) + S_{sp} \tag{6.7}$$

Here $U$ and $V$ are the velocity components in $x$ and $y$ directions. The density ($\rho$) and viscosity ($\mu$) of the gas mixture can be evaluated as (Reid *et al.*, 1987):

$$\rho = \frac{1}{\sum_{i=1}^{N} Y_i/\rho_i} \tag{6.8}$$

Table 6.1.   Values of $\sigma_i$ and $\varepsilon_{i,j}$ (compiled from ref. (Reid et al., 1987)).

|  | CO | CO$_2$ | H$_2$ | O$_2$ | CH$_4$ | N$_2$ | H$_2$O |
|---|---|---|---|---|---|---|---|
| $\sigma_i$ | 3.69 | 3.941 | 2.827 | 3.467 | 3.758 | 3.798 | 2.641 |
| $\varepsilon_i/k$ | 91.7 | 195.2 | 59.7 | 106.7 | 148.6 | 71.4 | 809.1 |

$$\mu = \sum_{i=1}^{n} \frac{y_i \mu_i}{\sum_{j=1}^{n} y_j \sqrt{\frac{M_j}{M_i}}} \tag{6.9}$$

where $M_i$ is molecular weight of species $i$ [kg kmol$^{-1}$], $\rho_i$ and $Y_i$ are the density and mass fraction of gas species $i$.

The effective diffusion coefficients $D_{i,m}^{eff}$ can be determined as:

$$\frac{1}{D_{i,m}^{eff}} = \begin{cases} \dfrac{\xi}{\varepsilon} \left( \dfrac{\sum_{j\neq i} \dfrac{X_j}{D_{ij}}}{1-X_i} + \dfrac{3}{2r_p} \sqrt{\dfrac{\pi M_i}{8RT}} \right), & \text{in porous electrodes} \\[3ex] \dfrac{\sum_{j\neq i} \dfrac{X_j}{D_{ij}}}{1-X_i}, & \text{in gas channels} \end{cases} \tag{6.10}$$

$$D_{ij} = \frac{0.0026 T^{1.5}}{P \sqrt{\dfrac{2 M_i M_j}{M_j + M_i}} \left( \dfrac{\sigma_i + \sigma_j}{2} \right)^2 \Omega_D} \tag{6.11}$$

$$\Omega_D = \frac{1.06036}{\left( \dfrac{k_b T}{\varepsilon_{i,j}} \right)^{0.1561}} + \frac{0.193}{\exp\left( 0.47635 \left( \dfrac{k_b T}{\varepsilon_{i,j}} \right) \right)} + \frac{1.03587}{\exp\left( 1.52996 \left( \dfrac{k_b T}{\varepsilon_{i,j}} \right) \right)} + \frac{1.76474}{3.89411 \left( \dfrac{k_b T}{\varepsilon_{i,j}} \right)} \tag{6.12}$$

$\xi$, $\varepsilon$, and $r_p$ are the tortuosity, porosity, and average pore radius of the porous electrodes. $D_{ij}$ is the binary diffusion coefficient of species $i$ and $j$. $\sigma_i$ is the mean characteristic length of species $i$. $\Omega_D$ is a dimensionless diffusion collision. $k_b$ is the Boltzmann's constant ($1.38066 \times 10^{-23}$ J K$^{-1}$). The values of $\sigma_i$ and $\varepsilon_{i,j}$ can be obtained from Reid et al. (1987) and are summarized in Table 6.1. $X_i$ is the molar fraction of specie $i$., which can be related to the mass fraction $Y_i$ by,

$$Y_i = X_i \left( \frac{M_i}{\sum_{i=1}^{N} X_i M_i} \right) \tag{6.13}$$

In the momentum equations (Equations (6.4) and (6.5)), the Darcy's law is used as source terms (Equations (6.14) and (6.15)):

$$S_x = -\frac{\mu U}{B_g} \tag{6.14}$$

$$S_y = -\frac{\mu V}{B_g} \tag{6.15}$$

$B_g$ is the permeability [m$^2$] measuring the capability of the porous media for fluid permeation. The use of Darcy's law as source terms allows the momentum equations to be applicable to both the gas channels and the porous media, as an infinitely large permeability can be used for the gas channel.

The source term ($S_T$, W m$^{-3}$) in the energy equation represents the heat generation/consumption by chemical reactions or electrochemical reactions. In the porous anode, the heat source term comes from heat generated by a WGSR and heat consumption by a DIR reaction. The heat generated by the electrochemical reaction, taking into account the irreversible overpotential losses, is applied evenly to the electrolyte. Therefore, the source term ST can be written as:

$$S_T = \begin{cases} R_{DIR}H_{DIR} + R_{WGSR}H_{WGSR}, & \text{in porous anode} \\ -\dfrac{J_{H_2}T\Delta S_{H_2} + J_{CO}T\Delta S_{CO}}{2FL} + \dfrac{J_{H_2}\eta_{t,H_2}}{L} + \dfrac{J_{CO}\eta_{t,CO}}{L}, & \text{in electrolyte} \end{cases} \tag{6.16}$$

$\Delta S_{H_2}$ and $\Delta S_{CO}$ are the entropy changes for the electrochemical reactions associated with H$_2$ fuel and CO fuel. $\eta_{t,H_2}$ and $\eta_{t,CO}$ refer to the total overpotentials for H$_2$ fuel and CO fuel. $R_{DIR}$ and $R_{WGSR}$ are the reaction rates (mol m$^{-3}$ s$^{-1}$) for DIR and WGSR, while $H_{DIR}$ and $H_{WGSR}$ are the corresponding reaction heat (J mol$^{-1}$). F is the Faraday constant (96485 C mol$^{-1}$). L is the thickness of the electrolyte (m). $J_{H_2}$ and $J_{CO}$ are the current densities produced from electrochemical oxidation of H$_2$ and CO, respectively.

The source term in the species equation ($S_{sp}$, kg m$^{-3}$s$^{-1}$) is caused by the chemical and electrochemical reactions. Taking H$_2$ as an example, the source term ($S_{H_2}$) can be written as (Ni, 2009; 2012; Ni and Leung, 2008; Reid *et al.*, 1987; Wang, 2004; Yuan *et al.*, 2003; Zheng *et al.*, 2013):

$$S_{H_2} = \begin{cases} 3R_{DIR}M_{H_2} + R_{WGSR}M_{H_2}, & \text{in porous anode} \\ -\dfrac{J_{H_2}M_{H_2}}{2F\Delta y}, & \text{at the anode-electrolyte interface} \end{cases} \tag{6.17}$$

where $\Delta y$ is the control volume width in y-direction (Fig. 6.1) at the anode-electrolyte interface. Similarly, for CO, the source term ($S_{H_2}$) can be written as:

$$S_{CO} = \begin{cases} R_{DIR}M_{CO} - R_{WGSR}M_{CO}, & \text{in porous anode} \\ -\dfrac{J_{CO}M_{CO}}{2F\Delta y}, & \text{at the anode-electrolyte interface} \end{cases} \tag{6.18}$$

CH$_4$ fuel only participates in the steam reforming reaction, while the direct electrochemical oxidation of CH$_4$ is not considered. Thus, the source term for CH$_4$ ($S_{CH_4}$) can be written as:

$$S_{CH_4} = -R_{DIR}M_{CH_4} \quad \text{in porous anode} \tag{6.19}$$

At the cathode side, oxygen molecules are involved in the electrochemical reaction at the TPB at the cathode-electrolyte interface. Thus, the source term for O$_2$ ($S_{O_2}$) can be written as:

$$S_{O_2} = -\dfrac{(J_{H_2} + J_{CO})M_{O_2}}{4F\Delta y} \quad \text{at the anode-electrolyte interface} \tag{6.20}$$

### 6.2.2 Electrochemical model

The current densities $J_{H_2}$ and $J_{CO}$ used in the CFD model can be obtained by an electrochemical model. In operation, the potential over the whole computational domain is considered to be uniform due to the use of a highly conductive interconnector along the SOFC. When the current density is zero, the SOFC potential is the highest and is called open-circuit voltage (OCV), or equilibrium potential. As the current density is increased, the potential of the SOFC is decreased due to various overpotential losses. The three overpotentials in the SOFC include the activation overpotential ($\eta_{act}$), the ohmic overpotential ($\eta_{ohmic}$), and the concentration overpotential ($\eta_{conc}$) (Ni and Leung, 2007). Thus, the current density ($J$) and potential ($V$) relationship can be established as (Ni, 2012):

$$V = E - \eta_{act,a} - \eta_{act,c} - \eta_{ohmic} \tag{6.21}$$

$$E_{H_2} = 1.253 - 0.00024516T + \frac{RT}{2F} \ln \left[ \frac{P^I_{H_2} \left( P^I_{O_2} \right)^{0.5}}{P^I_{H_2O}} \right] \tag{6.22}$$

$$E_{CO} = 1.46713 - 0.0004527T + \frac{RT}{2F} \ln \left[ \frac{P^I_{CO} \left( P^I_{O_2} \right)^{0.5}}{P^I_{CO_2}} \right] \tag{6.23}$$

$E$ is the OCV and the subscripts $H_2$ and CO represent the OCVs associated with $H_2$ and CO fuels; $T$ is temperature [K]. $R$ is the universal gas constant ($8.3145 \, J \, mol^{-1} \, K^{-1}$). $P^I$ used in Equations (6.22) and (6.23) are the partial pressure of gas species at the TPB. Therefore, the concentration overpotentials ($\eta_{conc}$) are included in $E$.

The ohmic overpotential ($\eta_{ohmic}$) is mainly caused by the resistance of the electrolyte to ion transport and can be determined with Ohm's law:

$$\eta_{ohmic} = JL \frac{1}{\sigma} \tag{6.24}$$

where $\sigma$ is the ionic conductivity ($\Omega^{-1} \, m^{-1}$) of the electrolyte and for YSZ it can be calculated as (Ferguson *et al.*, 1996),

$$\sigma = 3.34 \times 10^4 \exp\left( -\frac{10300}{T} \right) \tag{6.25}$$

The activation overpotential ($\eta_{act}$) represents the voltage loss involved in the electrochemical reactions. Based on experimental observations, the activation overpotentials for electrochemical oxidation of $H_2$ fuel ($\eta_{act,H_2,i}$) and CO fuel ($\eta_{act,CO,i}$) can be evaluated as:

$$\eta_{act,H_2,i} = \frac{RTJ_{H_2}}{n_{H_2} F J^0_{H_2,i}} \tag{6.26}$$

$$\eta_{act,CO,i} = \frac{RTJ_{CO}}{n_{CO} F J^0_{CO,i}} \tag{6.27}$$

$J^0_{H_2,i}$ and $J^0_{CO,i}$ are the exchange current densities [$A \, m^{-2}$] and represents the electrode's activity towards the electrochemical oxidation of $H_2$ and CO, respectively. Previous studies suggest that the values of $J^0_{H_2,a}$ and $J^0_{H_2,c}$ (exchange current density for the cathode) at 1073 K are $5300 \, A \, m^{-2}$ and $2000 \, A \, m^{-2}$, respectively (Chan and Xia, 2002). Thus, these values are adopted in the present study. In addition, experimental studies reveal that the rate of electrochemical oxidation of $H_2$ is about 2–3 times that of CO at a temperature of around 1073 K (Matsuzaki and Yasuda, 2000). Thus, $J^0_{CO,a}$ is assumed to be $J^0_{CO,a} = 0.2J^0_{H_2,a}$ or $J^0_{CO,a} = 0.6J^0_{H_2,a}$ in the present study.

### 6.2.3   *Chemical model*

The chemical model is developed to determine the rates of DIR ($R_{DIR}$, $mol \, m^{-3}s^{-1}$) and WGSR ($R_{WGSR}$, $mol \, m^{-3}s^{-1}$) as well as the corresponding reaction heat. The reaction rates of DIR and

WGSR can be determined as (Haberman and Young, 2004):

$$R_{\text{WGSR}} = k_{\text{sf}} \left( p_{H_2O}\, p_{CO} - \frac{p_{H_2}\, p_{CO_2}}{K_{\text{ps}}} \right) \tag{6.28}$$

$$k_{\text{sf}} = 0.0171 \exp\left( \frac{-103191}{RT} \right) [\text{mol m}^{-3}\,\text{Pa}^{-2}\,\text{s}^{-1}] \tag{6.29}$$

$$K_{\text{ps}} = \exp\left(-0.2935 Z^3 + 0.6351 Z^2 + 4.1788 Z + 0.3169\right) \tag{6.30}$$

$$Z = \frac{1000}{T\,(\text{K})} - 1 \tag{6.31}$$

$$R_{\text{DIR}} = k_{\text{rf}} \left( p_{CH_4}\, p_{H_2O} - \frac{p_{CO}\left(p_{H_2}\right)^3}{K_{\text{pr}}} \right) \tag{6.32}$$

$$k_{\text{rf}} = 2395 \exp\left( \frac{-231266}{RT} \right) \tag{6.33}$$

$$K_{\text{pr}} = 1.0267 \times 10^{10} \times \exp(-0.2513 Z^4 + 0.3665 Z^3$$
$$+ 0.5810 Z^2 - 27.134 Z + 3.277) \tag{6.34}$$

The reaction heat for endothermic DIR ($H_{\text{WGSR}}$, J mol$^{-1}$) and exothermic WGSR ($H_{\text{DIR}}$, J mol$^{-1}$) can be determined as (Chase, 1998; Ni, 2012):

$$H_{\text{DIR}} = -(206205.5 + 19.5175T) \tag{6.35}$$

$$H_{\text{WGSR}} = 45063 - 10.28T \tag{6.36}$$

## 6.3  NUMERICAL METHODOLOGIES

For the computational domain, the boundary conditions are set in this section. A velocity of 1 m s$^{-1}$ (or specified otherwise) is applied to the inlet ($x=0$) of the gas channels while zero velocity is specified at the inlet of the porous electrodes and the interconnectors. As the current single SOFC cell is one of the repeating cells in a planar SOFC stack, symmetric boundary conditions are applied to the bottom and top of the computational domain. That is, zero temperature gradient is applied to $y=0$ and $y=y_M$. At $x=x_L$, zero gradients for temperature, velocity, and mass fraction are applied to the gas channels, while zero velocity is applied to the solid part and the porous layer.

The governing equations are discretized and solved with the finite volume method (FVM). The equations used in the CFD model can be rewritten in a general form as (Patankar, 1980; Tao, 1988):

$$div(\rho \mathbf{U}\phi) = div(\Gamma_\phi\, \text{grad}\phi) + S_\phi \tag{6.37}$$

where $\phi$ is a general variable to be solved; $\Gamma_\phi$ and $S_\phi$ are general diffusion and source terms. Applying $\phi=1$, the above general transport equation is reduced to the continuity equation, in which the source term is caused by mass consumption or production from electrochemical reactions. Setting $\phi=U$ and $V$, the general equation becomes the momentum equations in $x$ and $y$ directions, respectively. Setting $\phi=T$, the general equation becomes the energy equation. With the general transport equation and the 2D grid for discretization (shown in Fig. 6.3), the governing equations can be converted to discretized forms with the same form:

$$a_P\phi_P = a_E\phi_E + a_W\phi_W + a_N\phi_N + a_S\phi_S + b \tag{6.38}$$

Here, central difference scheme and the upwind scheme are used to treat the diffusion and convection terms, respectively.

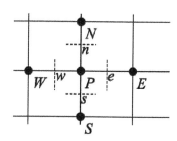

Figure 6.3.   Schematic of the 2D grid for discretization.

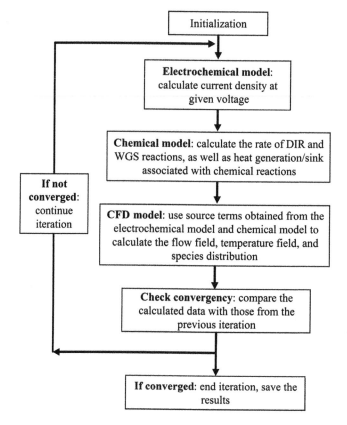

Figure 6.4.   Iteration scheme.

The coefficients for Equation (6.38) can be found in Patankar (1980). The pressure and velocity are treated with the SIMPLEC algorithm, in which the pressure correction $p'$ is done by solving the pressure correction equation (Equation (6.39)) which is derived based on the continuity equation (Tao, 1988). The TDMA-based iteration schemes are adopted to solve the discretized equations:

$$a_P p'_P = a_E p'_E + a_W p'_W + a_N p'_N + a_S p'_S + b \tag{6.39}$$

The iteration scheme is shown in Figure 6.4. First, initial data, such as initial temperature, gas composition, pressure etc., are applied to the whole computational domain. Then, the electrochemical model is solved to calculate the local current density at a given operating potential, followed by solving of the chemical model to calculate the reaction rates and reaction heat. Based

Table 6.2.   Typical values used in simulation.

| Parameter | Value |
|---|---|
| Operating temperature, $T$ [K] | 1073 |
| Operating pressure, $P$ [Pa] | 1.0 |
| Electrode porosity, $\varepsilon$ | 0.4 |
| Electrode tortuosity, $\xi$ | 3.0 |
| Average pore radius, $r_p$ [μm] | 0.5 |
| Anode-supported electrolyte: | |
|    Anode thickness $d_a$ [μm] | 500 |
|    Electrolyte thickness, $L$ [μm] | 100 |
|    Cathode thickness, $d_c$ [μm] | 100 |
| Height of gas flow channel [mm] | 1.0 |
| Length of the planar SOFC [mm] | 20 |
| Thickness of interconnector [mm] | 0.5 |
| Inlet velocity at anode: $U_0$ [m s$^{-1}$] | 1.0 |
| Inlet velocity at cathode: [m s$^{-1}$] | 3.0 |
| Cathode inlet gas molar ratio: $O_2/N_2$ | 0.21/0.79 |
| Anode inlet gas molar ratio: $H_2O/CH_4/H_2/CO_2/CO$ | 0.493/0.171/0.263/0.044/0.029 |
| SOFC operating potential [V] | 0.4 |
| Thermal conductivity of SOFC component [W m$^{-1}$K$^{-1}$]: | |
|    Anode | 11.0 |
|    Electrolyte | 2.7 |
|    Cathode | 6.0 |
|    Interconnect | 1.1 |

on the source terms calculated from the electrochemical model and the chemical model, the CFD model can be solved and the fluid field and temperature field can be updated. Then the program goes back to solve the electrochemical model again with the updated data. Computation is repeated until convergence is achieved. The program is written in an in-house code in FORTRAN. Before parametric simulations, initial simulations have been performed to compare the simulation results with the data from ref. (Yuan *et al.*, 2003), and the data from commercial software – FLUENT. Thus, the CFD code is well validated before use. Details of the code validation can be found from the previous publications (Ni, 2010).

## 6.4   RESULTS AND DISCUSSION

After rigorous code evaluation, parametric simulations are then performed to look at the coupled transport and reaction phenomena in SOFCs. The typical simulation conditions from the literature are adopted and summarized in Table 6.2.

### 6.4.1   *Base case*

Simulation is performed at an inlet temperature of 1073 K, operating potential of 0.5V, and $J^0_{CO} = 0.2 J^0_{H_2}$. Figure 6.5 shows the calculated reaction rates of DIR (a) and WGSR (b) in an SOFC.

   The DIR reaction rate is found to increase from about 12 mol m$^{-3}$ s$^{-1}$ at the inlet to be about 25 mol m$^{-3}$ s$^{-1}$ at the outlet (Fig. 6.5a). The calculated reaction rate is lower but of the same order as the results found by Lehnert *et al.* (2000), as the simulation temperature in their study (1123 K) was higher than in the present study. The increase in DIR reaction rate is mainly caused by an increase in the SOFC temperature along the flow channel (Fig. 6.6), which will be

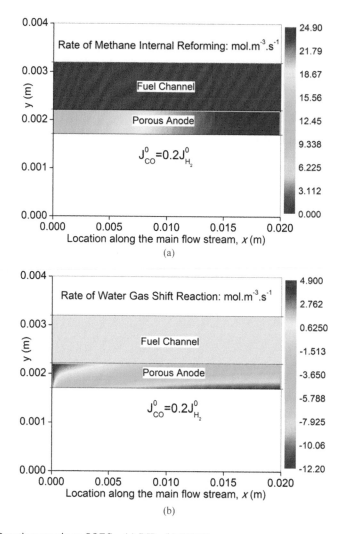

Figure 6.5.   Reaction rates in an SOFC – (a) DIR; (b) WGSR.

discussed later. For comparison, the WGSR reaction rate is positive near the inlet and negative in the downstream (Fig. 6.5b), especially near the anode-electrolyte interface. This is due to the electrochemical oxidation of CO at the TPB, which lowers the CO concentration and facilitates the reverse WGSR.

The temperature in the SOFC depends on heat generation and consumption due to chemical and electrochemical reactions. In an SOFC with internal reforming, the methane DIR is a strongly endothermic reaction which consumes heat. The WGSR and the electrochemical reactions are exothermic and generate heat. The irreversible overpotential losses also contribute to heat generation. The temperature in the SOFC increases from 073 K at the inlet to 109 K at the outlet (Fig. 6.6), indicating that the heat generation from the electrochemical reaction and the irreversible overpotential exceeds the heat consumption by endothermic DIR and reversed WGSR reactions.

Figure 6.7 shows the contours of $U/U_0$ in the SOFC. For both anode channel and cathode channel. The velocity distribution shows a clear development stage at the inlet and becomes fully developed in the downstream. The gas velocity in the porous electrodes is in general very low,

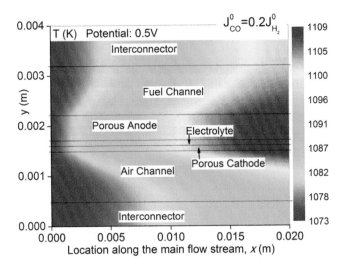

Figure 6.6. Temperature in the SOFC.

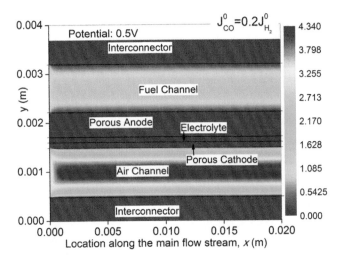

Figure 6.7. Velocity ratio $U/U_0$ contours in the SOFC.

indicating little effect from convection in the porous structure, which is consistent with previous studies (Yuan and Sunden, 2004).

The molar fractions of gas species are shown in Figure 6.8, taking CO, $H_2$, and $O_2$ as examples. As can be seen, the molar fractions of both CO and $H_2$ decrease along the gas flow channel. This indicates that the consumption of CO and $H_2$ by electrochemical oxidation is higher than the generation of these two gases from chemical reactions (DIR and WGSR) (Figs. 6.8a and b). The molar fraction of $O_2$ in the cathode also decreases apparently along the gas flow stream (Fig. 6.8c). In addition, it is found that the molar fraction of gas species near the electrode surface is obviously lower than that at the electrode-electrolyte interface, even for $O_2$ in a thin cathode layer. This is mainly caused by the resistance of the porous structure to gas transport.

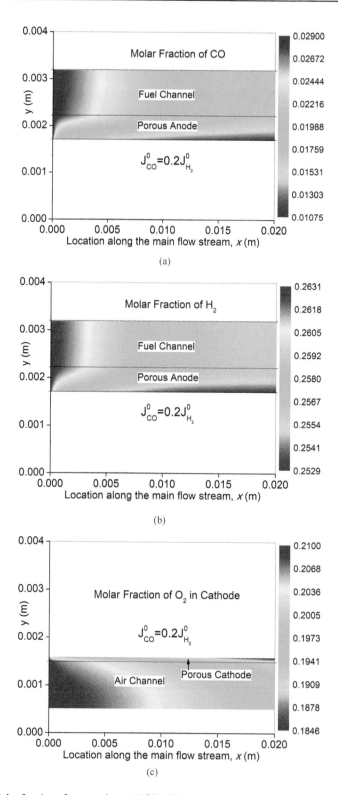

Figure 6.8.   Molar fraction of gas species – (a) CO; (b) $H_2$; and (c) $O_2$.

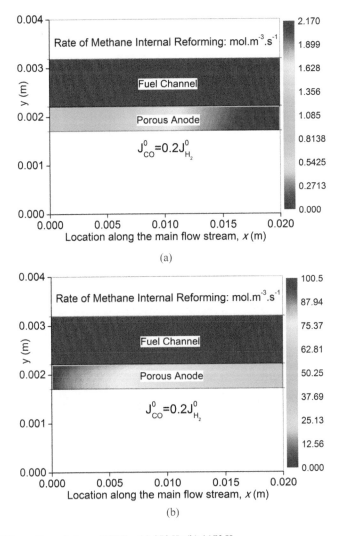

Figure 6.9.   DIR reaction rate in an SOFC – (a) 973 K; (b) 1173 K.

### 6.4.2   *Temperature effect*

To investigate the temperature effect, simulations are performed at an inlet temperature of 973 and 1173 K.

The rates of DIR reaction in an SOFC at 973 K and at 1173 K are compared in Figure 6.9. At an inlet temperature of 973 K, the rate of DIR reaction is very small (only 1–2 mol m$^{-3}$ s$^{-1}$) and increases along the gas flow stream (Fig. 6.9a). For comparison, at 1173 K, the DIR reaction rate at 1173 K is highest (about 100 mol m$^{-3}$ s$^{-1}$) at the inlet and decreases considerably in the downstream (Fig. 6.9b). At 973 K, the increase in DIR reaction rate is mainly caused by the increasing temperature along the cell (Fig. 6.10a), as the very low DIR reaction rate leads to negligible variation in CH$_4$ molar fraction. For comparison, at an inlet temperature of 1173 K, the DIR reaction rate is mainly caused by a large reduction in CH$_4$ molar fraction along the cell (Fig. 6.11), although the temperature is increased by about 10 K (Fig. 6.10b).

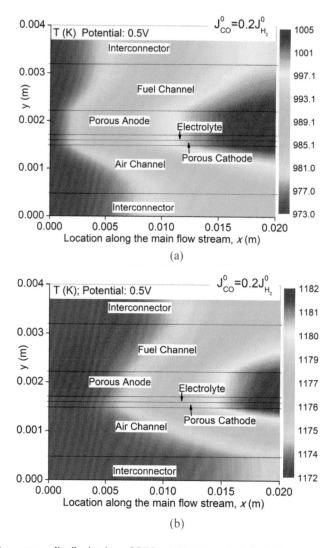

Figure 6.10.   Temperature distribution in an SOFC – (a) 973 K; and (b) 1173 K.

The reaction rates of WGSR in an SOFC are shown in Figure 6.12 at inlet temperatures of 973 K and 1173 K. The distribution pattern is found to be quite different from that of DIR reaction. At 973 K, the reaction rate is found positive at the inlet and negative in the downstream (Fig. 6.12a). This is because the electrochemical oxidation of CO decreases the CO molar fraction and facilitates the reversed WGSR, as can be seen from Figure 6.13a. For comparison, the reaction rate is positive in most of the anode layer from the inlet to the outlet while negative near the anode-electrolyte interface (Fig. 6.12b). This behavior is caused by a high rate of CO generation from the DIR reaction at a high temperature (Fig. 6.9b) which raises the CO molar fraction (Fig. 6.13b) and facilitates the forward WGSR.

### 6.4.3   *Operating potential effect*

To examine the operating potential effect, another simulation is performed at an operating potential of 0.8 V at an inlet temperature of 1073 K. It is found that, in contrast to the base case (Fig. 6.5a),

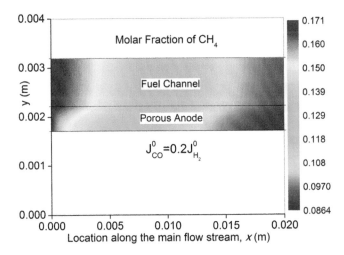

Figure 6.11.  Molar fraction of CH$_4$ in an SOFC at an inlet temperature of 1173 K.

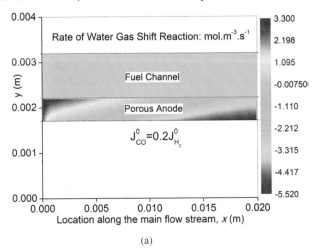

(a)

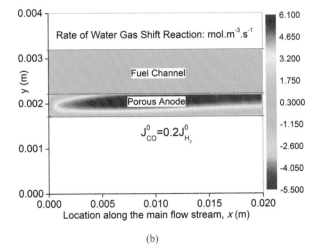

(b)

Figure 6.12.  Rate of WGSR in an SOFC – (a) 973 K; (b) 1173 K.

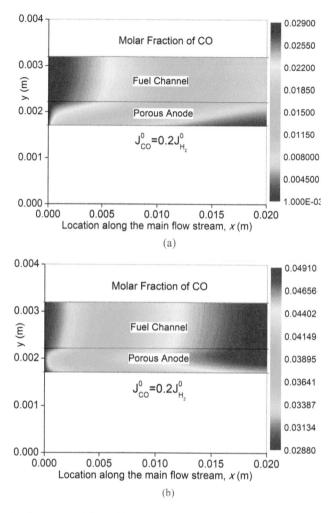

Figure 6.13.   Molar fraction of CO in an SOFC – (a) 973 K; (b) 1173 K.

the DIR reaction rate at 0.8V decreases along the flow channel (Fig. 6.14a). In addition, the rate of WGSR is in general positive and decreases along the channel (Fig. 6.14b), which is also different from the base case (Fig. 6.5b). The reduction in DIR reaction rate along the cell is mainly caused by the decreasing $CH_4$ molar fraction, as the temperature remains almost unchanged along the cell (Fig. 6.14c). This indicates that an SOFC can be almost thermally self-sustained at certain operating conditions, eliminating the need of heat supply (Shao *et al.*, 2005).

Compared with the base case, the total current density generated from the SOFC at 0.8 V is considerably lower (Fig. 6.15). A lower current density generates less heat as both the electrochemical heat generation and the heat generation from irreversible overpotential loss depend on the current density. This can explain why the base case has a higher temperature increase while the temperature increase is negligible at 0.8 V. Moreover, the current density associated with CO electrochemical oxidation is much lower at 0.8 V than that at 0.5 V (Fig. 6.15). Thus, less CO is consumed for power generation, which yields a higher CO molar fraction than the base case and can explain the positive WGSR in an SOFC at 0.8 V (Fig. 6.14b).

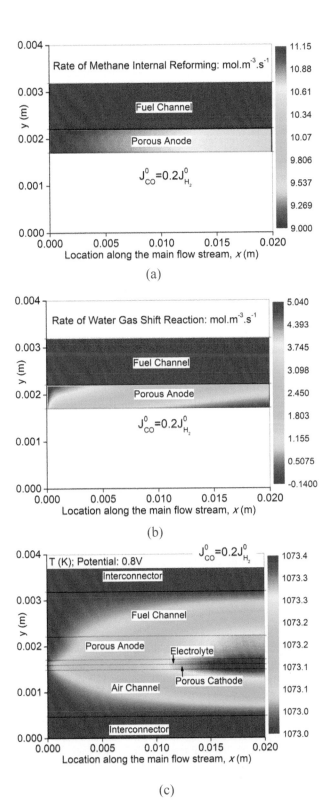

Figure 6.14.   SOFC at 0.8 V and 1073 K – (a) distribution of DIR reaction rate; (b) distribution of WGSR rate; and (c) temperature.

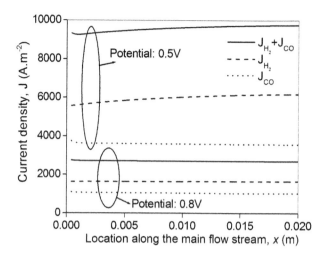

Figure 6.15.  Comparison of current density between the base case and the 0.8 V case.

### 6.4.4  *Effect of electrochemical oxidation rate of CO*

In the previous simulations, it is assumed that $J_{CO}^0 = 0.2 J_{H_2}^0$. To examine the effect of electrochemical oxidation rate of CO on an SOFC's performance, $J_{CO}^0 = 0.6 J_{H_2}^0$ is adopted in the simulation in this section. It is found that $J_{H_2}$ remains unchanged when $J_{CO}^0 = 0.2 J_{H_2}^0$ is changed to $J_{CO}^0 = 0.6 J_{H_2}^0$. However, $J_{CO}$ is increased considerably with $J_{CO}^0 = 0.6 J_{H_2}^0$ (Fig. 6.16a). As a result, the total current density with $J_{CO}^0 = 0.6 J_{H_2}^0$ is higher than that with $J_{CO}^0 = 0.2 J_{H_2}^0$, which in turn leads to more heat generation in the SOFC with $J_{CO}^0 = 0.6 J_{H_2}^0$. Therefore, the temperature increase in the SOFC with $J_{CO}^0 = 0.6 J_{H_2}^0$ is found larger than that with $J_{CO}^0 = 0.2 J_{H_2}^0$ (Fig. 6.16b and Fig. 6.10b).

Due to the higher temperature, the reaction rate of DIR in the SOFC with $J_{CO}^0 = 0.6 J_{H_2}^0$ is found to be slightly higher than that with $J_{CO}^0 = 0.2 J_{H_2}^0$ (Fig. 6.16c and Fig. 6.9b). Interestingly, at a high rate of electrochemical oxidation of CO ($J_{CO}^0 = 0.6 J_{H_2}^0$), the WGSR is found to be reversed (Fig. 6.16d), as the high consumption rate of CO force the reaction to be backward. Thus, from this section, it can be seen that the electrochemical reaction and chemical reactions are highly coupled and affect the SOFC's performance in a complicated manner.

### 6.5  CONCLUSIONS

This chapter provides an introduction to SOFC technology for stationary power generation. A 2D numerical model is developed by coupling the computational fluid dynamics (CFD) methodology with the electrochemical and chemical reaction kinetics. The rates of chemical reaction and electrochemical reaction as well as corresponding reaction heats are implemented in the CFD model as source terms.

Parametric simulations are performed to examine the effect of operating parameters on an SOFC's performance with pre-reformed methane gas mixture as a model fuel. It is found that the chemical reactions and the electrochemical reactions influence each other considerably. Obvious gas diffusion resistance by the porous electrodes is also observed, even for a thin cathode with a thickness of only 0.1 mm. By increasing the inlet temperature, the rate of DIR reaction is significantly improved, leading high $H_2$ and CO generation, which is beneficial for the SOFC's performance. Increasing the operating potential considerably decreases the current density and

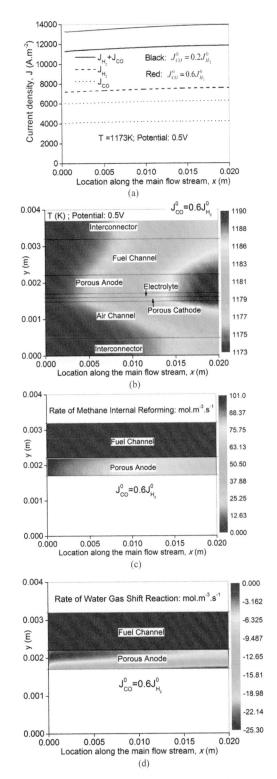

Figure 6.16. An SOFC with $J^0_{CO} = 0.6J^0_{H_2}$ – (a) current density compared with the $J^0_{CO} = 0.2J^0_{H_2}$ case; (b) temperature distribution; (c) DIR reaction rate; (d) WGSR rate.

thus heat generation in the cell. At certain operating conditions, a thermally self-sustained operation of the SOFC can be achieved. With a high rate of CO electrochemical oxidation, the total current density can be increased, leading to a higher temperature and a slightly higher rate of DIR reaction. However, WGSR tends to be reversed.

The results presented in this chapter offer some fundamental information on the coupled transport and reaction phenomena in a micro-SOFC running on hydrocarbon fuels. With further technology development, SOFCs are believed to play an important role in future clean energy conversion.

## ACKNOWLEDGEMENTS

The author thanks The Hong Kong Polytechnic University and the Hong Kong Research Grant Council (RGC) for funding support (Project Number: PolyU 5238/11E) for SOFC research.

## REFERENCES

Achenbach, E. & Riensche, E.: Methane/steam reforming kinetics for solid oxide fuel-cells. *J. Power Sources* 52 (1994), pp. 283–288.

Chan, S.H. & Xia, Z.T.: Polarization effects in electrolyte/electrode-supported solid oxide fuel-cells. *J. Appl. Electrochem.* 32 (2002), pp. 339–347.

Chase, M.W.: *NIST-JANAF thermochemical tables.* 4th edition, American Chemical Society, American Institute of Physics for the National Institute of Standards and Technology, 1998.

Ferguson, J.R., Fiard, J.M. & Herbin, R.: Three-dimensional numerical simulation for various geometries of solid oxide fuel-cells. *J. Power Sources* 58 (1996), pp. 109–122.

Haberman, B.A. & Young, J.B.: Three-dimensional simulation of chemically reacting gas flows in the porous support structure of an integrated-planar solid oxide fuel-cell. *Int. J. Heat Mass Transfer* 47 (2004), pp. 3617–3629.

Kawano, M., Toshiaki, T., Matsui, T., Kikuchi, R., Yoshida, H., Inagaki, T. & Eguchi, K.: Steam reforming on Ni-samaria-doped ceria cermet anode for practical size solid oxide fuel-cell at intermediate temperature. *J. Power Sources* 182 (2008), pp. 496–502.

Lehnert, W., Meusinger, J. & Thom, F.: Modeling of gas transport phenomena in SOFC anodes. *J. Power Sources* 87 (2000), pp. 57–63.

Matsuzaki, T. & Yasuda, I.: Electrochemical oxidation of $H_2$ and CO in a $H_2$-$H_2O$-CO-$CO_2$ system at the interface of a Ni-YSZ cermet electrode and YSZ electrolyte. *J. Electrochem. Soc.* 147 (2000), pp. 1630–1635.

Nahar, G. & Kendall, K.: Biodiesel formulations as fuel for internally reforming solid oxide fuel-cell. *Fuel Process. Technol.* 92 (2011), pp. 1345–1354.

Ni, M.: On the source terms of species equations in fuel-cell modelling. *Int. J. Hydrogen Energy* 34 (2009), pp. 9543–9544.

Ni, M.: 2D thermal-fluid modelling and parametric analysis of a planar solid oxide fuel-cell. *Energy Convers. Manage.* 51 (2010), pp. 714–721.

Ni, M.: Thermo-electrochemical modelling of ammonia-fueled solid oxide fuel-cells considering ammonia thermal decomposition in the anode. *Int. J. Hydrogen Energy* 36 (2011), pp. 3153–3166.

Ni, M.: Modeling of SOFC running on partially pre-reformed gas mixture. *Int. J. Hydrogen Energy* 37 (2012), pp. 1731–1745.

Ni, M., Leung, M.K.H. & Leung, D.Y.C.: Parametric study of solid oxide fuel-cell performance. *Energy Convers. Manage.* 48 (2007), pp. 1525–1535.

Ni, M., Leung, D.Y.C. & Leung, M.K.H.: Modeling of methane fed solid oxide fuel-cells: comparison between proton conducting electrolyte and oxygen ion conducting electrolyte. *J. Power Sources* 183 (2008), pp. 133–142.

Patankar, S.V.: *Numerical heat transfer and fluid flow.* McGraw-Hill, New York, NY, 1980.

Reid, R.C., Prausnitz, J.M. & Poling, B.E.: *The properties of gases and liquids.* 4th edition, McGraw-Hill Book Company, New York, NY, 1987.

Shao, Z.P., Haile, S.M., Ahn, J., Ronney, P.D., Zhan, Z.L. & Barnett, S.A.: A thermally self-sustained micro solid-oxide fuel-cell stack with high power density. *Nature* 435 (2005), pp. 795–798.

Shi, W.Y., Yi, B.L., Hou, M. & Shao, Z.G.: The effect $H_2S$ and CO mixtures on PEMFC performance. *Int. J. Hydrogen Energy* 32 (2007), pp. 4412–4417.

Singhal, S.C. & Kendall, K.: *High temperature solid oxide fuel-cells – Fundamentals, design and applications*. Elsevier, New York, NY, 2003.

Tao, W.Q.: *Numerical heat transfer*. Xi'an Jiaotong University Publishing, 1988.

Veyo, S.E., Shockling, L.A., Dederer, J.T., Gillett, J.E. & Lundber, W.L.: Tubular solid oxide fuel-cell/gas turbine hybrid cycle power systems: Status. *J. Eng. Gas Turb. Power* 124 (2002), pp. 845–849.

Wang, C.Y.: Fundamental models for fuel-cell engineering. *Chem. Rev.* 104 (2004), pp. 4727–4765.

Yang, Z.G., Zhang, J.L., Kintner-Meyer, M.C.W., Lu, X.C., Choi, D., Lemmon, J.P. & Liu, J.: Electrochemical energy storage for green grid. *Chem. Rev.* 111 (2011), pp. 3577–3613.

Yuan, J. & Sunden, B.: A numerical investigation of heat transfer and gas flow in proton exchange membrane fuel-cell ducts by a generalized extended Darcy model. *Int. J. Green Energy* 1 (2004), pp. 47–63.

Yuan, J.L, Rokni, M. & Sunden, B.: Three-dimensional computational analysis of gas and heat transport phenomena in ducts relevant for anode-supported solid oxide fuel-cells. *Int. J. Heat Mass Transfer* 46 (2003), pp. 809–821.

Zhao, T.S., Kreuer, K.D. & Nguyen, T.: *Advances in fuel-cells*. Elsevier, New York, NY, 2007.

Zheng, K.Q., Ni, M. & Sun, Q.: On the local thermal non-equilibrium SOFCs considering internal reforming and ammonia thermal cracking reaction. *Energy Technol.* 1 (2013), pp. 35–41.

# CHAPTER 7

## Nano-structural effect on SOFC durability

Yao Wang & Changrong Xia

## 7.1 INTRODUCTION

Recently, the interest in nano-technology has provided inspiration in the search for highly-performing and stable electrodes for solid oxide fuel cells (SOFCs). Positively to say, the decrease in the size of the electro-catalysts has greatly enhanced the cell performance due to the increase in three phase boundary (TPB) sites and the improved catalytic activities of nano-particles. The promotion has been reported in a number of combinations including perovskite/yttria stabilized zirconia (YSZ) cathodes (Chen *et al.*, 2008; Huang *et al.*, 2005), perovskite/doped ceria cathodes (Chen *et al.*, 2007; Jiang and Leng *et al.*, 2006; Wang, 2005; Wang *et al.*, 2012; Xu *et al.*, 2006; Zhang *et al.*, 2008) and Ni/electrolytes anodes (Ding *et al.*, 2008a,b; Jiang *et al.*, 2002; 2005; Liu *et al.*, 2011; Wang *et al.*, 2006; Zhu *et al.*, 2006). The unique structure with nano-particles seems also to be beneficial to cell stability, although this remains controversial. Taking a cathode derived by the impregnation method as an example: it is prepared by depositing the nano-catalysts into porous pre-sintered backbones, which are co-fired with the dense electrolyte layer at a high temperature. The backbones are strongly bonded with the electrolyte forming a robust incorporate body for tough operational conditions (Zhao, 2008a). The nano-sized catalysts are formed from the decomposition of the nitrate solutions at relatively low temperatures, under which the possible deleterious reaction between the catalysts and the electrolyte has been effectively suppressed, thus enlarging the selection alternatives for cathode materials. In the case of nano-anodes, derived by the impregnation technique, the nickel catalysts are free to suffer the support strength, and a little amount of nickel is introduced to reach the percolation limit for electron transfer (Suzuki *et al.*, 2005; Uchida *et al.*, 2003). Thus, the mechanical properties of the anode have been inevitably enhanced by having little influence from the micro-structure change of the nickel during the thermal or redox operation. However, with regard to thermodynamics, the nano-sized particles tend to grow up to release the free energy spontaneously by reducing the surface area. Grain coarsening could be observed in the nano-structured electrodes after a long isothermal test, especially at high temperatures, which was responsible for the performance degradation (Jiang *et al.*, 2009a,b,c; Suzuki *et al.*, 2009). Therefore, the use of the nano-structured electrode for practical application still needs to be carefully considered unless it definitely has a long-time stability with a reasonable degradation rate or a predictable performance as a function of operating time.

This chapter focuses on the nano-structured effects on electrode durability, i.e. cell durability. Fundamental aging mechanisms of the electrode components are also summarized. Long-time performances of the nano-structured electrodes are introduced to reveal the nano-size effects on electrode durability. Furthermore, models ever reported for the prediction of durability are included for better understanding the influence from the nano-size scale.

## 7.2 AGING MECHANISM OF SOFC ELECTRODES

### 7.2.1 *Aging mechanism of the anodes*

The SOFC anode is traditionally fabricated by mechanically mixing NiO with the electrolyte. Taking the Ni-YSZ cermet as an example, the mixture of NiO and YSZ powders is firstly co-pressed or tape-casted with a thin layer of YSZ electrolyte. The two green layers are then co-sintered at a high temperature in excess of 1300°C (Minh, 1993) to make the electrolyte layer dense and the anode/electrolyte interface well-connected. Subsequently, the NiO in the substrate is reduced in-situ to metallic Ni by exposure to $H_2$ when the operating temperature is above 450°C. The reduction process from NiO to Ni creates a number of pores for gas diffusion, and the Ni-YSZ cermet is finally obtained in the form of a dense YSZ electrolyte layer onto a porous Ni/YSZ anode substrate. In this composite structure, the YSZ particles serve as a pathway for oxygen ion transfer, while the metallic Ni particles act as the electronic conductor and catalyst for hydrogen oxidation. It is important that YSZ grains in the dense electrolyte should be well connected to the YSZ particles in the porous anode since this provides a conduction path for $O^{2-}$ ions. A volume of 30% Ni is required to form an interconnected Ni network, according to the percolation theory, which is also found to show the highest electronic conductivity by Dees *et al.* (1987). These conventional Ni-based cermets have been widely used as the SOFC anodes, but they still have several problems during the long-term test in the conditions of isothermal and other tough processes, such as thermal, redox cycling and/or in the presence of hydrocarbon and $H_2S$. Typical anode degradation exhibits decreased conductivity, increased ohmic or polarization resistance, or reduced cell potential which is mainly caused by grain coarsening and/or poisoning of the metallic nickel, as has been summarized in Table 7.1.

### 7.2.1.1   *Grain coarsening*
The anodic reaction can only occur near the TPB sites, which are defined as the sites where the electrolyte phase, the electron conducting phase, and the gas come together. The performance of the cermet anode is greatly enhanced by the electrolyte phase because of the increased TPB sites and this electrolyte phase extends the oxidation reaction zone from the physical interface of the anode and electrolyte to the anode bulk, about 10 μm beyond the interface (Brown *et al.*, 2000; Tanner *et al.*, 1997; Wang *et al.*, 2001). If there is a breakdown in connectivity in any one of the three phases, the reaction will not occur and the site is not electrochemical active anymore. The nickel has a low melting point compared with the ceramic phase, though the nickel particles are separated by the YSZ component. These conventional mechanical mixed structures may not be effective in preventing the coarsening of Ni particles, which occurs spontaneously to lower the free energy by reducing the surface area in a high-temperature sintering or a lengthy operation process. The agglomeration of nickel will reduce the TPBs by losing contact with the ceramic phases and consequently increasing the reactive polarization resistance. It will also reduce the electrical pathways by breaking the Ni-Ni contact and thus increase the ohmic resistance. For example, Figure 7.1 shows the impedance evolution for a Ni-YSZ anode in humidified hydrogen under a current loading density of 200 mA cm$^{-2}$ at different discharge time (Jiao *et al.*, 2012). The ohmic resistance increases from 0.75 Ω cm$^2$ to 0.98 Ω cm$^2$, and the polarization resistance from 0.54 Ω cm$^2$ to 1.82 Ω cm$^2$ after 250 h operation. Separated Ni, YSZ component and TPB networks are also analyzed with the reconstructed 3D images of Ni-YSZ anodes, Figure 7.2. It is clearly that the Ni particles experience certain morphological change, its specific surface area has decreased by 16.9% and 17.9% after 100 and 250 hours, and the percolation connectivity has reduced by 10% and 40%, which is consistent with the sintering mechanism of Ni coarsening. The reduction of both Ni surface area and Ni-YSZ inter-phases surface area corresponds to the decrease of the active TPB density, about 31% at 100 h and 57% at 250 h, which is further responsible for the increase of the ohmic and polarization resistance. Similar experimental results are also reported by Simwonis *et al.* (2000) a decrease of electrical conductivity from 3900 to 2600 S cm$^{-1}$ after exposure in Ar/4% $H_2$/3% $H_2O$ at 1000°C for 4000 h. The average Ni diameter was increased about 30% from 2 to 2.6 μm, while the YSZ and pore size distributions did not show any pronounced change. The evolution of electrical conductivity as a function of exposure time could be predicted by a concept of contiguity theory (Fan *et al.*, 1993; Fan, 1996), which suggests that the Ni continuity is interrupted by increased particle size. It was deduced that the Ni fraction contributed to the electronic conductivity when it was reduced from 23 to 20%. Furthermore, other results from the nickel sintering were exhibited as a decrease in gas permeability and a

Table 7.1. Anode degradation examples for Ni-YSZ cermets.

| Anode composition | Test conditions | Test time | Change of performance | Change of micro-structure | Reference |
|---|---|---|---|---|---|
| Ni-YSZ | 4% $H_2$ + Ar + 3% $H_2O$, 1000°C | 4000 h | $\sigma$: 3900 → 2600 S cm⁻¹ | $d_{Ni}$: 2 → 2.6 $\mu$m; $d_{pore}$: 1.6 → 1.47 $\mu$m | Simwonis et al. (2000) |
| Ni-TZ3Y | moist 10% $H_2$/90% $N_2$, 0.25 A cm⁻², 1000°C | 2500 h | $r_{\eta,loss}$: 21 mV/1000h; $r_{degradation}$: 2.7%/1000 h | $d_{Ni}$: 1.5 → 2.7 $\mu$m; $\varepsilon$: 25% → 19% | Jiang (2003) |
| Ni-YSZ | 10% $H_2$/90% $N_2$, 600°C | 30 h | $\sigma$: 1200 → 1000 S cm⁻¹ | Ni grain growth | Pihlatie et al. (2011) |
|  | 850°C | 160 h | $\sigma$: 1000 → 600 S cm⁻¹ |  |  |
|  | 10% $H_2$/90% $N_2$ + 3% $H_2O$, 850°C | 90 h | $\sigma$: 820 → 500 S cm⁻¹ | Ni grain growth and loss |  |
|  | 600°C | 160 h | $\sigma$: 1200 → 1100 S cm⁻¹ |  |  |
| Ni-YSZ | $H_2$ + 9% Ar + 4% $H_2O$, 1.9 A cm⁻², 850°C | 1500 h | $R_p$: 0.14 → 0.19 $\Omega$ cm² | $d_{Ni}$: 0.8 → 1.2 $\mu$m; A loss of Nickel | Hagen et al. (2006) |
| Ni-YSZ | $H_2$ + 3% $H_2O$, 0.3 A cm⁻², 850°C | 1800 h | $\Delta R_p$: 0.11 ± 0.01 $\Omega$ cm²/1000 h; $r_{degradation}$: 31 ± 3%/1000 h | Ni grain growth | Primdahl and Mogensen (2000) |
| Ni-YSZ | $H_2$ + 3% $H_2O$, 0.2 A cm⁻², 800°C | 250 h | $R_o$: 0.75 → 0.98 $\Omega$ cm²; $R_p$: 0.54 → 1.82 $\Omega$ cm² | $L_{TPB}$: 1.29 → 0.98 $\mu$m $\mu$m⁻³ | Jiao et al. (2012) |
| Ni-GDC (16 nm) | dry $H_2$/$N_2$, 560°C | 1600 h | $\sigma$: $10^5$ → $10^4$ S cm⁻¹ | Ni grain growth | Muecke et al. (2008) |
|  | 200–600°C cycling | 130 times | $\sigma$: $10^5$ → $10^3$ S cm⁻¹ | Ni grain growth, micro-cracks |  |
|  | Redox cycling | 25 cycles | $\sigma$: $2 \times 10^5$ → $4 \times 10^{-7}$ S cm⁻¹ | nickel segregation on the surface |  |
| Ni-TZ3Y | isothermal redox cycles, 850°C | 3 times | $CRS_{max}$, increased by 20 times | $\varepsilon$: 34% → 9% volumetric expansion of the Ni phase small Ni grain coarsening | Klemensø et al. (2005) |
| Ni-YSZ | RT–750°C thermal cycles, 800°C | 8 times | $\sigma$: 610 → 396 S cm⁻¹ | $d_{Ni}$: 0.86 → 1.74 $\mu$m; $\varepsilon$: 0.5% → 3.7% | Guan et al. (2011) |
| Ni-YSZ | isothermal redox cycles, 1000°C | 2 times | TEC: 12.4 → 10.33 × 10⁻⁶ K⁻¹ | Relative linear changes: 10.63% | Pihlatie et al. (2009a) |
| Ni-TZ3Y | isothermal redox cycles, 600°C | 3 times | $CRS_{max}$: 0 → 0.25 | $dL/L_o$: 0.31 | Pihlatie et al. (2009b) |
|  | 850°C |  | $CRS_{max}$: 0 → 1.55 | $dL/L_o$: 1.02 |  |
|  | 850°C, humid |  | $CRS_{max}$: 0 → 2.78 | $dL/L_o$: 0.97 |  |
|  | 1000°C |  | $CRS_{max}$: 0 → 3.25 |  |  |
| Ni-YSZ | isothermal redox cycles, 800°C | 3 times | $\eta_{cell}$: 0.57 → 0.47 V | Micro-cracks in the anode | Hatae et al. (2009) |
| Ni-SDC | 3.9% iso-octane + 96.1% air, 0.5 V, 600°C | 35 h | $P_{0.5V}$: 0.12 → 0 W cm⁻² | Heavy carbon deposition on Ni | Ding et al. (2008a) |
| Ni-YSZ | 0.6 V, 800°C $H_2$-$N_2$ + 50 ppm $H_2S$ | 120 h | i: 0.25 → 0.2 A cm⁻² | surface reconstruction of Ni | Zha et al. (2007) |
|  | 0.6 V, 800°C $H_2$-$N_2$ | 40 h | i: 0.2 → 0.24 A cm⁻² |  |  |
| Ni-YSZ | $H_2$ + 40 ppm $H_2S$ 0.4 A cm⁻², 700°C | 13 min | $\eta_{cell}$: 0.45 → 0 V |  | Kurokaw et al. (2007) |
| Ni-YSZ | $H_2$ + 50 ppm $H_2S$ 0.2 A cm⁻², 600°C | 15 h | $\eta_{cell}$: 0.5 → 0.2 V |  | Yun et al. (2011) |
| Ni-YSZ | moist $H_2$ + 1 ppm $H_2S$, 0.7 V, 800°C | 25 h | $ASR_{after}/ASR_{before}$: 1.4 |  | Marina et al. (2011) |

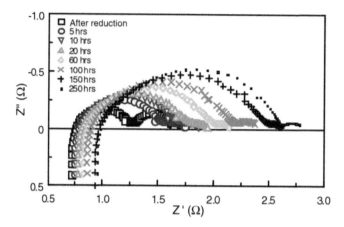

Figure 7.1.   AC impedance spectrum (under OCV) evolution at different discharge time (Jiao *et al.*, 2012).

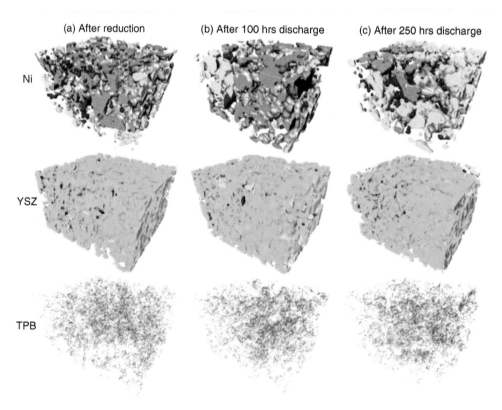

Figure 7.2.   Separated Ni, YSZ and TPB networks. Red: non-percolated particle, yellow: unknown-status particle (Jiao *et al.*, 2012).

reduction in mechanical strength (Jiang, 2003; Simwonis *et al.*, 2000), since the Ni components also served as strength supports in the conventional mixed anodes. The micro-structural changes were responsible for some local cracks in the anode body and delaminating defects at the interface between the anode and the electrolyte.

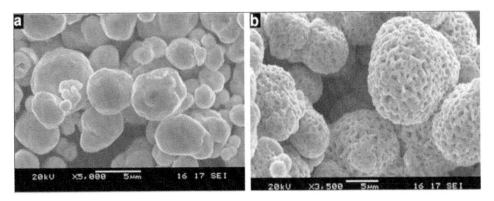

Figure 7.3.   SEM of (a) as received Ni and (b) fully oxidized Ni (NiO) particles. The secondary electron images were recorded using beam energy of 20 keV (Sarantaridis *et al.*, 2008).

### 7.2.1.2   *Redox cycling*

Before a SOFC operation, the NiO should undergo a reduction to Ni particles. As long as the fuel is supplied, the nickel remains in the reduced state. However, interruption of the fuel supply may occur accidently as a result of an error in the system control or intentionally upon system shutdown, when oxygen will diffuse to the anode causing the re-oxidation of Ni to NiO. Unfortunately, this reoxidation is not only associated with a volume expansion, but also with significant structural changes in the anode micro-structure, generating stresses in the anode. These stresses can exceed the stability limit of the components, potentially promoting crack growth, which leads to degradation of the SOFC or complete failure (Klemensø *et al.*, 2005; Malzbender *et al.*, 2005; Pihlatie *et al.*, 2009a,b; Waldbillig, 2005; Wood *et al.*, 2006). Klemensø *et al.* (2006) have observed micro-structural changes in situ during the redox cycling of nickel-3YSZ anodes. Rounding of the nickel grains was seen to proceed simultaneously with reduction which is associated with increased porosity and shrinkage, while grain growth was observed during the re-oxidation by expansion of the particles into the surrounding voids, which results in macro cracks by cumulative damage of residual stress. These micro-structural observations are consistent with the model of the redox mechanism (Klemensø *et al.*, 2006). Sarantaridis *et al.* (2008) have investigated the change in physical and mechanical properties of Ni-YSZ composites during the redox cycling at 800°C. Grain growth and formation of pore were observed (Fig. 7.3). Increased elastic modulus has been reported with the degree of the Ni oxidation. Some slight warping was observed after the intermediate steps of interrupted oxidation.

The anode-supported cell is more sensitive to the dimension stability of the anode, while thermo-mechanical calculations indicate that the anode substrate should not expand or shrink more than 0.2% in order to avoid cracking of the thin electrolyte layer (Malzbender *et al.*, 2005; Sarantaridis *et al.*, 2007). However, the first complete reoxidation of the typical Ni/YSZ anode substrate causes a linear expansion of about 1%, and subsequent redox cycles result in even more irreversible linear expansion due to the reorganization of Ni particles (Klemensø *et al.*, 2005; Malzbender *et al.*, 2005).

### 7.2.1.3   *Coking and sulfur poison*

The main problem preventing direct oxidation of hydrogen is the coke formation, because Ni is also an excellent catalyst for synthesizing carbon in addition to steam reforming. For example, a thermodynamic calculation predicts that carbon can be formed at 800°C and 1 atm unless the $H_2O:CH_4$ ratio is higher than 1.0 (Gorte *et al.*, 2003). The carbon covers the surface of Ni leading to a rapid reduction in electrochemical active sites, TPBs. Furthermore, carbon dissolution into the bulk metal can produce very high pressure causing materials to fracture. For example, the Ni cermet fractured completely after exposure to 100% $CH_4$ at 800°C for 1.5 h (Kim *et al.*, 2001; 2002).

Similar to coking poison, degradation of the electrodes are known to be caused by sulfur compounds, which are mixed as impurities or added purposely in the fuel. The nickel-based anodes have very limited tolerance to $H_2S$: significant poisoning was observed if the $H_2S$ concentration was above 5–10 ppm at 950–1000°C (Matsuzaki and Yasuda, 2000; Zha *et al.*, 2007). $H_2S$ is quickly dissociated into hydrogen and elemental sulfur, which strongly adsorbs on the nickel surface and blocks the active sites for the further oxidation of the fuel, leading to increased anode polarization. Nevertheless, the process is reversible when $H_2S$ stays in a concentration below the critical value at high temperatures within a short period of exposure. Once the nickel sulfides are formed, the poison effects are irreparable, leading to a permanent decreased performance.

### 7.2.2   *Aging mechanism of cathodes*

A cathode is the place where oxygen molecular is reduced to oxygen ions with the combination of electrons from the external circuit. Cathodic kinetics generally dominate the performance of the whole cell because of the high activation energy (often $> 1.5$ eV (Clarke and Levi, 2003)) for $O_2$ reduction compared with that for fuel oxidation. Consequently, cathode development with high performance and stability is rather critical for SOFCs, especially operated at intermediate temperatures. Similar to the anode, performance degradation related to the cathode is mostly caused by the following effects: (i) coarsening of the micro-structure due to sintering; (ii) decomposition of the cathode material; (iii) chemical reaction with electrolyte to form insulating phases at interfaces; (iv) spallation of the cathode. Any of them would result in a decrease of electronic or ionic conductivity, electrochemical active sites, or the porosity.

Strontium-doped lanthanum manganate (LSM) is one of the most classical cathode materials and the LSM and LSM-YSZ composite electrodes have been widely studied in the literature. It is generally accepted that LSM is chemically compatible with YSZ, its thermal expansion coefficient matches electrolytes, and it is stable under SOFC fabrication and operational conditions. However, the durability of the composite cathode still needs to be considered for the practical application of a SOFC, in which there has been unfortunately found some degradation during a lengthy operation. For example, Liu *et al.* (2009) have studied the LSM-YSZ/YSZ interface of anode-supported SOFCs, which were subjected to 1500 h testing at 750°C under electrical loading of $0.75 A\,cm^{-2}$. The LSM coverage and the TPB length were estimated to be reduced by 50 and 30% when the cell was tested in air, while the values correspond to 10 and 4% for the cell tested in oxygen. The loss of the LSM coverage and TPB length was possibly due to the formation of insulating zirconate phases, which present locally and preferably in LSM/YSZ electrolyte contact areas. Jérgensen *et al.* (2000) have also studied the durability of the LSM-YSZ composite cathodes. The electrode overpotential exceeded 100% of the initial value after 2000 h test keeping at constant $-300\,mA\,cm^{-2}$ at 1000°C in air. Pore formation near the cathode/electrolyte interface was observed and considered to be responsible for the cathode degradation. Performance of LSM-YSZ composite cathodes after lengthy test in other literatures are listed in Table 7.2, where cathodic degradation occurs due to various reasons, including grain growth, chemical reactions, and micro-structure changes induced by thermal, chemical and/or electrical attacks such as high current density, high local gradients in both electric potential and oxygen partial pressure.

Co-based perovskites are also developed as the cathode materials for intermediate temperature SOFCs based on ceria electrolytes. However, rapid cathode degradation could be induced by the large mismatch in the thermal expansion coefficients between the electro-catalyst and the electrolyte. The difference in the thermal behavior results in a strain which may break the connectivity of the interfaces, corresponding to a reduction of TPBs, further causing a decreased cathode performance. The misfit stain is proportional to the thermal expansion coefficient (TEC) difference between the two phases. Performance degradation becomes serious for larger thermal stress if the TEC of the electro-catalyst is much higher than that of the electrolyte. These common Co-based perovskites are unfortunately to have large TEC ($La_{0.6}Sr_{0.4}CoO_{3-\delta}$ (LSC): $23 \times 10^{-6}\,K^{-1}$ (Tsipis

Table 7.2.   Cathode degradation for LSM-YSZ composite electrodes.

| Test conditions | Test time | Change of performance | Change of composition | Change of micro-structure | Reference |
|---|---|---|---|---|---|
| 0.30 A cm², 1000°C | 2000 h | $\eta_{cathode}$: 0.025 → 0.08 V<br>$R_o$: 1.3 → 2.0 Ω cm²<br>$R_p$: 0.1 → 0.3 Ω cm² | Mn migration towards the grain boundaries of the YSZ electrolyte | Pore formation at the LSM-YSZ/YSZ interface<br>Densification of LSM-YSZ cathode | Jørgensen et al. (2000) |
| 0.75 A cm², 750°C | 1500 h | $\eta_{cell}$: 0.65 → 0.5 V<br>$\Delta ASR$: 0.15 Ω cm² | LaZrO₃ formation at the LSM/YSZ interface<br>Mn dissolution from LSM to the sintering neck of YSZ | Flattening and reduction of TPBs | Hagen et al. (2008) |
| 800°C | 500 h | $P_{0.7V}$: 0.62 → 0.32 W cm⁻²<br>$R_o$: 0.21 → 0.30 Ω cm²<br>$R_p$: 0.35 → 0.52 Ω cm² | Delamination defects at the LSM-YSZ/YSZ interface | Delamination defects at the LSM-YSZ/YSZ Interface<br>Coarsening and shrinkage of LSM Reduction of TPBs | Song et al. (2008) |
| 0.5 A cm², 800°C | 19000 h | $\eta_{cell}$: 3.3 → 3.05 V | Mn enhancement at the grain boundaries of YSZ | | Malzbendera et al. (2012) |
| 0.75 A cm², 750°C | 1500 h | $\eta_{cell}$: 0.63 → 0.5 V<br>$R_p$: 0.47 → 0.89 Ω cm² | | a small gap (<500 nm) between LSM-YSZ and YSZ interface | Hagen et al. (2006) |
| 0.75 A cm², 750°C | 1500 h | $R_{p,after}/R_{p,before} = 1.9$ | LaZrO₃ formation at the LSM/YSZ interface | LSM coverage decreased by 50%<br>Flattening of TPBs | Liu et al. (2009) |
| 800°C | 80 h | $R_p$: 0.5 → 0.9 Ω cm² | | YSZ phase contiguity: 1 → 0.44 | Song et al. (2009) |

and Kharton, 2009; Zhao *et al.*, 2008a), $Sm_{0.5}Sr_{0.5}CoO_{3-\delta}$ (SSC): $19.6 \times 10^{-6}$ K$^{-1}$ (Zhang and Xia, 2010), $La_{0.6}Sr_{0.4}Co_{0.2}Fe_{0.8}O_{3-\delta}$ (LSCF): $14.8 \times 10^{-6}$ K$^{-1}$ (Tietz *et al.*, 2006a; Tsipis and Kharton, 2008; Zhang and Xia, 2010), almost twice that of the doped-ceria electrolyte (SDC, GDC: $11$–$12 \times 10^{-6}$ K$^{-1}$ (Tsipis and Kharton, 2008; Zhao *et al.*, 2008a). According to Zhang's model (Zhang and Xia, 2010), performance degradation could be quantitatively predicted. For example, the polarization resistance of a LSC-SDC composite electrode increases from 1.2 to 1.5 $\Omega$ cm$^2$ after seven RT-800°C thermal cycles. The performance is further decreased to 2.4 $\Omega$ cm$^2$ after 25 thermal cycles.

## 7.3    STABILITY OF NANO-STRUCTURED ELECTRODES

### 7.3.1    *Fabrication and electrochemical properties of nano-structured electrodes*

A nano-structured electrode has an ideal micro-structure which exhibits high electrochemical performance due to high catalytic activities and rich reaction sites of the nano-catalysts with large specific area. The nano-sized electrodes are often prepared by two means. One method is using a simple mixture of the two phases in nano-size (Song *et al.*, 2008), or a nano-composite in the form of a surface coagglomeration like core-shell (Sato *et al.*, 2010), which is prepared by co-precipitation method using nano-sized particles as seed crystals. A colloidal suspension with nano-particles (such as YSZ) is mixed with the metal nitrate solution. The mixture is then added into $NH_4HCO_3$ to nucleate the constituent metal carbonates onto the surface of dispersed YSZ particles. The nano-composite particles are finally obtained after heating the precursor. The nano-composites exhibit excellent performance both in electrochemical properties and durability due to the improved phase contiguity and interfacial coherence. The other method, which is called as impregnation/infiltration (Jiang *et al.*, 2010), is a popular technique used to fabricate the nano-structured electrodes. The impregnation technique involves depositing nano-particles into a porous backbone, which is prepared in advance by co-firing with the electrolyte at high temperature, forming a strong and well-connected incorporation. Either the strong bonding between the backbones and the electrolyte or the good connection of the skeleton improves the mechanical properties and the stability of the interface. The nano-particles with the desired phase are deposited by thermal decomposition of the nitrate precursors at a temperature often much lower than that required for the traditional composite. The reaction between the cathode and the electrolyte could thus be avoided, enlarging the alternatives of electrodes with high electrochemical performance but poor chemical stability at high temperatures. The nano-scale characteristics of particles are beneficial to achieving high catalytic activity and large reaction sites. Many nano-structured electrodes have been reported to perform well with a much lower polarization resistance and/or higher power output. The promotion factors are listed in Table 7.3, which shows 2.3–78 for the $O_2$ reduction reaction, 1.6–25 for the $H_2$ oxidation reaction.

### 7.3.2    *Models about nano-structured effects on stability*

Although nano-structured electrodes have demonstrated high electrochemical performance, their long-term stability still needs to be considered for commercial applications, which require a cell life up to 50,000 h with very small degradation rates. At present, the nano-structure effects on long-term stability are complicated and have not reached an agreement due to the contradictory results, which are sensitive to particle size, electrode structure and/or other factors such as temperature, current density and thermal cycles. In principle, grain coarsening could be greatly promoted by reducing the particle to nanometer level since the nano-sized particles are proved to have significant surface diffusivity due to the relatively low melting points. It occurs spontaneously to reduce the free energy by reducing the surface area, which leads to a decreased cell performance, i.e. an increased polarization resistance or ohmic resistance due to the reduction of the particle connection and surface area. On the contrary, grain growth could be blocked

Table 7.3. Performance and promotion factors of various nano-structured electrodes.

| Impregnated nano-particles | Scaffold | Performance | Promotion factors, $f_p$ | Reference |
|---|---|---|---|---|
| **Cathode of SOFC** | | | | |
| PrBaCo$_2$O$_5$ (30 wt%) | SDC | $R_E = 0.08\,\Omega\,cm^2$ @600°C | | Wang et al. (2012) |
| SDC | Sr$_2$Fe$_{1.5}$Mo$_{0.5}$O$_6$ | $R_E = 0.11\,\Omega\,cm^2$ @750°C | | Zhang et al. (2011) |
| SSC | Sc$_{0.1}$Zr$_{0.89}$Ce$_{0.01}$O$_{2-x}$ | $R_E = 0.29\,\Omega\,cm^2$ @600°C | | Zhang et al. (2010) |
| LSM | YDB | $P = 1.13\,W\,cm^2$ @750°C | | Jiang et al. (2009c) |
| Ag | BaCe$_{0.8}$Sm$_{0.2}$O$_{2.9}$ | $R_E = 0.11\,\Omega\,cm^2$ @600°C | | Wu et al. (2010) |
| LSC | SDC | $P = 0.82\,W\,cm^2$ @750°C | | Zhao et al. (2009) |
| SSC | BaCe$_{0.8}$Sm$_{0.2}$O$_{2.9}$ | $R_E = 0.21\,\Omega\,cm^2$ @600°C | | Wu et al. (2009) |
| YDB (50 wt%) | LSM | $R_E = 1.08\,\Omega\,cm^2$ @600°C | | Jiang et al. (2008b) |
| GDC (5.8 mg cm$^{-2}$) | LSM | $R_E = 0.21\,\Omega\,cm^2$ @700°C | 56 for O$_2$ reduction | Jiang and Wang (2005) |
| Pd (1.8 mg cm$^{-2}$) | LSM/YSZ | $R_E = 0.9\,\Omega\,cm^2$ @600°C | 78 for O$_2$ reduction | Liang et al. (2008) |
| Pd (1.2 mg cm$^{-2}$) | LSCF | $R_E = 2.9\,\Omega\,cm^2$ @600°C | 1.9 for O$_2$ reduction | Chen et al. (2009) |
| GDC (1.5 mg cm$^{-2}$) | LSCF | $R_E = 1.6\,\Omega\,cm^2$ @600°C | 3.4 for O$_2$ reduction | Chen et al. (2009) |
| LSM ($\sim$2 mg cm$^{-2}$) | YSZ | $R_E = 1.6\,\Omega\,cm^2$ @600°C | 44 for O$_2$ reduction | Liang et al. (2008) |
| LSCF (1.1 mg cm$^{-2}$) | YSZ | $R_E = 0.54\,\Omega\,cm^2$ @600°C | | Chen et al. (2008) |
| LSCF (12.5 vol%) | GDC | $R_E = 0.25\,\Omega\,cm^2$ @600°C | 14 for O$_2$ reduction | Shah and Barnett (2008) |
| LSC (30 vol%) | YSZ | $P = 2.1\,W\,cm^2$ @800°C | | Armstrong and Rich (2006) |
| LSC (55 wt%) | SDC | $R_E = 0.36\,\Omega\,cm^2$ @600°C | | Zhao et al. (2009) |
| Ag | LSCF/GDC | $P = 0.98\,W\,cm^2$ @800°C | 3.3 for H$_2$/air | Liu et al. (2007) |
| Pd (1.4 mg cm$^{-2}$) | YSZ | $R_E = 0.22\,\Omega\,cm^2$ @700°C | | Liang et al. (2008) |
| BSCF (1.8 mg cm$^{-2}$) | LSM | $R_E = 1.3\,\Omega\,cm^2$ @700°C | 12 for O$_2$ reduction | Ai et al. (2010) |
| SSC | LSM/YSZ | $R_E = 8.5\,\Omega\,cm^2$ @600°C | 2.3 for O$_2$ reduction | Lu et al. (2006) |
| YSB | LSM | $R_E = 0.14\,\Omega\,cm^2$ @700°C | | Jiang et al. (2009b) |
| **Anode of SOFC** | | | | |
| SDC | Ni/SDC | $P = 0.75\,W\,cm^2$ @600°C | | Liu et al. (2011) |
| CeO$_2$ | Ni/SDC | $P = 0.77\,W\,cm^2$ @600°C | | Liu et al. (2011) |
| SDC (30 wt%) | Ni/SDC | $P = 0.72\,W\,cm^2$ @600°C | | Ding et al. (2008b) |
| SDC (20 mg cm$^{-2}$) | Ni/SDC | $P = 0.57\,W\,cm^2$ @600°C | | Zhu et al. (2006) |
| SDC (4 mg cm$^{-2}$) | Ni/YSZ | $R_E = 0.24\,\Omega\,cm^2$ @800°C | 7.3 for H$_2$ reduction | Jiang et al. (2002) |
| GDC (4 mg cm$^{-2}$) | LSCM | $R_E = 0.44\,\Omega\,cm^2$ @800°C | 26 for CH$_4$ reduction | Jiang et al. (2006) |
| GDC (4 mg cm$^{-2}$) | LSCM | $R_E = 0.12\,\Omega\,cm^2$ @800°C | 20 for H$_2$ reduction | Jiang et al. (2006) |
| GDC (1.42 mg cm$^{-2}$) | Ni | $R_E = 1.29\,\Omega\,cm^2$ @800°C | 25 for CH$_4$ reduction | Wang (2006) |
| Pd | LSCM/YSZ | $R_E = 0.88\,\Omega\,cm^2$ @800°C | 1.5 for H$_2$ reduction | Jiang et al. (2008a) |
| Pd | LCCM/YSZ | $R_E = 1.1\,\Omega\,cm^2$ @750°C | 6.5 for CH$_4$ reduction | Babaei et al. (2010) |
| Pd | Ni/GDC | $R_E = 0.6\,\Omega\,cm^2$ @700°C | 5 for H$_2$ reduction | Babaei et al. (2009) |
| Pd+CeO$_2$ | LSCM/YSZ | | 5 for H$_2$/air | Ye et al. (2008) |

in the unique structure of the nano-composites, which have high homogeneity that presents strong interaction, requiring higher energy to break the binding, making the motion of the nano-particles difficult and suppressing grain coarsening. In addition, the desired impregnated phases are formed at a relatively low temperature, which seems to minimize the potential of chemical reaction during the sintering process and enlarge the material selection for electrodes. However, nano-sized particles are also proved to be so active that the temperature requiring for chemical reaction of the interface also decreases, leading to the formation of an inert phase during the cell operation. Hence, the stability of the nano-structured electrodes is still a matter of considerable controversy.

### 7.3.2.1  *Nano-size effects on isothermal grain growth*

As mentioned above, the cell performance will be decreased by the grain growth, which results in increased ohmic and polarization resistances. Grain growth kinetics has been reported as being quite different for particles with the grain size in the micron- and nano-regime. According to the classical migration kinetics deduced by Burke and Turnbull (Burke, 1949; Burke and Tumbull, 1952), the isothermal grain growth for polycrystalline materials is given as:

$$d^n - d_0^n = k_n(t - t_0) \tag{7.1}$$

where $d$ denotes the average grain size at the time $t$, $d_0$ the initial grain size at time $t_0$, and $k_n$ is given by:

$$k_n = k_0 \cdot e^{-\frac{Q}{kT}} \propto D \tag{7.2}$$

where $k_0$ is a material constant, $Q$ the activation energy, $k$ the Boltzmann constant and $T$ the temperature in Kelvin, the term is proportional to the diffusion coefficient $D$. The grain growth exponent ranges from $n = 2$ up to $n = 10$, which depends on the temperature within a given poly-crystalline system (Rupp, 2006). The particles tend to grow continually with the annealing time, and the characteristic is commonly observed for micro-grains at higher temperatures. However, when the particles sizes are reduced to the nano-regime, the grain coarsening sometimes no longer obey the classical parabolic laws but a self-limited grain relaxation:

$$d - d_0 = (d_L - d_0)\left(1 - e^{-\frac{t}{\tau}}\right) \tag{7.3}$$

where $d_L$ is the limited grain size and $\tau$ is the relaxation time. The grain growth could occur by a relaxation of the micro-strain. The process is limited by defects in the grain boundaries, which exist in a high density in nano-crystalline. The nano-particles grow to a certain limited size in the first dwell, then suddenly slow down and maintain a limited level, finally resulting in a stable micro-structure regardless of the annealing time. However, the relaxation process only applies at low temperatures since the micro-strain of the nano-crystalline might fully relax if the temperature is sufficiently high to activate volume diffusion in addition to the grain boundaries' diffusion. In that case, the common grain growth kinetics dominates and results in a continual coarsening with time.

### 7.3.2.2  *Nano-structured effects on durability against thermal cycle*

The thermal cycle model has been proposed by Zhang *et al.* (2010) and matches well with the experimental data, Figure 7.4a. The performance will also be decreased by the mismatch of the two phases, which result in an increased polarization resistance due to the break of the interface between the electronic and ionic phases. When TEC of the electrolyte ($\alpha_i$) and the electro-catalyst ($\alpha_j$) are different, a misfit strain will be resulted from a temperature change, $\Delta T$:

$$\Delta \varepsilon = (\alpha_j - \alpha_i)\Delta T \tag{7.4}$$

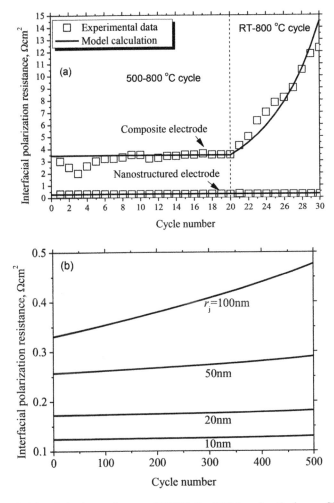

Figure 7.4. Interfacial polarization resistances (600°C) for LSC-based cathodes on SDC electrolytes. (a) Comparison between model prediction and experimental result for LSC-SDC composite cathode and nano-structured cathode in thermal cycle processes ($\Phi_i = \Phi_j = 0.5$, $\Phi_g = 0.41$, $r_i = 1\,\mu m$, and $r_j = 1\,\mu m$ for composite cathode; $r_i = 1\,\mu m$ and $r_j = 50\,nm$ for nano-structured cathode). (b) Effects of nano-particle size for LSC–SDC nano-structured cathode under thermal cycle treatment ($\Phi_i = \Phi_j = 0.5$, $\Phi_g = 0.41$, $r_i = 0.5\,\mu m$, and RT-800°C cycle) (Zhang and Xia, 2010).

At a stress balance state, the imposition of this misfit strain results in a couple of equal and opposite stresses, $\sigma_i$ in the electrolyte phase and $\sigma_j$ in the electro-catalyst phase, holding $\sigma_i = -\sigma_j = \sigma$:

$$\Delta\varepsilon = \sigma\left(\frac{1}{\tilde{E}_i} + \frac{1}{\tilde{E}_j}\right) \tag{7.5}$$

where $\tilde{E}_i$ and $\tilde{E}_j$ are the effective modulus of electrolyte and electro-catalyst phases. It is the stresses that break the interface, resulting in the reduction of TPB and the electrochemical degradation. Fortunately, the interface may survive, the survival probability, $P_{s,i-j}$, under positive

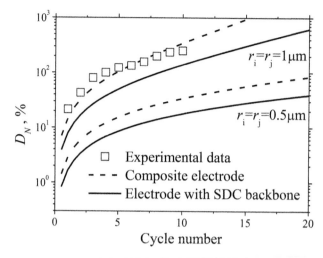

Figure 7.5.    Degradation rate, $D_N$, of LSC-SDC electrode with SDC backbone structure (the same structure as the nano-structured electrode) and composite LSC-SDC electrode as a function of RT-800°C cycle number (Zhang and Xia, 2010).

stress $\sigma$ can be estimated by the Weibull weakest-link theory (Bazant, 2004; Paramonv and Andersons, 2007):

$$P_{s,i-j}(\sigma) = \exp\left(-r_{i-j}^2\left(\frac{\sigma}{\Sigma_{i-j}}\right)^{m_{i-j}}\right) \tag{7.6}$$

where $\Sigma_{i-j}$ and $m_{i-j}$ are Weibull parameters corresponding to the interface between $i$ and $j$ phase, $r_{i-j}$ is the smaller particle radius of $i$ and $j$ phase, for instance, the size of the nano-particles in the case of impregnated electrodes. It should be suggested that higher survival probability is associated with smaller grain size, Figure 7.4. Figure 7.4b shows that durability increases substantially with the decrease of the impregnated particle size from 100 to 10 nm. The resistance has increased from 0.33 to 0.48 $\Omega$ cm$^2$ after 500 cycles with the degradation rate of 45% when the size is 100 nm; while only slight resistance increases from 0.12 to 0.13 $\Omega$ cm$^2$ with a much low degradation rate of 8.3% for 10 nm. Remarkable durable properties are also predicted for nano-structured electrodes, which have low degradation rates compared with the conventional composite, Figure 7.5, since the effect of the $i$–$i$ interface breaking on the continuity of the oxygen-ion conduction network is not as severe as that for composite cathodes in which the networks are formed by percolated electrolyte particles in randomly packing systems. The prediction has also been proven by experimental results from a SDC-LSC impregnated cathode, Figure 7.4a, Ni-YSZ anode, and the micro-tubular single cell (Zhang *et al.*, 2010).

## 7.4    LONG-TERM PERFORMANCE OF NANO-STRUCTURED ELECTRODES

### 7.4.1    *Anodes*

#### 7.4.1.1    *Enhanced interfacial stabilities of nano-structured anodes*
Nano-structured anodes are effective in improving mechanical stability by setting metal catalysts free from strength supports. The nano-structure is usually formed by depositing nano-particles into a porous pre-sintered backbone, which is co-fired with the dense electrolyte at a high temperature to achieve a good connectivity of the oxide phase, ensuring enough mechanical strength. The porous backbones are often fabricated with pore formers. NiO (Ni) is used as the pore former due to the well-developed anode-supported SOFC fabrication method. Taking the Ni-YSZ-based

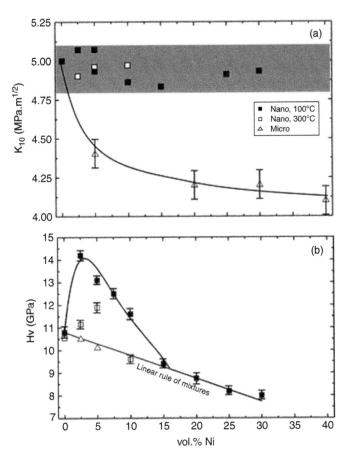

Figure 7.6.   Toughness (a) and Vickers hardness (b) of $ZrO_2$/Ni composites for increasing volume (López-Esteban *et al.*, 2006).

anode as an example, a NiO-YSZ composite, with a thin YSZ electrolyte, is first prepared in a normal ceramic process. After reduction of NiO, the Ni can be leached out of the substrate, using boiling nitric acid, leaving a porous YSZ network supporting the dense YSZ electrolyte layer. Normal pore formers, such as starch, graphite, carbon, and PMMA are also used. The YSZ powders are mixed with these pore formers using binders, then the mixtures are co-pressed or tape-casted with the YSZ electrolyte and sintered at high temperatures. During the sintering process, the binders and pore formers are burned out, forming dense YSZ electrolytes on porous YSZ backbones. As the backbone has the same material as the electrolyte, an additional effect should be mentioned: that residue stress caused by different coefficient of thermal expansion between the electrolyte and the anode substrate could be obviously reduced, which also leads to an improved stability of the interface.

Ni particles are formed in nano-size by the decomposition of nickel nitrate solution. Some unique interface properties are reported when the particle sizes are in nano-level (López-Esteban *et al.*, 2006). Toughness, as shown in Figure 7.6a, remains nearly constant for nano-composites while it decreases significantly in the case of micro-composites. Hardness as a function of Ni contents presented in Figure 7.6b also behaves quite differently. In micro-level (1–2 μm), a linear softening with increasing nickel content is observed, which corresponds to the rule of the mixture. However, in nano-size (10–100 nm), the hardness is at its highest with the Ni content at

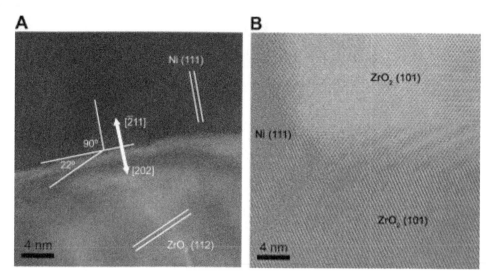

Figure 7.7.   HRTEM micrographs of zirconia/nickel interfaces found in (A) a nano-composite and (B) a micro-composite. In micrograph (A), from the angular orientation of (111) planes in Ni and (112) planes in $ZrO_2$ with respect to the interface, it has been assigned as $ZrO_2(002)/Ni(110)$ (Moya *et al.*, 2007).

2.5–5 vol% but decreases linearly when the concentration is above 15 vol%. The nano-composite hardness *versus* the nickel content is in good agreement with the theoretical model proposed by Pecharroman *et al.* (2004). The improved mechanical properties for the interface of nano-particles can be explained by HRTEM micrographs (Moya *et al.*, 2007). Nickel nano-particles in Figure 7.7a present a faceted aspect, with curved lines flanking sharp interfaces, and a good lattice matching between Ni and $ZrO_2$ taking into account a 45° rotation in one of the crystal. The density functional theory (DFT) calculation (Beltrán *et al.*, 2003; Pecharromán *et al.*, 2005) predicts that it needs 5000–5700 mJ m$^{-2}$ to separate the $ZrO_2(001)/Ni(001)$ interface. The energy is similar to the separation work of bulk Ni. Conversely, the $ZrO_2(011)/Ni(001)$ interface of the micro-composite shown in Figure 7.7b presented a separation work of only 900 mJ m$^{-2}$, which could be related to the poor atomic match explaining the decrease in the toughness.

Similar results about the enhanced interface force were found subsequently by Kim *et al.* (2006). They reported an improved durability of SOFC anodes against thermal cycling. As shown in Figure 7.8, the electrical conductivity of the anode, made from NiO-YSZ mixtures, decreases continuously with the progress of thermal cycling. After twenty thermal cycles, the electrical conductivity decreases from 1208.8 to 635.4 S cm$^{-1}$ (47% reduction). However, the electrical conductivity of the anode made from NiO-YSZ nano-composites shows little change during the twenty thermal cycling tests (10% reduction). These results exactly correspond with the crystal plane orientation variations of Ni-YSZ anodes depending on particle size. Sato *et al.* (2009) have introduced another long-stable nano-structured Ni-YSZ anode interdispersed with 160 nm NiO and 75 nm YSZ, which has a good interface connection. The polarization potential at 700°C under the constant current density of 0.2 A cm$^{-2}$ has maintained at 0.9 V for more than 900 h due to the spontaneous formation of a stable hetero-interface between Ni and YSZ grains suppressing the rearrangement and coalescence of Ni during the operation. NiO/YSZ composite anode (Fukui *et al.*, 2003) fabricated with nano-sized NiO grains partially or fully covered with fine YSZ grains also shows superior stability with constant anode polarization for 7200 h, which suggests that the Ni coarsening could be effectively prevented by nano-sized YSZ phase. In addition, it is interesting to observe that the interface reaction between anode support and dispersed Ni particles has been enhanced during the long-term operation. A depth of about 20–30 nm of the bottom part

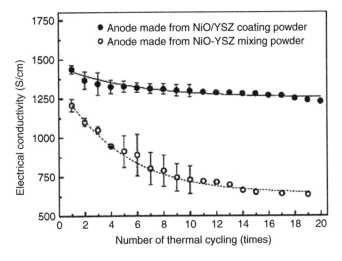

Figure 7.8.  Electrical conductivity variations of anodes depending on thermal cycling (Kim *et al.*, 2006).

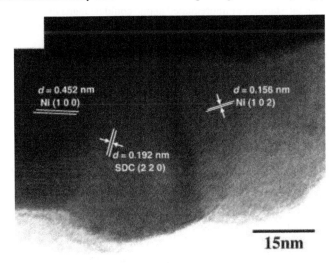

Figure 7.9.  High resolution TEM image of Ni-dispersed SDC anode after the long-term (1100 h) operation in SOFC (Suzuki *et al.*, 2006).

of the nickel (Fig. 7.9) is embedded into SDC and some small SDC particles are sandwiched between Ni particles after a long time test (Suzuki *et al.*, 2006). Such a structure can suppress the disconnection in an electronic network and can maintain the effective reaction area by keeping the number of active sites. The coating of thin a SDC layer at the bottom part of a pre-anchored Ni particle can also enhance the reaction zone, because $H_2$ can reach the Ni site through the thin SDC layer. Therefore, the unique interface reactions could improve the mechanical and electronic properties of the nano-structured anode by spontaneously forming a strong bonding with crystal orienting, leading further to high stability during the long-term tests of SOFCs.

### 7.4.1.2  *Durability of nano-structured anodes against redox cycle*
Ni infiltration is also used as a possible solution to the redox problem of SOFC. No dimensional changes are recorded after one redox cycle of an infiltrated 16 wt % Ni/YSZ composite has taken place. These cermets are thus dimensionally redox stable, which is attributed to the properties

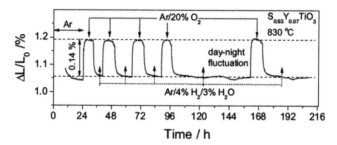

Figure 7.10.    Chemical expansion behavior of $Sr_{0.93}Y_{0.07}TiO_{3-d}$ ceramic (sintered in Ar/4% $H_2$ at 1 300°C for 10 h and then annealed in Ar at 1 100 °C for 3 h, porosity 15%) between Ar/4% $H_2$/3% $H_2O$ and Ar/20% $O_2$ at 830°C (Fu and Tietz, 2008).

of the micro-structure that is produced by infiltration. Nevertheless, Ni impregnation is a time-consuming issue and it sometimes needs at least ten cycles of impregnation-decomposition steps (Tucker *et al.*, 2007). Conductivity degradations attributed to Ni coarsening are still reported for Ni impregnated anodes in the literatures (Busawon *et al.*, 2008; Klemens *et al.*, 2010). Further improvements about the stability of impregnated anode could be made by the optimization of the micro-structure, introducing sintering inhibitors (Klemens *et al.*, 2010) and/or substituting other oxide materials.

Recently, perovskite-based electronic conductors with better redox stability have been reported as the anode, which are structurally stable under both oxidizing and reducing conditions for a large tolerance of oxygen, experiencing a slight dimensional change. Unfortunately, their electrochemical performance is usually very poor: a small amount of metallic catalyst is still needed to improve the performance to an acceptable level. A special structure, with nano-metal particles impregnating ceramic backbones, has been designed, in which the detrimental effect associated with the re-oxidation can be avoided since any dimensional change of the metals can be tolerated inside the pore structure of the anode and the rigid ceramic framework will remain intact. For example, a 5% Ni-impregnated $Sr_{0.93}Y_{0.07}TiO_{3-d}$ (SYT)/YSZ anode (Fu and Tietz, 2008) has shown a polarization resistance of 0.21 $\Omega\,cm^2$ at 800°C in wet Ar/5% $H_2$. The chemical linear expansion of SYT is 0.14% upon redox cycling between Ar/4% $H_2$/3% $H_2O$ and Ar/20% $O_2$ at 830°C (Fig. 7.10). In addition, the dimensional change is reversible during the five cycles. High stability is also observed for the anode with a similar structure by depositing the nano-sized Co-Fe alloy into $K_2NiF_4$-type structured $Pr_{0.8}Sr_{1.2}(Co,Fe)_{0.8}Nb_{0.2}O_{4+\delta}$ (K-PSCFN) matrix (Yang *et al.*, 2012). The cell output remains at 0.8 W $cm^{-2}$ and is not affected by the redox cycling of the anode during a total of twenty-six cyclic tests.

### 7.4.1.3    *Durability of nano-structured anodes against coking and sulfur poisoning*
Some progress has been made in improving the hydrocarbon and sulfur tolerance by altering their composition and micro-structure. One effective way is to prevent the direct contact with Ni and the fuel gas by introducing highly-performed catalyst, such as ceria-based oxides (Yun *et al.*, 2011; Zhu *et al.*, 2006). Kurokawa *et al.* (2007) have checked the stability of infiltrated and non-infiltrated cells. They found that under a constant current density of 0.4 A $cm^{-2}$ at 700°C, the non-infiltrated cell with the traditional Ni-YSZ anode deteriorated rapidly as its voltage quickly dropped from 0.45 V to 0 immediately after operating in 40 ppm $H_2S$ for 13 h. In contrast, the ceria infiltrated cell performed for 500 h with a rapid drop of voltage by 170 mV to 0.6 V in the initial 20 h, then there was a slow decrease from 0.6 V to 0.56 V during the following 278 h, finally maintaining this level for the remaining 200 h. Moreover, the voltage can return back quickly in about 12 min to 0.75 V, which is only 20 mV lower than the initial operating voltage when the testing environment was switched back from the 40 ppm $H_2S$ to $H_2$ (Fig. 7.11). It is possible that an infiltration of ceria nano-particulate layer prevents the deleterious formation of

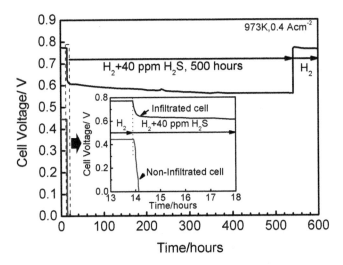

Figure 7.11. Cell voltage as a function of time for cells exposed to 40 ppm H₂S at 700°C (Kurokaw *et al.*, 2007).

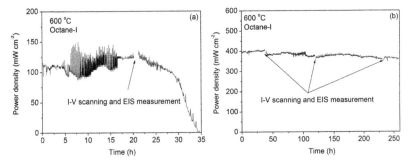

Figure 7.12. Power density evolution with operation time for (a) conventional anodes, (b) SDC coated Ni-SDC anode (Ding *et al.*, 2008a).

$Ni_3S_2$, increasing sulfur tolerance of Ni-YSZ drastically. Furthermore, the thin ceria coating layer was also helped in reducing Ni coarsening. The doped ceria coated Ni-SDC anode also helps to hinder carbon deposition. For example, Zhu *et al.* (2006) have reported that the 25 mg cm$^{-2}$ SDC impregnated nickel anode could be operated at 600°C for 50 h without obvious degradation when methane is directly used as the fuel. Ding *et al.* (2008a) have further found that this modified anode could improve the anode stability even using fuels with higher C/H ratio, such as iso-octane. Figure 7.12 shows that, when there is a power density evolution within a timed operation, the output at 600°C drops rapidly from initial 100 to 0 mW cm$^{-2}$ within 35 h for the conventional anode because of the heavy carbon deposition on the Ni catalysts as can be observed in Figure 7.13a. The degradation is remarkably delayed for the nano-structured anode. Its power density is initially about 387 mW cm$^{-2}$ and remains at 360 mW cm$^{-2}$ after more than 260 h of operation. The improved tolerance against carbon formation is attributed to the enhanced electrochemical activity of the anode for the continuous oxygen-ion conduction with the addition of nano-sized SDCs, which would help to remove possible deposited carbon.

Another approach to solving the problems is replacing the Ni in the anode with a much less catalytically active metal, such as Cu, or metal alloy (Kim *et al.*, 2001; Park *et al.*, 2000). Cu is an excellent electronic conductor but a poor catalyst for CC bond activation. Therefore, carbon formation, which troubles the use of Ni cermet, is avoided bythe replacement of Cu.

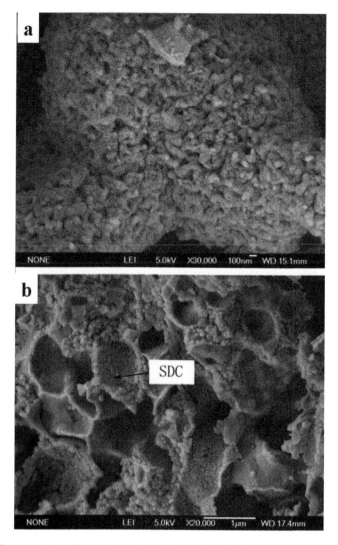

Figure 7.13.   Cross-sectional SEM micrographs of the anodes after lengthy testing, (a) the conventional anode, (b) nano-SDC coated anode (Ding *et al.*, 2008a).

However, copper and its oxides have low melting temperatures. They are not allowed to fabricate the Cu cermets like conventional Ni/YSZ anodes, which require high sintered temperatures in excess of 1300°C in order to achieve good connectivity of YSZ components. More seriously, Cu ions are also observed to migrate into YSZ following high temperature calcination. A novel fabrication method is developed by impregnating copper nitrate solution into a pre-sintered YSZ scaffold. Well-connected pathways are established by nano-sized Cu particles, which are obtained by the decomposition at 450°C in reduced conditions. Ceria is also added for its high activity in hydrocarbon oxidation and its high ionic conductivity. The nano-structured Cu-ceria anode has exhibited very high sulfur tolerance and can operate on fuels with sulfur contents up to 400 ppm without significant loss in performance (He *et al.*, 2005). It has also showed the direct electrochemical oxidation of various hydrocarbons ($C_4H_{10}$, $CH_4$) (Ding *et al.*, 2008a; Park *et al.*, 2000; Zhu *et al.*, 2006).

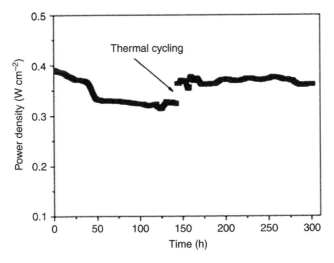

Figure 7.14.   Power output at 0.7 V and 700°C as a function of time for the cell with LSM infiltrated YSB backbone as the cathode (Jiang *et al.*, 2009a).

### 7.4.2   *Cathodes*

#### 7.4.2.1   *LSM*

LSMs are novel cathodes used on various electrolytes for their good thermal and chemical compatibility. However, the oxygen reduction on LSM cathodes only occurs near the TPBs at the electrode-electrolyte interface due to its low ionic conductivity. Performance of the LSM-based cathodes could be improved by using a nano-structure which provides large active sites for electrochemical reactions. In addition, the nano-structured LSM cathodes exhibits long-term stability because of the unique nano-sized effects, as mentioned above. For instance, Jiang *et al.* (2009a) have investigated the durable performance of the LSM impregnated yttria-stabilized bismuth oxide (YSB) cathodes. The power density declines from 0.39 to 0.33 W cm$^{-2}$ in the initial 50 h, but is then gradually stabilized in the subsequent 100 h, after that the cell is operated under a 700°C-to-room temperature thermal cycle and the power density has increased to 0.36 W cm$^{-2}$ and maintained relatively stable for over 150 h (Fig. 7.14). From the SEM images shown in Figure 7.15b, the initial power decline might be attributed to the coarsening of LSM, which is likely to be inevitable because of the high sinterability of small particles. However, the particle growth seems to be self-limited due to the size effect, and contact between the nano-sized LSM and the YSB pore walls tends to increase after a 300 h operation, as a result of the secondary sintering process. Song *et al.* (2009) have investigated the durable performance of a well-engineered nano-composite, where fine LSM and YSZ particles were mixed in the form of co-agglomeration. The polarization resistance of the traditional LSM-YSZ composite was increased by 463% after 80 h of cyclic current stress, while it remained unchanged for the novel nano-structured composite cathodes due to the improved phase contiguity and interfacial coherence, which was also proven to be beneficial to suppress the micro-structural damage, including the delamination defects at the cathode/electrolyte interface, or void in the local cathode. This is consistent with the model prediction that smaller particles correspond to higher mechanical stability (Zhang and Xia, 2010). The volume fraction of the conjugated phase plays a critical role in the optimization of the cathode micro-structure for maintaining electrode durability, while still exhibiting good cell performance. Theoretically, the percolation threshold depends on the particle size ratio as well as the thermal history. Different ratios of surface-conjugated YSZ to YSZ core particles have been investigated with regard to the composition effects. It has been shown that a long-term performance with

Figure 7.15.    Cross-sectional images of (a) as-prepared LSM infiltrated YSB backbone, and (b) LSM infiltrated YSB backbone cathode after 300 h testing (Jiang *et al.*, 2009a).

500 h stable power export of about 549 mW cm$^{-2}$ during the accelerated life time test was maintained for the firm YSZ skeleton, fabricated by using a higher content of conjugated YSZ, which is suitable for mechanical and electrochemical long-term performance. Another homogeneous LSM-YSZ nano-composite, fabricated by a co-precipitation method using 3 nm YSZ particles as seed crystals, isalso reported to present high long-term stability (Sato *et al.*, 2010). No significant performance degradation of the cathode has been observed during an operation at 700°C for 1000 h under a constant current density of 0.2 A cm$^{-2}$.

### 7.4.2.2    *LSC*

LSCs are attractive cathode materials for intermediate temperature SOFCs in terms of its high catalytic activity and surface exchange coefficient. However, the material also has a large coefficient of thermal expansion, which is about $23 \times 10^{-6}$ K$^{-1}$, almost twice that of typical electrolytes, such as YSZ and doped ceria electrolytes ($11$–$12 \times 10^{-6}$ K$^{-1}$). The TEC mismatch between the electrolyte and cathode will result in a thermal stress, which is responsible for delamination at the

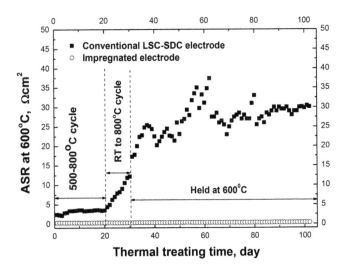

Figure 7.16.   Area specific resistance (ASR) at 600°C for the impregnated electrode and the conventional LSC–SDC electrode upon thermal treatment (Zhao *et al.*, 2008a).

cathode/electrolyte interface, and/or cracking of the electrolyte during the operation (Chen *et al.*, 2005). In addition, LSC reacts readily with YSZ to form insulated $La_2Zr_2O_7$ and $SrZrO_3$ phases at the temperature above 1000°C (Künga *et al.*, 2010; Yamamoto *et al.*, 1987). Thus, the material could not be used directly with YSZ at high temperatures.

Much progress has been made to adjust the thermal compatibility between the LSC electrode and the electrolyte both in compositions by Fe and/or Ni doping or in micro-structure. A specific architecture of LSC-based composite are designed by depositing nano-LSC particles into a porous matrix, which is strongly bonded with the electrolyte by co-sintering at a high temperature. Thermal stress is greatly released since the electrolyte is usually made from the same material as the matrix, which dominates the thermal properties of the infiltrated cathodes (Huang *et al.*, 2004). For example, the TEC of 55 wt% LSC impregnated YSZ is $12.6 \times 10^{-6} K^{-1}$, which is close to that of the YSZ electrolyte and much lower than that of pure LSC. Thus, a high resistance to thermal shock is expected for the impregnated cathodes. Zhao *et al.* (2008a) have investigated the stability of the impregnated cathode upon thermal treatments by measuring the change of the area's specific resistance for symmetrical cells. Properties of the conventional LSC-SDC composite cathode were also included for comparison. The polarization resistance at 600°C of the conventional composite electrode was increased from the initial 2.42 $\Omega$ cm$^2$ to 3.5 $\Omega$ cm$^2$ after twenty 500–800°C thermal cycles and further to 12.2 $\Omega$ cm$^2$ after ten room-temperature-to-800°C thermal cycles, finally reaching up to about 30 $\Omega$ cm$^2$ after more than 2000 h isothermal treatment at 600°C. However, the resistance of the impregnated cathode was maintained at 0.25 $\Omega$ cm$^2$ over one hundred days and no obvious degradation was observed. The results indicate that excellent thermal cycle durability can be achieved using nano-structured composite cathodes with strong incorporate substrates.

Additionally, the stability of the impregnated cathode is greatly related to the nano-particles, which are sensitive to the chemical compatibility and grain coarsening. According to Peters' thermo-dynamical modeling for nano-particles (Peters, 2008), a chemical reaction between nano-sized LSCs and YSZ would occur when the temperature was down to 500°C, much lower than the 900°C needed for conventional powder (Labrincha *et al.*, 1993), as evidenced by XRD. Figure 7.17 represents the impedance spectra as a function of time for 30 wt% LSCs in YSZ at 700°C in air (Hjalmarsson *et al.*, 2012; Huang *et al.*, 2004). The total polarization resistance increases from 2.57 to 2.73 $\Omega$ cm$^2$ during the 250 h test, corresponding to 0.12 $\Omega$ cm$^2$ for ohmic

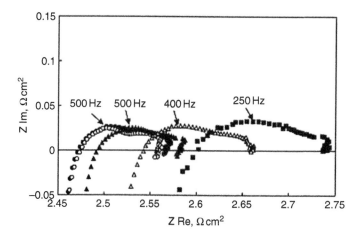

Figure 7.17.    Impedance spectra from symmetric cells, measured in air at 700°C as a function of time, using 30 wt% LSCo in YSZ as electrodes. The measurements were taken after the following times: •2, ∘ 5.5, ▲ 24, △ 72, and ■ 250 h (Huang *et al.*, 2004).

resistance and $0.04\,\Omega\,cm^2$ for ASR. It should be noticed that the increment of ohmic resistance plays the dominant role in the cell degradation, which is attributed to the formation of insulating phases (such as $SrZrO_3$) at the interface between Co-based perovskites and YSZ matrix. Similar results are also reported by Hjalmarsson *et al.* (2012), who have investigated the electrochemical performance and stability of nano-sized LSCs. Therefore, a layer of dense doped-ceria is required as a barrier, preventing solid-state reactions, if the cell is to be operated at high temperature (Künga *et al.*, 2010). Besides chemical compatibility, grain growth, and a degree of phase formation are also needed to be considered for a nano-structured composite electrode. The stability changes a lot with different operating temperatures even with the same cathode system, since the positive effect of LSC formation could be counteracted by grain coarsening, resulting in the confused relationship between cell stability and operating temperature.

### 7.4.2.3   *LSCF*

Mixed ionic and electronic conducted perovskites $La_{0.6}Sr_{0.4}Co_{0.2}Fe_{0.8}O_{3-d}$ have been extensively investigated as cathode materials for intermediate temperature SOFCs, which have successfully been developed giving reproducibly a power output of $1.0–1.2\,W\,cm^{-2}$ at 800°C and 0.7 V with hydrogen as the fuel (Mai *et al.*, 2005). These high power densities allow a reduction in operating temperature of about 100°C by maintaining the performance at the same level with LSM cathodes. However, a higher loss in performance has been reported when the cell with the conventional LSCF cathode revealed a degradation of about 0.5∼1.5%/1000 h during the 3000 h long-term operation at 750°C with the current load of $0.5\,A\,cm^{-2}$ (Tietz *et al.*, 2006b). Further investigations were conducted to reveal the possible reason for this and thus improve its long-term performance. For example, Shah *et al.* (2008) reported a stable nano-structured cathode by infiltration of LSCF into GDC backbones. The polarization resistance at 600°C was initially as low as $0.24\,\Omega\,cm^2$, decreased to $0.15\,\Omega\,cm^2$ after 75 h, and then was maintained at this level for more than 200 h. SEM images have demonstrated no obvious changes in micro-structures. However, the stability of the nano-structured cathode becomes a problem when it is operated at higher temperatures. Figure 7.18 shows the time dependence of the polarization resistance at different aging temperatures for the nano-structured LSCF electrodes. The polarization resistance increased dramatically at higher temperatures due to the structural changes.

A mathematical model was developed to predict the interfacial polarization resistance, $R_p$. According to the coarsening theory, the characteristic length $L$ evolves with time $t$ in the form of

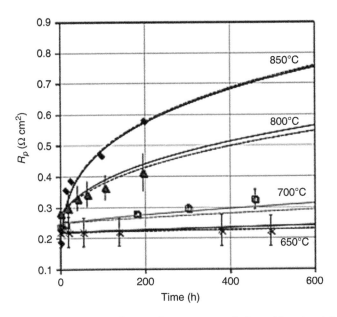

Figure 7.18.  The experimental data for different aging temperatures is shown (discrete points) along with the results from the fit of Equation (7.8), using $n = 4$ (continues lines). The dashed lines represent the fit from $Q = 2.86$ eV and $B = 1 \times 10^6$ $(\Omega\,cm^2)^4\,s^{-1}$, and the solid lines represents the fit from $Q = 2.52$ eV and $B = 3 \times 10^4$ $(\Omega\,cm^2)^4\,s^{-1}$ (Shah *et al.*, 2011).

a power law as Equation (7.1). Assuming that the atoms diffuse along the surface, $n$ has a value of 4, and resistivity of a specific interface, $R_s$, is a constant, $R_p$ could be explained in the following equation:

$$R_p \propto R_s \cdot L \qquad (7.7)$$

Then, the time-dependent polarization resistance with the micro-structure could be written in the form by combining Equations (7.1), (7.2) and (7.7) as:

$$R_p = (B \exp(-Q/kT)t + R_{p0}^n)^{1/n} \qquad (7.8)$$

where $R_{p0}$ is the initial polarization resistance, $B$ is the fitting parameter. Equation (7.8) could be used to predict the degradation rate over longer times, which is not easy to discover in experiments. The resistance is predicted to increase from a starting value of 0.2 to 0.28–0.35 after 40,000 h. Nonetheless, the above predictions are only valid if we assume that the cell degradation is caused by the grain growth. A number of other degradation mechanisms are possible such as Sr segregation to the surface of LSCF (Simner *et al.*, 2006) or insulating zirconate phase formation when using YSZ electrolytes (Mai *et al.*, 2006).

### 7.4.2.4  SSC
SSCs are also mixconducted perovskites by substituting the A-site La with Sm. The doping leads to great enhancement of the surface exchange kinetics, which has been proved to dominate the oxygen reduction process. The surface exchange coefficient of SSC is higher than that of the LSC, implying a faster surface exchange process. However, like other Co-based perovskites, it suffers from the drawback of the chemical reaction with YSZ and the thermal expansion coefficient mismatch with the electrolytes.

Novel nano-structured cathodes with high performance and stability are fabricated with SSC. For example, the polarization resistance at 700°C was reduced from 0.103 $\Omega\,cm^2$ for a blank porous LSCF electrode to 0.071 $\Omega\,cm^2$ with the impregnation of a thin LSCF layer, due to an

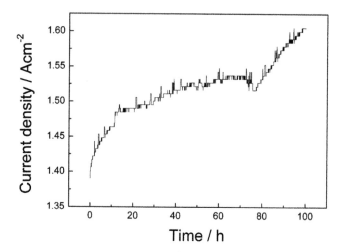

Figure 7.19.   Current density *versus* operating time with a constant voltage output of 0.5 V (Zhao *et al.*, 2008b).

increase in surface area from the LSCF nano-particles. It continually decreased to $0.036 \, \Omega \, cm^2$ with SSC nano-particles for an enhancement of the surface catalytic properties. In particular, the coating of SSC maintained the stability of LSCF cathodes over a 100 h operation with no evidence of the particles agglomerating (Lou *et al.*, 2009). As shown with SEM images (Zhao *et al.*, 2008b), nano-network structured cathodes consisting of SSC nano-wires in the form of aggregation of 5–8 nano-beads with the diameter of 50 nm also showed the highest performance among the SSC materials ever reported. The interfacial polarization resistance is only $0.21 \, \Omega \, cm^2$ at 500°C, and the peak power density of an anode-supported cell with 10 μm thick SDC electrolytes has reached a high value of $0.44 \, W \, cm^{-2}$ because of the high surface area, strong particle-particle connection and well developed porosity for oxygen reduction. Furthermore, it is also remarkable to notice that the current density under a constant voltage of 0.5 V at 600°C increased from 1.40 to $1.60 \, A \, cm^{-2}$ in 100 h, Figure 7.19, indicating no degradation in fuel cell operation. The increased performance for the nano-structured cathode is probably due to the cathode micro-structure's evolution, Figure 7.20, with the well-connected SSC nano-wires to strengthen the ionic and electronic conducting path or the higher porosity to optimize the gas diffusion, which has been proven to be accelerated from the reduction of the resistance of the low frequency in the impedance spectra, Figure 7.21.

## 7.5   SUMMARY

The aging of SOFC electrodes contributes significantly to the decrease of long-term performance. In the case of the conventional Ni-based anodes, the degradation is mainly caused by the micro-structure change during the thermal and redox cycling or the catalyst poisoning in the presence of hydrocarbon or $H_2S$. While the cathode durability is limited by the chemical incompatibility at the interface, the residue stress resulted from the TEC mismatch and grain growth, or the composition change related to the test conditions, including the temperature and current densities. The nano-structured electrodes are desirable alternatives for intermediate temperature SOFCs because of the enhanced electrochemical performance from the increased TPB sites. The unique structure is also beneficial to the cell's durability due to the strong bonding between nano-particles and the flexible fabrication, which could be easily designed to avoid high-temperature treatments of components with low melting points, or set the catalysts free in the composites. Although the

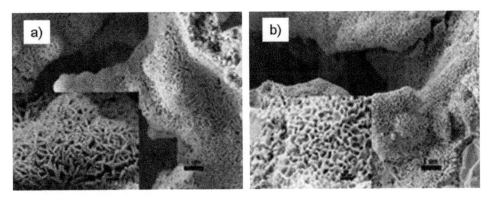

Figure 7.20.   Cross-sectional micro-structure views for the cathode, (a) as-prepared (b) after the durability test (Zhao *et al.*, 2008b).

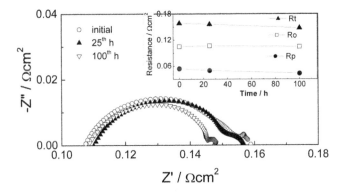

Figure 7.21.   Impedance spectra measured at 600°C at different testing time (Zhao *et al.*, 2008b).

nano-sized structure is not stable in terms of thermal-dynamics, the degradation could be delayed by controlling the proper range of grain size and reasonable operation conditions. At their best, the nano-structured electrodes have great potential for practical application, whilst providing enhanced electrochemical and durable properties.

## ACKNOWLEDGEMENTS

Financial support from the Ministry of Science and Technology of China (2012CB215403) is greatly appreciated.

## REFERENCES

Ai, N., Jiang, S.P., Lu, Z., Chen, K.F. & Su, W.H.: Nanostructured $(Ba,Sr)(Co,Fe)O_{3-\delta}$ impregnated $(La,Sr)MnO_3$ cathode for intermediate-temperature solid oxide fuel-cells. *J. Electroch. Soc.* 157 (2010), pp. B1033–B1039.
Armstrong, T.J. & Rich, J.G.: Anode-supported solid oxide fuel-cells with $La_{0.6}Sr_{0.4}CoO_{3-\delta}$–$Zr_{0.84}Y_{0.16}O_{2-\delta}$ composite cathodes fabricated by an infiltration method. *J. Electrochem. Soc.* 153 (2006), pp. A515–A520.
Babaei, A., Jiang, S.P. & Li, J.: Electrocatalytic promotion of palladium nanoparticles on hydrogen oxidation on Ni/GDC anodes of SOFCs via spillover. *J. Electrochem. Soc.* 156 (2009), pp. B1022–B1029.

Babaei, A., Zhang, L., Tan, S.L. & Jiang. S.P.: Pd-promoted (La, Ca)(Cr,Mn)O$_3$/GDC anode for hydrogen and methane oxidation reactions of solid oxide fuel-cells. *Solid State Ionics* 181 (2010), pp. 1221–1228.

Bazant, Z.P.: Probability distribution of energetic-statistical size effect in quasibrittle fracture. *Probabilist. Eng. Mech.*19 (2004), pp. 307–319.

Beltrán, J.I., Gallego, S., Cerdá, J., Moya, J.S. & Muòoz, M.C.: Bond formation at the Ni/ZrO$_2$ interface. *Phys. Rev.* B 68 (2003), pp. 075401–075405.

Burke, J.E.: Some factors affecting the rate of grain growth in metals. *Transactions of the American Institute of Mining, Metallurgical and Petroleum Engineers* 180 (1949), pp. 73–91.

Burke, J.E. & Turnbull, D.: Recrystallization and grain growth. *Prog. Met. Phys.* 3 (1952), pp. 220–224.

Brown, M., Primdahl, S. & Mogensen, M.: Structure/performance relations for Ni/yttria-stabilized zirconia anodes for solid oxide fuel-cells. *J. Electrochem. Soc.* 147 (2000), pp. A475– A485.

Busawon, A.N., Sarantaridis, D. & Atkinson, A.: Ni infiltration as a possible solution to the redox problem of SOFC anodes. *Electrochem. Solid-State Lett.* 11 (2008), pp. B186–B189.

Chen, F.L., Chen, J., Cheng, J.L., Jiang, S.P., He, T.M., Pu, J. & Li. J.: Novel nano-structured Pd+yttrium doped ZrO$_2$ cathodes for intermediate temperature solid oxide fuel-cells. *Electrochem. Commun.* 10 (2008), pp. 42–46.

Chen, J., Liang, F.L., Liu, L.N., Jiang, S.P., Chi, B., Pu, J. & Li, J.: Nanostructured (La,Sr)(Co,Fe)O$_3$+YSZ composite cathodes for intermediate temperature solid oxide fuel-cells. *J. Power Sources* 183 (2008), pp. 586–589.

Chen, J., Liang, F.L., Chi, B., Pu, J., Jiang, S.P. & Jian, L.: Palladium and ceria infiltrated La$_{0.8}$Sr$_{0.2}$Co$_{0.5}$Fe$_{0.5}$O$_{3-\delta}$ cathodes of solid oxide fuel-cells. *J. Power Sources* 194 (2009), pp. 275–280.

Chen, K.F., Lü, Z., Ai, N., Chen, X.J., Hua, J.Y., Huang, X.Q. & Su, W.H.: Effect of SDC-impregnated LSM cathodes on the performance of anode-supported YSZ films for SOFCs. *J. Power Sources* 167 (2007), pp. 84–89.

Chen, X., Yu, J. & Adler, S.B.: Thermal and chemical expansion of Sr-doped lanthanum cobalt oxide (La$_{1-x}$Sr$_x$CoO$_{3-\delta}$). Chemistry of Materials 17 (2005), pp. 4537–4546.

Clarke, D.R. & Levi, C.G.: Materials design for the next generation thermal barrier coating. *Annual Rev. Mater. Res.* 33 (2003), pp. 361–382.

Dees, D.W., Claar, T.D., Easler, T.E., Fee, D.C. & Mrazek, F.C.: Conductivity of porous Ni/ZrO$_2$-Y$_2$O$_3$ cermets. *J. Electrochem. Soc.* 134 (1987), pp. 2141–2146.

Ding, D., Liu, Z.B., Li, L. & Xia, C.R.: An octane-fueled low temperature solid oxide fuel-cell with Ru-free anodes. *Electrochem. Commun.* 10 (2008a), pp. 1295–1298.

Ding, D., Zhu, W., Gao, J.F. & Xia, C.R.: High performance electrolyte-coated anodes for low-temperature solid oxide fuel-cells: Model and experiments. *J. Power Sources* 179 (2008b), pp. 177–185.

Fan, Z.: A microstructural approach to the effective transport properties of multiphase composites. *Philosoph. Mag.* A 73 (1996), pp. 1663–1684.

Fan, Z., Miodownik, A.P. & Tsakiropoulos, P.: Microstructural characterisation of two phase materials. *Mater. Sci. Technol.* 12 (1993), pp. 1094–1100.

Fu, Q.X. & Tietz, F.: Ceramic-based anode materials for improved redox cycling of solid oxide fuel-cells. *Fuel-Cells* 8 (2008), pp. 283–293.

Fu, Q.X., Tietz, F., Sebold, D., Tao, S.W. & Irvine, J.T.S.: An efficient ceramic-based anode for solid oxide fuel-cells. *J. Power Sources* 171 (2007), pp. 663–669.

Fukui, T., Ohara, S., Naito, M. & Nogi, K.: Performance and stability of SOFC anode fabricated from NiO/YSZ composite particles. *J. Europ. Ceramic Soc.* 23 (2003), pp. 2963–2967.

Gorte, R.J. & Vohs, J.M.: Novel SOFC anodes for the direct electrochemical oxidation of hydrocarbons. *J. Catal.* 216 (2003), pp. 477–486.

Guan, Y., Gong, Y.H. & Li, W.J.: Quantitative analysis of micro structural and conductivity evolution of Ni-YSZ anodes during thermal cycling based on nano-computed tomography. *J. Power Sources* 196 (2011), pp. 10,601–10,605.

Hagen, A., Barfod, R., Hendriksen, P.V., Liu, Y.-L. & Ramousse, S.: Degradation of anode supported SOFCs as a function of temperature and current load. *J. Electrochem. Soc.* 153 (2006), pp. A1165–A1171.

Hagen, A., Liu, Y.L., Barfod, R. & Hendriksen. P.V.: Assessment of the cathode contribution to the degradation of anode-supported solid oxide fuel-cells. *J. Electrochem. Soc.* 155 (2008), pp. B1047–B1052.

Hatae, T., Matsuzaki, Y., Yamashita, S. & Yamazakib, Y.: Initial damage to anode microstructure caused by partial redox cycles during electrochemical oxidation. *J. Electrochem. Soc.* 156 (2009), pp. B609–B613.

He, H., Gorte, R.J. & Vohs, J.M.: Highly sulfur tolerant Cu-ceria anodes for SOFCs. *Electrochem. Solid-State Lett.* 8 (2005), pp. A279–A280.

Hjalmarsson, P., Hallinder, J. & Mogensen, M.: Electrochemical performance and stability of nano-particulate and bi-continuous $La_{1-x}Sr_xCoO_3$ and $Ce_{0.9}Gd_{0.1}O_{1.95}$ composite electrodes. *J. Solid State Electrochem.* 16 (2012), pp. 2759–2766.

Huang, Y.Y., Ahn, K., Vohs, J.M. & Gorte, R.J.: Characterization of Sr-doped $LaCoO_3$-YSZ composites prepared by impregnation methods. *J. Electrochem. Soc.* 151 (2004), pp. A1592–A1597.

Huang, Y.Y., Vohs, J.M. & Gorte, R.J.: Characterization of LSM-YSZ composites prepared by impregnation methods. *J. Electrochem. Soc.* 152 (2005), pp. A1347–A1353.

Jiang, S.P.: Sintering behavior of $Ni/Y_2O_3$-$ZrO_2$ cermet electrodes of solid oxide fuel-cells. *J. Mater. Sci.* 38 (2003), pp. 3775–3782.

Jiang, S.P. & Wang, W.: Fabrication and performance of GDC-impregnated $(La,Sr)MnO_3$ cathodes for intermediate temperature solid oxide fuel-cells. *J. Electrochem. Soc.* 152 (2005), pp. A1398–A1408.

Jiang, S.P., Duan, Y.Y. & Love, J.G.: Fabrication of high-performance $NiO/Y_2O_3$-$ZrO_2$ cermet anodes of solid oxide fuel-cells by ion impregnation. *J. Electrochem. Soc.* 149 (2002), pp. A1175–A1183.

Jiang, S.P., Zhang, S., Zhen, Y.D. & Wang, W.: Fabrication and performance of impregnated Ni anodes of solid oxide fuel-cells. *J. Amer. Ceramic Soc.* 88 (2005), pp. 1779–1785.

Jiang, S.P., Chen, X.J., Chan, S.H. & Kwok, J.T.: GDC-impregnated $(La_{0.75}Sr_{0.25})(Cr_{0.5}Mn_{0.5})O_3$ anodes for direct utilization of methane in solid oxide fuel-cells. *J. Electrochem. Soc.* 153 (2006), pp. A850–A856.

Jiang, S.P., Ye, Y.M., He, T.M. & Ho, S.B.: Nanostructured palladium-$La_{0.75}Sr_{0.25}Cr_{0.5}Mn_{0.5}O_3/Y_2O_3$-$ZrO_2$ composite anodes for direct methane and ethanol solid oxide fuel-cells. *J. Power Sources* 185 (2008a), pp. 179–182.

Jiang, Z.Y., Zhang, L., Feng, K. & Xia, C.R.: Nanoscale bismuth oxide impregnated $(La,Sr)MnO_3$ cathodes for intermediate-temperature solid oxide fuel-cells. *J. Power Sources* 185 (2008b), pp. 40–48.

Jiang, Z.Y., Xia, C.R., Zhao, F. & Chen, F.L.: $La_{0.85}Sr_{0.15}MnO_{3-\delta}$ infiltrated $Y_{0.5}Bi_{1.5}O_3$ cathodes for intermediate-temperature solid oxide fuel-cells. *Electrochem. Solid-State Lett.* 12 (2009a), pp. B91–B93.

Jiang, Z.Y., Lei, Z.W., Ding, B., Xia, C.R., Zhao, F. & Chen, F.L.: Electrochemical characteristics of solid oxide fuel-cell cathodes prepared by infiltrating $(La,Sr)MnO_3$ nanoparticles into yttria-stabilized bismuth oxide backbones. *Int. J. Hydrogen Energy* 35 (2009b), pp. 8322–8330.

Jiang, Z.Y., Zhang, L., Cai, L.L. & Xia, C.R.: Bismuth oxide-coated $(La,Sr)MnO_3$ cathodes for intermediate temperature solid oxide fuel-cells with yttria-stabilized zirconia electrolytes. *Electrochim. Acta* 54 (2009c), pp. 3059–3065.

Jiang, Z.Y., Xia, C.R. & Chen, F.L.: Nano-structured composite cathodes for intermediate-temperature solid oxide fuel-cells via an infiltration/impregnation technique. *Electrochim. Acta* 55 (2010), pp. 3595–3605.

Jiao, Z.J., Shikazono, N. & Kasagi, N.: Quantitative characterization of SOFC nickel-YSZ anode microstructure degradation based on focused-ion-beam 3D-reconstruction technique. *J. Electrochem. Soc.* 159 (2012), pp. B285–B291.

Jørgensen, M.J., Holtappels, P. & Appel, C.C.: Durability test of SOFC cathodes. *J. Appl. Electrochem.* 30 (2000), pp. 411–418.

Kim, H., Park, S., Vohs, J.M. & Gorte, R.J.: Direct oxidation of liluid fuels in a solid oxide fuel-cell. *J. Electrochem. Soc.* 148 (2001), pp. A693–A695.

Kim, H., Lu, C., Worrell, W.L., Vohs, J.M. & Gorte, R.J.: Cu-Ni cermet anodes for direct oxidation of methane in solid-oxide fuel-cells. *J. Electrochem. Soc.* 149 (2002), pp. A247–A250.

Kim, S.D., Moon, H., Hyun, S.H., Moon, J., Kim, J. & Lee, H.W.: Performance and durability of Ni-coated YSZ anodes for intermediate temperature solid oxide fuel cells. *Solid State Ionics* 177 (2006), pp. 931–938.

Klemensø, T., Appel, C.C. & Mogensen, M.: In situ observations of microstructural changes in SOFC anodes during redox cycling. *Electrochem. Solid-State Lett.* 9 (2006), pp. A403–A407.

Klemensø, T., Chung, C., Larsen, P.H. & Mogensen, M.: The mechanism behind redox instability of anodes in high-temperature SOFCs. *J. Electrochem. Soc.* 152 (2005), pp. A2186–A2192.

Klemens, T., Thydén, K., Chen, M. & Wang, H.J.: Stability of Ni–yttria stabilized zirconia anodes based on Ni-impregnation. *J. Power Sources* 195 (2010), pp. 7295–7301.

Kurokaw, H., Sholklapper, T.Z., Jacobson, C.P., Jonghe, L.C.D. & Visco, S.J.: Ceria nanocoating for sulfur tolerant Ni-based anodes of solid oxide fuel-cells. *Electrochem. Solid-State Lett.* 10 (2007), pp. B135–B138.

Küngas, R., Bidrawn, F., Vohs, J.M. & Gorte, R.J.: Doped-ceria diffusion barriers prepared by infiltration for solid oxide fuel-cells. *Electrochem. Solid-State Lett.* 13 (2010), pp. B87–B90.

Labrincha, J.A., Frade, J.R. & Marques, F.M.B.: $La_2Zr_2O_7$ formed at ceramic electrode/YSZ contacts. *J. Mater. Sci.* 28 (1993), pp. 3809–3815.

Leng, Y.J., Chan, S.H., Khor, K.A. & Jiang, S.P.: $(La_{0.8}Sr_{0.2})_{0.9}MnO_3-Gd_{0.2}Ce_{0.8}O_{1.9}$ composite cathodes prepared from $(Gd,Ce)(NO_3)_x$ modified $(La_{0.8}Sr_{0.2})_{0.9}MnO_3$ for intermediate-temperature solid oxide fuel-cells. *J. Solid State Electrochem.* 10 (2006), pp. 339–347.

Liang, F.L., Chen, J., Jiang, S.P., Chi, B., Pu, J. & Jian, L.: Development of nanostructured and palladium promoted (La,Sr)MnO3-based cathodes for intermediate-temperature SOFCs. *Electrochem. Solid-State Lett.* 11 (2008), pp. B213–B216.

Liu, Y., Mori, M., Funahashi, Y., Fujishiro, Y. & Hirano, A.: Development of micro-tubular SOFCs with an improved performance via nano-Ag impregnation for intermediate temperature operation. *Electrochem. Commun.* 9 (2007), pp. 1918–1923.

Liu, Y.L., Hagen, A., Barford, R., Chen, M., Wang, H.J., Poulsen, F.W. & Hendriksen, P.V.: Microstructural studies on degradation of interface between LSM–YSZ cathode and YSZ electrolyte in SOFCs. *Solid State Ionics* 180 (2009), pp. 1298–1304.

Liu, Z.B., Ding, D., Liu, B.B., Guo, W.W., Wang, W.D. & Xia, C.R.: Effect of impregnation phases on the performance of Ni-based anodes for low temperature solid oxide fuel-cells. *J. Power Sources* 196 (2011), pp. 8561–8567.

López-Esteban, S., Rodriguez-Suarez, T., Esteban-Betegón, F., Pecharromán, C. & Moya, J.S.: Mechanical properties and interfaces of zirconia/nickel in micro- and nanocomposites. *J. Mater. Sci.* 41 (2006), pp. 5194–5199.

Lou, X.Y., Wang, S.Z., Liu, Z., Yang, L. & Liu, M.L.: Improving $La_{0.6}Sr_{0.4}Co_{0.2}Fe_{0.8}O_{3-\delta}$ cathode performance by infiltration of a $Sm_{0.5}Sr_{0.5}CoO_{3-\delta}$ coating. *Solid State Ionics* 180 (2009), pp. 1285–1289.

Lu, C., Sholklapper, T.Z., Jacobson, C.P., Visco, S.J. & De Jonghe, L.C.: LSM-YSZ cathodes with reaction-infiltrated nanoparticles. *J. Electrochem. Soc.* 153 (2006), pp. A1115–A1119.

Mai, A., Haanappel, V.A.C., Uhlenbruck, S., Tietz, F. & Stöver, D.: Ferrite-based perovskites as cathode materials for anode-supported solid oxide fuel-cells: Part I. Variation of composition. *Solid State Ionics* 176 (2005), pp. 1341–1350.

Mai, A., Becker, M., Assenmacher, W., Tietz, F., Hathiramani, D., Ivers-Tiffee, E., Stöver, D. & Mader, W.: Time-dependent performance of mixed-conducting SOFC cathodes. *Solid State Ionics* 177 (2006), pp. 1965–1968.

Malzbender, J., Wessel, E., Steinbrech, R.W. & Singheiser, L.: Reduction and re-oxidation of anodes for solid oxide fuel-cells. *Solid State Ionics* 176 (2005), pp. 2201–2203.

Malzbendera, J., Batfalsky, P., Vaßen, R., Shemet, V. & Tietz, F.: Component interactions after long-term operation of an SOFC stack with LSM cathode. *J. Power Sources* 201 (2012), pp. 196–203.

Marina, O.A., Coyle, C.A., Engelhard, M.H. & Pederson, L.R.: Mitigation of sulfur poisoning of Ni/Zirconia SOFC anodes by antimony and tin. *J. Electrochem. Soc.* 158 (2011), pp. B424–B429.

Matsuzaki, Y. & Yasuda, I.: The poisoning effect of sulfur-containing impurity gas on a SOFC anode: Part I. Dependence on temperature, time, and impurity concentration. *Solid State Ionics* 132 (2000), pp. 261–269.

Minh, N.Q.: Ceramic fuel-cells. *J. Amer. Ceramic Soc.* 76 (1993), pp. 563–588.

Moya, J.S., Lopez-Esteban, S. & Pecharromán, C.: The challenge of ceramic/metal microcomposites and nanocomposites. *Prog. Mater. Sci.* 52 (2007), pp. 1017–1090.

Muecke, U.P., Graf, S., Rhyner, U. & Gauckler, L.J.: Microstructure and electrical conductivity of nanocrystalline nickeland nickel oxide/gadolinia-doped ceria thin films. *Acta Mater.* 56 (2008), pp. 677–687.

Paramonov, Y. & Andersons, J.: A family of weakest link models for fiber strength distribution. *Composites A Appl. Sci. Manufact.* 38 (2007), pp. 1227–1233.

Park, S., Vohs, J.M. & Gorte, R.J.: Direct oxidation of hydrocarbons in a solid-oxide fuel-cell. *Nature* 404 (2000), pp. 265–267.

Pecharromán, C., Esteban-Betegón, F., Bartolomé, J.F., Richter, G. & Moya. J.S.: Theoretical model of hardening in zirconia-nickel nanoparticle composites. *Nano Lett.* 4 (2004), pp. 747–751.

Pecharromán, C., Beltrán, J.I., Esteban-Betegón, F., López-Esteban, S., Bartolomé, J.F., Muòoz, M.C. & Moya, J.S.: Zirconia/nickel interfaces in micro and naocomposites. *Z. Metallk.* 96 (2005), pp. 507–513.

Peters. C.: *Grain-size effects in nanoscaled electrolyte and cathode thin films for SOFC.* Karlsruhe Scientific Publishing, Karlsruhe, Germany, 2008.

Pihlatie, M.H., Kaiser, A., Mogensen, M. &Chen, M.: Electrical conductivity of Ni–YSZ composites: Degradation due to Ni particle growth. *Solid State Ionics* 189 (2011), pp. 82–90.

Pihlatie, M., Kaiser, A., Larsen, P.H. & Mogensen, M.: Dimensional behavior of Ni–YSZ composites during redox cycling. *J. Electrochem. Soc.* 156 (2009a), pp. B322–B329.

Pihlatie, M., Ramos, T. & Kaiser, A.: Testing and improving the redox stability of Ni-based solid oxide fuel-cells. *J. Power Sources* 193 (2009b), pp. 322–330.

Primdahl, S. & Mogensen, M.: Durability and thermal cycling of Ni/YSZ cermet anodes for solid oxide fuel-cells. *J. Appl. Electrochem.* 30 (2000), pp. 247–257.

Rupp, J.L.M., Infortuna, A. & Gauckler, L.J.: Microstrain and self-limited grain growth in nanocrystalline ceria ceramics. *Acta Mater.* 54 (2006), pp. 1721–1730.

Sarantaridis, D. & Atkinson, A.: Redox cycling of Ni-based solid oxide fuel-cell anodes: A review. *Fuel-Cells* 7 (2007), pp. 246–258.

Sarantaridis, D., Chater, R.J. & Atkinson, A.: Changes in physical and mechanical properties of SOFC Ni–YSZ composites caused by redox cycling. *J. Electrochem. Soc.* 155 (2008), pp. B467–B472.

Sato, K., Abe, H., Misono, T., Murata, K., Fukui, T. & Naito, M.: Enhanced electrochemical activity and long-term stability of Ni–YSZ anode derived from NiO–YSZ interdispersed composite particles. *J. Europ. Ceramic Soc.* 29 (2009), pp. 1119–1124.

Sato, K., Kinoshita, T. & Abe, H.: Performance and durability of nanostructured $(La_{0.85}Sr_{0.15})_{0.98}MnO_3$/yttria-stabilized zirconia cathodes for intermediate-temperature solid oxide fuel-cells. *J. Power Sources* 195 (2010), pp. 4114–4118.

Shah, M. & Barnett, S.A.: Solid oxide fuel-cell cathodes by infiltration of $La_{0.6}Sr_{0.4}Co_{0.2}Fe_{0.8}O_{3-\delta}$ into Gd-doped ceria. *Solid State Ionics* 179 (2008), pp. 2059–2064.

Shah, M., Voorhees, P.W., Barnett, S.A.: Time-dependent performance changes in LSCF-infiltrated SOFC cathodes: The role of nano-particle coarsening. *Solid State Ionics* 187 (2011), pp. 64–67.

Simner, S.P., Anderson, M.D., Engelhard, M.H. & Stevenson, J.W.: Degradation mechanisms of La-Sr-Co-Fe-O₃ SOFC cathodes. *Electrochem. Solid-State Lett.* 9 (10) (2006), pp. A478–A481.

Simwonis, D., Tietz, F. & Stöver. D.: Nickel coarsening in annealed Ni/8YSZ anode substrates for solid oxide fuel-cells. *Solid State Ionics* 132 (2000), pp. 241–251.

Song, H.S., Hyun, S.H., Kim, J., Lee, H-W. & Moon, J.: A nanocomposite material for highly durable solid oxide fuel-cell cathodes. *J. Mater. Chem.* 18 (2008), pp. 1087–1092.

Song, H.S., Lee, S., Hyun, S.H., Kim, J. & Moon, J.: Compositional influence of LSM-YSZ composite cathodes on improved performance and durability of solid oxide fuel-cells. *J. Power Sources* 187 (2009), pp. 25–31.

Suzuki, S., Uchida, H. & Watanabe, M.: Microstructural analyses of ceria-based anode with highly dispersed Ni electrocatalysts for medium-temperature solid oxide fuel-cells. *Electrochem.* 73 (2005), pp. 128–134.

Suzuki, S., Uchida, H. & Watanabe, M.: Interaction of samaria-doped ceria anode with highly dispersed Ni catalysts in a medium-temperature solid oxide fuel-cell during long-term operation. *Solid State Ionics* 177 (2006), pp. 359–365.

Tanner, C.W., Fung, K.-Z. & Virkar, A.V.: The effect of porous composite electrode structure on solid oxide fuel-cell performance. *J. Electrochem. Soc.* 144 (1997), pp. 21–30.

Tietz, F., Arul Raj, I., Zahid, M. & Stöver, D.: Electrical conductivity and thermal expansion of $La_{0.8}Sr_{0.2}(Mn,Fe,Co)O_{3-\delta}$ perovskites. *Solid State Ionics* 177 (2006a), pp. 1753–1756.

Tietz, F., Haanappel, V.A.C., Mai, A., Mertens, J. & Stöver, D.: Performance of LSCF cathodes in cell tests. *J. Power Sources* 156 (2006b), pp. 20–22.

Tsipis, E.V. & Kharton, V.V.: Electrode materials and reaction mechanisms in solid oxide fuel-cells: a brief review. I. Performance-determining factors. *J. Solid State Electrochem.* 12 (2008), pp. 1039–1060.

Tucker, M.C., Lau, G.Y., Jacobson, C.P., DeJonghe, L.C. & Visco, S.J.: Performance of metal-supported SOFCs with infiltrated electrodes. *J. Power Sources* 171 (2007), pp. 477–482.

Uchida, H., Suzuki, S. & Watanabe, M.: High performance electrode for medium-temperature solid oxide fuel-cells. *Electrochem. Solid-State Lett.* 6 (2003), pp. A174–A177.

Waldbillig, D., Wood, A. & Ivey, D.G.: Electrochemical and microstructural characterization of the redox tolerance of solid oxide fuel-cell anodes. *J. Power Sources* 145 (2005), pp. 206–215.

Wang, W., Jiang, S.P., Tok, A.I.Y. & Luo, L.: GDC-impregnated Ni anodes for direct utilization of methane in solid oxide fuel-cells. *J. Power Sources* 159 (2006), pp. 68–72.

Wang, X.G., Nakagawa, N. & Kato, K.: Anodic polarization related to the ionic conductivity of zirconia at Ni-zirconia/zirconia electrodes. *J. Electrochem. Soc.* 148 (2001), pp. A565–A569.

Wang, Y., Zhang, H., Chen, F.L. & Xia, C.R.: Electrochemical characteristics of nano-structured $PrBaCo_2O_{5+x}$ cathodes fabricated with ion impregnation process. *J. Power Sources* 203 (2012), pp. 34–41.

Wood, A., Pastula, M., Waldbillig, D. & Ivey, D.G.: Initial testing of solutions to redox problems with anode-supported SOFC. *J. Electrochem. Soc.* 153 (2006), pp. A1929–A1934.

Wu, T.Z., Zhao, Y.Q., Peng, R.R. & Xia, C.R.: Nano-sized $Sm_{0.5}Sr_{0.5}CoO_{3-\delta}$ as the cathode for solid oxide fuel-cells with proton-conducting electrolytes of $BaCe_{0.8}Sm_{0.2}O_{2.9}$. *Electrochim. Acta* 54 (2009), pp. 4888–4892.

Wu, T.Z., Rao, Y.Y., Peng, R.R. & Xia, C.R.: Fabrication and evaluation of Ag-impregnated $BaCe_{0.8}Sm_{0.2}O_{2.9}$ composite cathodes for proton conducting solid oxide fuel-cells. *J. Power Sources* 195 (2010), pp. 5508–5513.

Xu, X.Y., Jiang, Z.Y., Fan, X. & Xia, C.R.: LSM-SDC electrodes fabricated with an ion-impregnating process for SOFCs with doped ceria electrolytes. *Solid State Ionics* 177 (2006), pp. 2113–2117.

Yamamoto, O., Takeda, Y., Kanno, R. & Noda, M.: Perovskite-type oxides as oxygen electrodes for high temperature oxide fuel-cells. *Solid State Ionics* 22 (1987), pp. 241–246.

Yang, C.H., Yang, Z.B., Jin, C., Xiao, G.L., Chen, F.L. & Han, M.F.: Sulfur-tolerant redox-reversible anode material for direct hydrocarbon solid oxide fuel-cells. *Adv. Mater.* 24 (2012), pp. 1439–1443.

Ye, Y.M., He, T.M., Li, Y., Tang, E.H., Reitz, T.L. & Jiang, S.P.: Pd-promoted $La_{0.75}Sr_{0.25}Cr_{0.5}Mn_{0.5}O_3$/YSZ composite anodes for direct utilization of methane in SOFCs. *J. Electrochem. Soc.* 155 (2008), pp. B811–B818.

Yun, J.W., Yoon, S.P., Park, S., Kim, H.S. & Nam, S.W.: Analysis of the regenerative $H_2S$ poisoning mechanism in $Ce_{0.8}Sm_{0.2}O_3$-coated Ni/YSZ anodes for intermediate temperature solid oxide fuel-cells. *Int. J. Hydrogen Energy* 36 (2011), pp. 787–796.

Zha, S.W., Cheng, Z. & Liu, M.L.: Sulfur poisoning and regeneration of Ni-based anodes in solid oxide fuel-cells. *J. Electrochem. Soc.* 154 (2007), pp. B201–B206.

Zhang, H., Zhao, F., Chen, F.L. & Xia, C.R.: Nano-structured $Sm_{0.5}Sr_{0.5}CoO_{3-\delta}$ electrodes for intermediate-temperature SOFCs with zirconia electrolytes. *Solid State Ionics* 192 (2010), pp. 591–594.

Zhang, L., Zhao, F., Peng, R.R. & Xia, C.R.: Effect of firing temperature on the performance of LSM-SDC cathodes prepared with an ion-impregnation method. *Solid State Ionics* 179 (2008), pp. 1553–1556.

Zhang, L., Liu, Y.Q., Zhang, Y.X., Xiao, G.L., Chen, F.L. & Xia, C.R.: Enhancement in surface exchange coefficient and electrochemical performance of $Sr_2Fe_{1.5}Mo_{0.5}O_6$ electrodes by $Ce_{0.8}Sm_{0.2}O_{1.9}$ nanoparticles. *Electrochem. Commun.* 13 (2011), pp. 711–713.

Zhang, Y.X. & Xia, C.R.: A durability model for solid oxide fuel-cell electrodes in thermal cycle processes. *J. Power Sources* 195 (2010), pp. 6611–6618.

Zhao, F., Peng, R.R. & Xia, C.R.: A $La_{0.6}Sr_{0.4}CoO_{3-\delta}$ based electrode with high durability for intermediate temperature solid oxide fuel-cells. *Mater. Res. Bull.* 43 (2008a), pp. 370–376.

Zhao, F., Wang, Z.Y., Liu, M.F., Zhang, L., Xia, C.R. & Chen, F.L.: Novel nano-network cathodes for solid oxide fuel-cells. *J. Power Sources* 185 (2008b), pp. 13–18.

Zhao, F., Zhang, L, Jiang, Z.Y., Xia, C.R. & Chen, F.L.: A high performance intermediate-temperature solid oxide fuel-cell using impregnated $La_{0.6}Sr_{0.4}CoO_{3-\delta}$ cathode. *J. Alloy. Compd.* 487 (2009), pp. 781–785.

Zhu, W., Xia, C.R., Fan, J., Peng, R.R. & Meng, G.Y.: Ceria coated Ni as anodes for direct utilization of methane in low-temperature solid oxide fuel-cells. *J. Power Sources* 160 (2006), pp. 897–902.

# CHAPTER 8

## Micro- and nano-technologies for microbial fuel cells

Hao Ren & Junseok Chae

### 8.1 INTRODUCTION

An energy crisis and global warming are among the foremost critical issues that we face today. According to the United Kingdom Energy Research Centre, both oil and gas production will peak within the next twenty years (Bentley, 2002; Sorrell *et al.*, 2009; 2010), and afterwards our society will suffer from post-peak reduction. Global warming, possibly caused by the greenhouse effect of $CO_2$ accumulation from massive fossil energy consumption since the industrial revolution, has been a worldwide issue because it changes global climate, boosts the sea level, etc. Thus, the search for and utilization of carbon-neutral, renewable, and sustainable energy is increasingly critical for the sustainability of our planet.

Many carbon-neutral, renewable, and sustainable energy sources are available, including hydroenergy, solar energy, geothermal energy, wind energy, and bioenergy. Among them bioenergy is an attractive alternative to fossil energy, which had been the main energy source in history before the industrial revolution. Bioenergy is abundant and easy to access. Biomass in wastewater contains approximately $1.5 \times 10^{11}$ kWh of potential energy, the equivalent of 17 GW, and agricultural practice could produce $1.34 \times 10^{12}$ kg of biomass, equivalent to a power of 600 GW, which could provide 120% of the annual electricity generation in the United States of America, based on a report that wastewater treatment consumes 15 GW, which is currently only 3% of US annual electricity generation. (Logan, 2004; Logan and Rabaey, 2012; McCarty *et al.*, 2011; Perlack *et al.*, 2005).

Microbial fuel cells (MFCs), which convert biomass directly to electricity with the catalytic activity of some specific bacteria, named exoelectrogenic or anode respiring bacteria (ARB), have attracted worldwide attention, due to direct electricity conversion, and high conversion efficiency. During the past ten years, more than two thousand papers have been published in this area (Xie *et al.*, 2012). Of these MFCs, most are in macro- or meso-scale forms that scientists aim to utilize for wastewater treatment and renewable energy production, bioremediation, and/or power supplies for remote sensors in harsh conditions. Another interesting approach is, via miniaturization of a MFC, to scale the MFC to be sub-millimeter form factor, a small size power supply or lab-on-a-chip device for scientific research on exoelectrogens. Since the first micro-scale MFC study was published in 2002 (Chiao *et al.*, 2002), many research groups have reported on micro-scale MFCs, aiming to explore the fundamental phenomena of exoelectrogens, as well as to enhance the power performance in micro-scale MFCs (Chiao *et al.*, 2002; 2006; Choi and Chae, 2012; Choi *et al.*, 2011; Inoue 2012; Mink *et al.*, 2012; Qian *et al.*, 2009; Ren and Chae, 2012; Ren *et al.*, 2012; 2013).

By applying micro-fabrication and micro-fluidic techniques, the advantages of economical mass production and large surface-area-to-volume ratio drive a micro-scale MFC towards a potential candidate in the miniaturized energy converter. However, previously reported micro-scale MFCs face overcoming a critical challenge; their low power density. The areal and volumetric power densities have improved by more than $10^4$ and $10^5$ fold in the past decade, and a highest areal power density of 7.72 W m$^{-2}$ and volumetric power density of 3320 W m$^{-3}$ have been reported (Ren *et al.*, 2013; 2015). Yet those areal and volumetric power densities are still about a few fold to four orders of magnitudes lower than those of a conventional Lithium ion battery – a golden standard of portable power source (Pikul *et al.*, 2013). Thus, enhancing the power density of micro-scale MFCs is much needed.

Besides small size power supplies, miniaturization enables MFCs-on-a-chip, suitable for scientific studies, including elucidating the mechanism of extracellular electron transfer (EET) of exoelectrogens, screening the electricity generation capability of individual exoelectrogen, etc. Applying miniaturized MFCs in these research venues allows for the integration of multiple MFCs on one chip for parallel analysis, thus enhancing the efficiency of the analyses, thanks to batch fabrication features. Miniaturized MFC-on-a-chip devices also find application in biosensors for toxic chemical detection, such as formaldehyde (Davila *et al.*, 2011).

## 8.2    ELECTRICITY GENERATION FUNDAMENTAL

### 8.2.1    *Electron transfer of exoelectrogens*

Figure 8.1 is a schematic illustration of the operational principle of an MFC. Electron transport from cells to the anode of the MFC is currently on active research, and the following steps are commonly adopted among researchers: cells break down organic substances to generate NADH through the tricarboxylic acid cycle (TCA) or glycolysis, specific types of bacteria are able to transfer electrons stored in NADH outside their body through a series of electron shuttles, such as coenzyme Q, ubiquinone, cytochrome c, etc., and then electrons outside their body are transported through electron shuttles, such as cytochrome or flavins, etc., to the anode. The series of electron transport is named extracellular electron transport (EET). To date, various species of bacteria have been found to have the capability of EET, including the *Geobacter* species (*Geobacter sulfurreducens*, *Geobacter metallireducens*), the *Shewanella* species (*Shewanella oneidensis* MR-1, *Shewanella putrefacians* IR-1, *Shewanella oneidensis* DSP10), the *Pseudomonas* species (*Pseudomonas aeruginosa* KRP1), *Rhodopseudomonas palustris* DX-1, *Saccharomyces cerevisiae*, *Escherichia coli*, etc. Of them, the *Geobacter* and *Shewanella* species are the most often researched.

To date, two EET mechanisms have been found. One is indirect EET, relying on electron shuttles, which can be either self-secreted or externally added. Electron shuttles are redox-driven and have two states: oxidized and reduced states. Driven by redox, the oxidized shuttles become reduced via the acquisition of electrons at the exoelectrogen outer membrane. The reduced shuttles are then driven to the anode and release electrons to the anode and become oxidized. This process repeats to transfer electrons from the exoelectrogens to the anode. For instance, *Shewanella oneidensis* MR-1 is reported to mainly rely on electron shuttles named flavins (Marsili *et al.*, 2008).

The other mechanism is direct EET, mainly involving the *Geobacter* species which, however, is still largely open to debate. Years ago, it was reported that the direct EET mechanism involves the nano-meter scale pili secreted by exoelectrogens, and both *in situ* and *ex situ* measurements indicate the pili are metallic-like (Malvankar *et al.*, 2011). According to this mechanism, electrons produced by the exoelectrogen can be transferred to the anode directly through the conductive pili. Recently some researchers reported that EET relies on the electron super-exchange (electron hopping) between the adjacent cytochrome c, and this process is driven by the redox of cytochrome c (Snider *et al.*, 2012). The cytochrome c is immobilized on the exoelectrogen outer membrane and inside the exoelectrogen. The reduced cytochrome c, via the respiration of exoelectrogen, can pass electrons to adjacent oxidized cytochrome c. After multiple electron exchanges, electrons are transferred to the anode from the exoelectrogen. With regards to how electrons are transferred from inside the exoelectrogen to the cytochrome c immobilized on the outer membrane still remains to be discovered.

Once electrons reach an anode, they flow to a cathode, oxidized by an electron acceptor, such as oxygen, ferricyanide, etc., driven by a high potential at the cathode. By connecting a load between

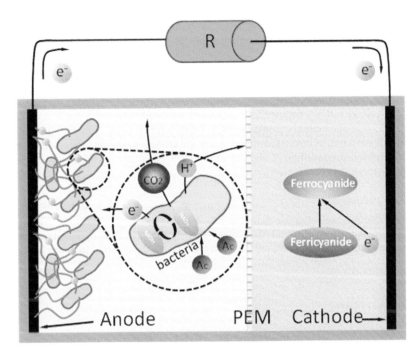

Figure 8.1.   Illustration of a typical two-chamber microbial fuel cell (MFC): specific bacteria species in the anode chamber, named exoelectrogenic or anode-respiring bacteria (ARB), break down organic substrates, i.e., acetate, to produce electrons, protons, and $CO_2$. The electrons pass through an external resistor to be reduced at the cathode while protons pass through the proton exchange membrane (PEM) from the anode to the cathode chamber.

the anode and the cathode, these electrons power the load and thus electricity is harvested through the microbial metabolism.

### 8.2.2   *Voltage generation*

The voltage of the fuel cell is determined by the electrochemical potential at the anode and the cathode. Electrochemical potential is determined by the Nernst equation:

$$E_0' = E_0 - \frac{RT}{nF} \ln \frac{[\text{products}]^p}{[\text{reactants}]^r} \tag{8.1}$$

where $E_0$ represents the standard anode or cathode potentials [V], $n$ is the number of moles of electrons transferred in the cell reaction, and $[\text{products}]^p/[\text{reactants}]^r$ is the reaction quotient which is the ratio of the activities of the products divided by the reactants raised to their respective stoichiometric coefficient. The Nernst equation is used to calculate the anode and cathode potential, and the total cell potential is calculated by:

$$E_{\text{EMF}}' = E_{\text{cathode}}' - E_{\text{anode}}' \tag{8.2}$$

For an MFC using acetate as an organic substance, and ferricyanide as a catholyte, the half-reaction in the anode and the cathode compartment can be written as:

$$\text{Anode}: \quad CH_3COO^- + 2H_2O \rightarrow 2CO_2 + 8e^- + 7H^+ \tag{8.3}$$

$$\text{Cathode}: \quad Fe(CN)_6^{3+} + e^- \rightarrow Fe(CN)_6^{2+} \tag{8.4}$$

$E_0$ for anode and cathode are $-0.284$ and $0.361$ V, respectively (Logan, 2008). Through these formula, with reaction quotient measured, the total cell potential can be calculated.

### 8.2.3    Parameter for MFC characterization

#### 8.2.3.1    Open circuit voltage ($E_{OCV}$)

Open circuit voltage (OCV), denoted by $E_{OCV}$, is the voltage between the anode and the cathode when an MFC is at open circuit, similar with conventional fuel cells. Note the OCV is different from total cell potential, as discussed above; generally OCV is lower than the total cell potential, due to different sources resulting in overpotentials. OCV is usually measured under open circuit by a multi-meter or data acquisition system.

#### 8.2.3.2    Areal/volumetric current density ($i_{max, areal}$, $i_{max, volumetric}$) and areal/volumetric power density ($p_{max, areal}$, $p_{max, volumetric}$)

Assuming $I$, $R_i$, and $R_e$ are current [A], internal resistance [$\Omega$], and external load resistance of an MFC [$\Omega$]. According to Ohm's law, current flowing through the external load can be written as:

$$I = \frac{U}{R_e + R_i} \tag{8.5}$$

Output power of MFC can be written as:

$$P = I^2 R_e = E_{OCV}^2 R_e / (R_i + R_e)^2 \tag{8.6}$$

When $R_i = R_e$, the maximum power can be obtained from:

$$P_{max} = E_{OCV}^2 / 4R_i \tag{8.7}$$

In this case, the maximum current can be calculated as:

$$I_{max} = E_{OCV} / 2R_i \tag{8.8}$$

In order to evaluate the performance of the MFC, the maximum current and power output is normalized to unit electrode area or unit of chamber volume, to get areal or volumetric current/power densities:

$$i_{max, areal} = I_{max} / A, \; p_{max, areal} = P_{max} / A \tag{8.9}$$

$$i_{max, volumetric} = I_{max} / V, \; p_{max, volumetric} = P_{max} / V \tag{8.10}$$

Here, $i_{max, areal}$, $i_{max, volumetric}$, $p_{max, areal}$, $p_{max, volumetric}$ are the areal current density, volumetric current density, areal power density, and volumetric power density, respectively.

#### 8.2.3.3    Internal resistance ($R_i$) and areal resistivity ($r_i$)

Internal resistance is a critical parameter and it can be calculated from the electrical characterization of an MFC. A polarization curve measurement is a popular method to extract $R_i$. A multi-meter or data acquisition system (DAQ) can be used to measure a MFC output voltage while a series of external resistors are used to measure the corresponding voltages. Then current density versus voltage and current density versus power density is plotted. An example of a polarization curve, which often contains two plots, is shown in Figure 8.2. In (a), at low current density, activation resistance dominates, and the voltage across external load drops fast with the increase of current density. When the current density increases to an intermediate range, the voltage versus current density profile becomes approximately linear, indicating that ohmic resistance dominates. As the current density further increases, again the voltage drops quickly; this is because the concentration resistance dominates.

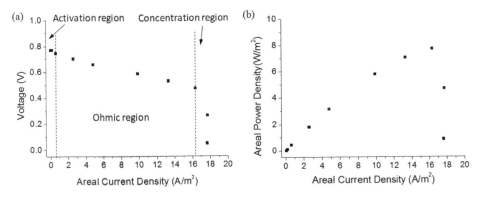

Figure 8.2. Schematic of the voltage *versus* current density of a micro-scale MFC with an ultramicroelectrode (Ren *et al.*, 2013); (a) voltage *versus* current density chart, (b) areal power density *versus* areal current density chart. In (a) three distinct zones exist, representing activation resistance, ohmic resistance, and concentration resistance zones, respectively.

By linearly fitting the ohmic region in the polarization curve, the slope is approximately equal to the internal resistance of ohmic region, $R_i$. As shown in Equation (8.6), minimizing internal resistance yields a high power density MFC.

It is often difficult to compare internal resistance of different sizes of MFCs side by side. For a macro/meso-scale MFC, the internal resistance can be as low as several to 10 s of $\Omega$, while most miniaturized MFCs have internal resistance on the order of $k\Omega$, because generally a larger size electrode results in a smaller internal resistance. Electrical resistance can be shown as:

$$R = \frac{\rho l}{A} \qquad (8.11)$$

Here $R$ [$\Omega$] is the electrical resistance, $\rho$ is the resistivity [$\Omega\,m$], $l$ is the length [m], and $A$ is the effective area of reaction occurrence [$m^2$]. $R$ is directly proportional to $1/A$, thus it is fair to normalize the resistance at a given reaction area. Areal resistivity, $r_i$ [$\Omega\,m^2$], denotes the normalized internal resistance at a unit of the electrode area.

#### 8.2.3.4 Efficiency – Coulombic efficiency (CE) and energy conversion efficiency (EE)

Two efficiency parameters are often used to evaluate MFCs: Coulombic efficiency (CE) and energy conversion efficiency (EE).

##### 8.2.3.4.1 Coulombic efficiency (CE)

Coulombic efficiency is the ratio of coulombs harvested by the anode of an MFC via decomposing organic substrates by the exoelectrogen, to the theoretical maximum coulombs converted from all organic substrates:

$$CE = \frac{C_P}{C_T} \times 100\% \qquad (8.12)$$

where $C_P$ is the total coulombs calculated by integrating the current over the time for substrate consumption [C], and $C_T$ is the maximum possible coulombs of the substrate $C_T = V \times b \times N_A \times e \times mol_{substrate}$ [C]. $V$ is the volume of the anode chamber [$m^3$], $b$ is the number of moles of electrons produced by oxidation of the substrate ($b = 8$ mol e$^-$/mol acetate), $N_A$ is Avogadro's number ($6.023 \times 10^{23}$ molecules/mol), $e$ is the electron charge ($1.6 \times 10^{-19}$ C/electron), and $mol_{substrate}$ is the moles of acetate oxidized. The theoretical $CE$ can reach 100%, and the actual $CE$ can vary largely from 0.1% to more than 90%. Miniaturized MFCs often show low $CE$, primarily due to electron loss in the anode compartment, such as oxygen leakage.

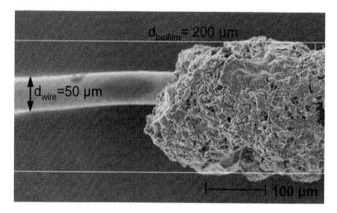

Figure 8.3.   SEM image of a biofilm on a platinum wire with a diameter of 50 μm; biofilm thickness was determined to be 75 μm (Pocaznoi *et al.*, 2012).

#### 8.2.3.4.2   *Energy conversion efficiency (EE)*

Energy conversion efficiency is the ratio of the total energy harvested by an MFC to the maximum possible energy that the biomass can produce (standard molar enthalpy (heat) of biomass):

$$EE = \frac{E_P}{E_T} \times 100\% \tag{8.13}$$

where $E_P$ is the total energy calculated by integrating the power output over the time for substrate consumption, and $E_T$ is the maximum possible energy of the biomass $E_T = V \times c \times \Delta_f H°$. $V$ is the volume of the anode chamber [m$^3$], c is the concentration of biomass in the anode chamber ($c = 25$ mol m$^{-3}$, sodium acetate), and $\Delta_f H°$ is standard molar enthalpy (heat) of formation (708.8 kJ mol$^{-1}$ for sodium acetate).

#### 8.2.3.5   *Biofilm morphology*

The formation of biofilm on an anode has an important impact on current/power generation capability. As a result, biofilm morphology, including the uniformity and thickness, are often characterized in studying MFCs. Advanced imaging techniques are deployed for the characterization, including scanning electron microscopy (SEM) and confocal laser scanning microscopy (CLSM).

Scanning electron microscopy (SEM) can achieve the resolution of a sub-micrometer, thus it is often used to visualize a microbial biofilm as well as an individual exoelectrogen. Figure 8.3 illustrates an SEM image of a biofilm taken in a platinum wire with a diameter of 50 μm. The biofilm was determined to be 75 μm (Pocaznoi *et al.*, 2012). *In situ* biofilm characterization cannot be performed by SEM, because focused electron beams need to bombard the biofilm in order to generate secondary electrons for imaging. As a result, the biofilm needs to be fixed before imaging. The fixation process includes fixing the biofilm in 2% glutaraldehyde solution overnight and then dehydrating it by serial 10 min transfers through 50, 70, 90, and 100% ethanol. Due to the low conductivity of fixed biofilm, a layer of Au is deposited on top of the biofilm prior to imaging.

Confocal laser scanning microscopy (CLSM) is a powerful tool to perform *in situ* measurement of density and thickness of biofilm. The unique capability of CLSM is to acquire in-focus images of biofilm at a given depth. By acquiring multiple images with varying depth of focus through the biofilm, a 3D topological image of the biofilm can be reconstructed. Figure 8.4 shows the CLSM image of a biofilm (Richter *et al.*, 2008).

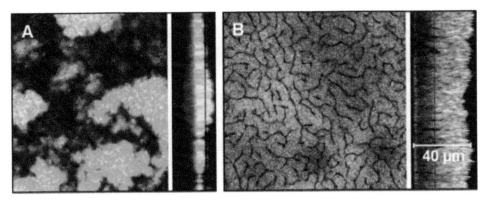

Figure 8.4.   Biofilm images obtained using confocal laser scanning microscopy (CLSM). *Geobacter* forms
a biofilm in a macroscale MFC (a) 10 days after the start-up 12 μm thick and (b) 18 days after
the start-up, 40 μm thick. CLSM is capable of measuring the thickness of density of the biofilm
(Richter *et al.*, 2008).

## 8.3   PRIOR ART OF MINIATURIZED MFCS

A miniaturized MFC was first reported in 2002 by Chiao *et al.* (2002) by applying baker's
yeast, *Saccharomyces cerevisiae* to break down glucose and generate electricity. The MFC had an
electrode with an area of 0.07 cm$^2$, and generated a power density of 5.72 nW m$^{-2}$. Afterwards, the
miniaturized MFC was optimized by creating micro-fluidic channels in the anode and cathode
chambers, and the areal and volumetric power densities enhanced significantly to 23 μW m$^{-2}$
and 0.276 W m$^{-3}$, which are more than three orders of magnitudes higher than the first prototype
(Chiao *et al.*, 2006). Later, by fabricating micro-pillars on a PDMS substrate using soft lithography,
which effectively increased the surface-area-to-volume ratio, their power densities were further
enhanced to 4 mW m$^{-2}$ and 40 W m$^{-3}$, respectively (Siu and Chiao, 2008).

A miniaturized MFC with a volume of 1.5 μL which utilized *Shewanella oneidensis* MR-1
was presented by Qian *et al.* (2009) recording 1.5 mW m$^{-2}$ and 15 W m$^{-3}$. The MFC was later
optimized by implementing micro-fluidic channels and soft lithography to have a volume of 4 μL,
producing 6.25 mW m$^{-2}$ and 62.5 W m$^{-3}$ (Qian *et al.*, 2011).

Parra and Lin (2009) reported the first miniaturized MFC utilizing *Geobacter* species, and a
power density of 0.12 W m$^{-2}$ and 0.34 W m$^{-3}$ was presented with a plain gold electrode. Later
they improved the MFC by using Carbon nanotubes (CNT) as electrode material because they
have a large surface-area-to-volume ratio and are shown to be biocompatible with microbes
(Inoue, 2012). With CNTs as electrodes, a power density of 73.8 mW m$^{-2}$ and 16.4 W m$^{-3}$ was
achieved. Mink *et al.* (2012) also adopted a CNT-based anode which deployed vertically-grown
multi-walled CNTs. Ni silicide was applied to reduce the contact resistance, yielding a power
density of 20 mW m$^{-2}$ and 392 W m$^{-3}$.

By adding L-cysteine inside the anode chamber to successfully mitigate oxygen intrusion, Choi *et al.* (2011) constructed a miniaturized MFC producing a power density of
47 mW m$^{-2}$/2333 W m$^{-3}$. The same group presented the first miniaturized MFC array in a series
stack to mark a power output of 100 μW and a power density of 0.33 W m$^{-2}$ and 667 W m$^{-3}$ (Choi
and Chae, 2012). The open circuit voltage of the MFC array was reported to be as high as 2.47 V.

By improving the mass transfer of substrate into exoelectrogens, a 100 μL MFC was presented
by Ren and Chae (Ren and Chae, 2012). A power density of 0.83 W m$^{-2}$ and 3320 W m$^{-3}$ was
reported. Mass transfer was further improved by implementing an ultramicroelectrode (UME), and
a record high power density of 7.72 W m$^{-2}$, regardless of the size of the MFCs, was achieved (Ren
and Chae 2012; Ren *et al.* 2014). Furthermore, by successfully mitigating the oxygen intrusion

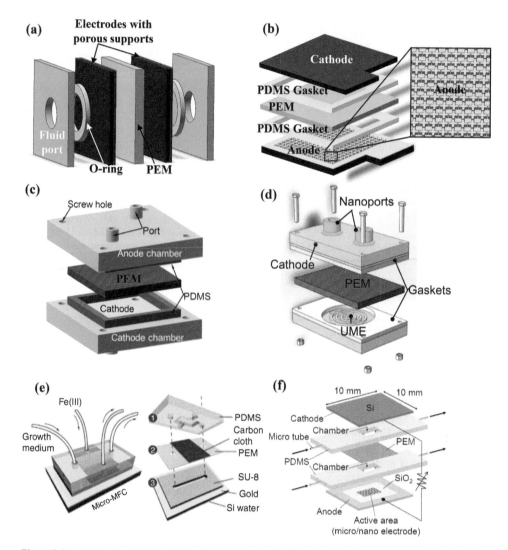

Figure 8.5.   Miniaturized MFCs: (a) the first miniaturized MFC, using *Saccharomyces cerevisiae*, by Chiao
*et al.* (2002); (b) a miniaturized MFC presented by Siu and Chiao (2008), using *Saccharomyces cerevisiae*; (c) a miniaturized MFC presented by Choi *et al.* (2011) using *Geobacter
sulfurreducens* mixed culture; (d) a miniaturized MFC with an ultramicroelectrode anode, using
*Geobacter sulfurreducens* mixed culture, presented by Ren *et al.* 2013; (e) a miniaturized MFC
presented by Qian *et al.*, which used *Shewanella* as the exoelectrogen; (f) a miniaturized MFC
presented by Inoue *et al.* (2012), using *Geobacter sulfurreducens*.

into the anode chamber, a high CE of 79% was reported. Figure 8.5 illustrates the various types
of miniaturized MFCs. Table 8.1 summaries the specifications of miniaturized MFCs.

Figure 8.6 shows a comparison between the power density and coulombic efficiency of a
miniaturized MFC in comparison with those of macro/meso-scale MFCs. Nowadays, miniaturized MFCs have achieved both higher power density and coulombic efficiency compared with
macro/meso-scale MFCs.

Table 8.1.  Specifications of miniaturized MFCs.

| Reference | Volume (total) [μL] | Anode size [cm²]/ material | Inoculum | $P_{max, areal}$ [W m⁻²] | $P_{max, volumetric}$ [W m⁻³] | CE [%] | $r_i$ [Ω cm²] | SAV [m⁻¹] |
|---|---|---|---|---|---|---|---|---|
| Choi et al. (2011a) | 9 | 2.25/gold | Geobacter sulfurreducens mixed | 0.047 | 2333 | 31 | 22.5 k | 50000 |
| Choi and Chae (2012) | 100 (per cell) | 1/gold | Geobacter sulfurreducens mixed | 0.33 | 667 | NA | 32 k | 4000 |
| Inoue et al. (2011) | 80 | 0.24/carbon nanotube | Geobacter sulfurreducens mixed | 0.0738** | 16.4** | NA | NA | 589.5 |
| Parra et al. (2009) | 550 | 0.01/gold | Geobacter sulfurreducens mixed | 0.12* | 0.343* | NA | 3.3 k | 2.9 |
| Qian et al. (2011) | 8 | 0.4/gold | Shewanella oneidensis MR-1 | 0.0625* | 62.5 | NA | 6.4 k | 10000 |
| Qian et al. (2009) | 3 | 0.15/gold | Shewanella oneidensis MR-1 | 0.0015 | 15.3 | 2.8 | 36 k | 10000 |
| Siu and Chiao (2008) | 30 | 1.2/gold | Saccharomyces cerevisiae | 0.000424 | 4.24* | 14.7 | 3 k | 10000 |
| Biffinger et al. (2006) | 25*** | 0.45/carbon/pt ink | Shewanella oneidensis DSP10 | 0.006 | 10 | NA | 3.375 k | 1666 |
| Chen et al. (2011) | 25 | 0.126/gold | Shewanella oneidensis MR-1 | 0.029 | 14.6* | NA | 76.86 k | 504 |
| Mink et al. (2012) | 3 | 0.25/carbon nanotube | Geobacter sulfurreducens mixed | 0.02 | 392 | NA | NA | 19600 |
| Ren and Chae (2012) | 200 | 4/gold | Geobacter sulfurreducens mixed | 0.83 | 3320 | 79 | 5.6 k | 400 |
| Ren et al. (2013) | 39.3 | 0.186/gold, ultramicroelectrode | Geobacter sulfurreducens mixed | 7.72 | 3658 | 70.2 | 167 | 400 |

*: Calculated based on the reported data.

**: Calculated based on the enhancement of current outputs by the CNT electrode 205%, with regard to the flat-reference electrode.

***: The volume of the device.

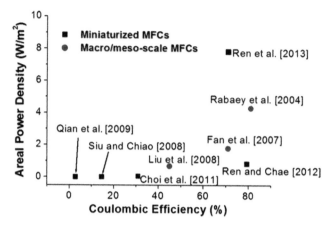

Figure 8.6.    A comparison of areal power density and CE of this work with those of existing macro/meso (Rabaey *et al.*, 2004) and miniaturized MFCs (Choi *et al.*, 2011; Qian *et al.*, 2009; Ren and Chae, 2012; Ren *et al.*, 2013; Siu and Chiao, 2008).

## 8.4    PROMISES AND FUTURE WORK OF MINIATURIZED MFCS

### 8.4.1    *Promises*

In the past ten years, miniaturized MFCs have demonstrated remarkable success. For instance, the current/power densities of them have improved by $10^4$ fold, making them attractive for active research. Advantages of miniaturized MFCs include their small size, ease of forming arrays, high aspect ratio, and fast mass transfer, etc.

In the past half-century, micro-/nano-fabrication technology has revolutionized the life of the human being. This technology has been employed in the micro-electronics and micro-electro-mechanical systems (MEMS) industry, with numerous commercially available products to better the everyday life of humans, such as micro-processors, memorys, liquid crystal displays, micro-sensors (accelerometers and gyroscopes), etc. Batch fabrication of micro-/nano-fabrication technology lowers the cost by manufacturing multiple devices simultaneously. In addition to the batch fabrication capability, the very fine feature sizes, down to nano-meter scale, of micro-/nano-fabrication technology are attractive for exploring the fundamental study of exoelectrogen characterization. Attractive materials, such as carbon nanotubes, graphene, conductive nano-wires, etc., have become available for micro-/nano-fabrication technology and these have potential to improve the current/power harvesting.

The very first outcome of miniaturization is the small feature size, which allows for less building materials to be used, less space to be taken, and less electrolytes required for the operation, etc. This results in substantially less expense for MFC research than for macro-scale research: a practical issue that most researchers may consider. Along with the small feature size, batch fabrication allows ease of forming arrays. An MFC reactor array is useful for screening exoelectrogens and can be connected in a series or parallel to boost the voltage of the current.

Miniaturized MFCs have a higher surface-area-to-volume ratio than a macro-scale MFC, which results in higher volumetric current/power density. The internal resistance is directly proportional to $1/A$:

$$R_i \propto \frac{1}{A} \tag{8.14}$$

The areal and volumetric current and power density can be written as:

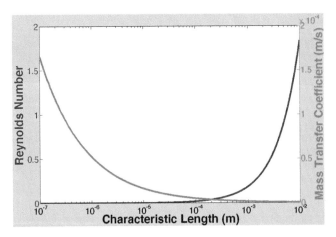

Figure 8.7. Reynolds number and mass transfer coefficient *versus* the characteristic length (arrows indicate the corresponding curve for two *y*-axis parameters).

$$i_{\text{max, areal}} = \frac{I_{\text{max}}}{A} = \frac{E_{\text{OCV}}}{2R_i \cdot A}, \quad p_{\text{max, areal}} = \frac{P_{\text{max}}}{A} = \frac{E_{\text{OCV}}^2}{4R_i \cdot A} \tag{8.15}$$

$$i_{\text{max, volumetric}} = \frac{I_{\text{max}}}{V} = \frac{E_{\text{OCV}}}{2R_i \cdot V} = \frac{E_{\text{OCV}}}{2R_i \cdot A} \cdot SAV$$

$$p_{\text{max, volumetric}} = \frac{P_{\text{max}}}{V} = \frac{E_{\text{OCV}^2}}{4R_i \cdot V} = \frac{E_{\text{OCV}^2}}{4R_i \cdot A} \cdot SAV \tag{8.16}$$

Assuming the areal resistivity ($R_i \cdot A$) and open circuit voltage ($E_{\text{OCV}}$) remain constant, the areal current and power densities remain constant as the device dimensions are scaled, while the volumetric areal and power densities increase linearly with *SAV* as the device dimensions are scaled.

Miniaturized MFCs benefit from high mass transfer, thanks to the smaller characteristic length, which results in higher current and power densities. Mass transfer coefficient in the anode chamber can be written as:

$$k_c = 0.664 \text{Re}^{1/2} \, Sc^{1/3} \left(\frac{D}{L_s}\right) \tag{8.17}$$

where $k_c$, $Re$, $S_c$, $D$, and $L_s$ are the mass transfer coefficient, Reynold's number, Schmidt number, diffusivity, and the characteristic length. $Re$ can be written as: $Re = \rho v L / \mu$, where $\rho$ is the specific density of the fluid, $v$ is the linear velocity of the fluid, $\mu$ is the viscosity of the fluid, and $L$ is the characteristic length.

Assuming all the substrate flux is converted to electricity, the maximum current and power can be written as:

$$I_{\text{max}} = k_c \cdot b \cdot A \cdot e \cdot CE \cdot c \tag{8.18}$$

$$P_{\text{max}} = k_c \cdot b \cdot A \cdot e \cdot CE \cdot c \cdot E \tag{8.19}$$

Figure 8.7 shows the Reynold's number and mass transfer coefficient as a function of characteristic length. As the characteristic length reduces, the Reynolds number, which is a dimensionless number that gives a measure of the ratio of inertial forces to viscous forces, decreases and mass transfer coefficient increases. Consequently, the areal current and power density increases.

### 8.4.2    Future work

Miniaturized MFCs have demonstrated plenty of promise and indeed much progress has prevailed for such a short period of time; yet, miniaturized MFCs face many different challenges to overcome, including further enhancing current and power density, applying air-cathodes to replace potassium ferricyanide, and autonomous operation.

#### 8.4.2.1    Further enhancing current and power density

Enhancing current/power density requires lowering internal resistance and areal resistivity. The reported internal resistance and areal resistivity of miniaturized MFCs are still rather high, in the order of $k\Omega$ and $100\,s\,\Omega\,cm^2$, respectively. The internal resistance is composed of three parts: ohmic resistance, activation resistance and concentration resistance.

*Ohmic resistance*

Ohmic resistance can be divided into resistance of anode $R_a$, cathode $R_c$, electrolyte $R_e$ and ion exchange membrane $R_m$. For simplicity of comparison, all the resistances are normalized to the surface area of the anode to be areal resistivity:

$$r_i = r_a + r_c + r_e + r_m \tag{8.20}$$

where $r_a, r_c, r_e$ and $r_m$ are the areal resistivity of the anode, cathode, electrolyte and the membrane, respectively. $r_a$ is composed of the areal resistivity of the anode itself and of the electron generation and transfer from exoelectrogen to anode. $r_a$ is a function of the anode's material properties, the population of exoelectrogens, and the mechanism of electron transfer from exoelectrogen to anode, etc. $r_c$ is the areal resistivity of the cathode and of ferricyanide ions transferring from the bulk solution to the vicinity of the cathode and reduced electrons, which is mainly determined by the material properties. $r_m$ is the areal resistivity of the ion exchange membrane, which is mainly determined by the properties of electrolytes across the membrane, such as pH, concentration, and substance. $r_e$ is determined by the distance between two electrodes:

$$r_e = \frac{l}{K} \tag{8.21}$$

where $l$ is the distance between the two electrodes, and $K$ is the specific conductivity between two electrodes.

Let us assume the height of both anode and cathode chambers is $100\,\mu m$, and the conductivity for both anolyte and catholyte is $0.017\,S\,cm^{-1}$. The calculated $r_e$ is $1.2\,\Omega\,cm^2$. $r_m$, resistivity of the ion exchange membrane (Nafion, 117 for instance) is approximately $10\,\Omega\,cm^2$ in phosphate buffer medium. $r_c$, resistivity of the cathode is also low when the current density is low enough that the concentration loss is not a dominating factor, which is the case in the ohmic region. As a result, the $r_a$, anode resistivity dominates the total resistivity in miniaturized MFCs.

The anode resistivity, $r_a$, is attributed to electrode resistivity, and to the resistivity associated with the electron transfer from the exoelectrogen to the anode (it can also be seen as the contact resistance. Resistance of the biofilm is also incorporated into the electron transfer resistivity). For a conductive anode, such as gold, CNT, carbon, etc., the electrode resistivity is negligible. Consequently, the main source of the anode resistivity comes from the electron transfer from the exoelectrogen to the anode.

Because the anode resistivity, $r_a$, dominates the overall resistivity, having a biofilm of high catalytic capability is essential. Materials with higher SAV, lower contact resistance, and better biocompatibility, which result in a larger exoelectrogen population, help to reduce the high areal resistivity. 3D structured materials with a high surfacearea-tovolume ratio, such as CNT and graphene, have been adopted in MFCs, and a significant power density enhancement has been achieved. Figure 8.8 shows examples of such efforts: CNT and graphene-based anodes presented by Inoue *et al.* (2012) (micro-scale MFC) and Xie *et al.* (2012) (macro-scale MFC).

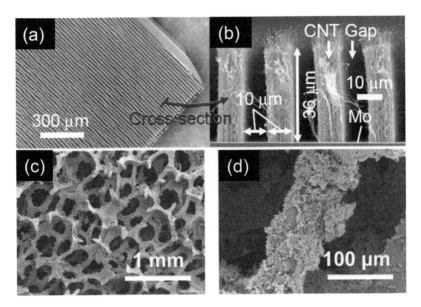

Figure 8.8.   (a, b) SEM images of biofilm on vertically aligned CNT forest electrodes presented by Inoue *et al.* (2012): (a) top view (b) cross section view (Inoue, 2012); (c, d) SEM images of biofilm on graphene-sponge composite at different scales ((Xie *et al.*, 2012).

In the future, other 3D electrodes, such as electrodes made from 3D printing, electrospun nano-wire, 3D electrodes of conductive sponge, etc., can be used to fully take up the volume of the anode chamber, offering a larger surface area to volume ratio.

*Activation and concentration resistance*
Activation resistance mainly dominates at low current density and concentration resistance mainly dominates at high current density, as shown in Figure 8.2. As discussed in detail (Ren *et al.*, 2012), increasing the surface area to volume ratio and applying a catalyst onto the anode and cathode, helps reduce the activation resistance. In order to reduce the concentration resistance, enhancing the mass transfer into the biofilm, including increasing substrate and buffer concentration, increasing flow rate, etc., can be applied.

### 8.4.2.2   *Applying air-cathodes to replace potassium ferricyanide*
Almost all miniaturized MFCs employ potassium ferricyanide as an electron acceptor in the cathode which is toxic and environmentally unfriendly. As an alternative, oxygen is the ultimate electron acceptor, because it is renewable and has a higher standard potential than ferricyanide (0.805 V *versus* 0.361 V). However, several challenges remain against the use of oxygen as an electron acceptor. The first challenge is its high overpotential. Even with platinum decorated cathode, overpotential can be as high as 0.3–0.4 V, thus the OCV of MFCs with oxygen as an electron acceptor becomes low and often less than 0.8 V.

Another challenge are the low reaction kinetics of oxygen. Performance of air-cathode MFCs is mostly limited by the air-cathode, which degrades the performance of the MFC. Fan *et al.* (2007) changed the size of the air-cathode, and reported that as anode/cathode area ratio reduces from 1:1 to 1:7, the areal power density increased from 1.04 W m$^{-2}$ to 6.71 W m$^{-2}$ (normalized to the anode area). Thus, enhancing the reaction kinetics of oxygen is critical.

A further critical challenge in miniaturized MFCs is that most research utilizes an electrically driven peristaltic or syringe pump to drive the electrolyte into the corresponding chambers, and more often than not, the electricity generated is far less than the electricity consumed by these pumps.

### 8.4.2.3   *Reducing the cost of MFCs*

The cost of miniaturized MFCs is still high. The cost of building one micro-scale MFC from our group is ~30 US$ (Ren and Chae, 2012), which can be reduced substantially. Of the total cost, the proton exchange membrane takes one third of the total cost, thus acquiring inexpensive proton exchange membranes is critical. Therefore, developing low-cost ion exchange membranes, and having comparable performances to existing exchange membranes, is attractive for affordable miniaturized MFCs.

## 8.5   CONCLUSION

This chapter reviewed the promises and challenges of micro-/nano-technologies on a microbial fuel cell, which is a carbon-neutral, renewable energy converter. Through micro-/nano-technologies, its power density may be significantly enhanced, which enables it to be a promising candidate for a portable power converter. After briefly introducing the technological backgrounds and experimental methods, we reviewed the significant enhancement in current and power ($10^3$ and $10^4$) in the micro-scale MFCs. The performance may continue to improve upon further research on MFCs, and the impact of having high power miniaturized MFCs may be significant in sustainable energy developments.

## REFERENCES

Bentley, R.W.: Global oil & gas depletion: an overview. *Energy Policy* 30:3 (2002), pp. 189–205.

Chiao, M., Lam, K.B., Su, Y. & Lin, L.: A miniaturized microbial fuel-cell. In: *Solid-State Sensor, Actuator and Microsystems Workshop*. Hilton Head Island, SC, 2002.

Chiao, M., Lam. K.B. & Lin, L.: Micromachined microbial and photosynthetic fuel-cells. *J. Micromech. Microeng.* 16:12 (2006), pp. 2547–2553.

Choi, S. & Chae, J.: An array of microliter-sized microbial fuel-cells generating 100 μW of power. *Sensor. Actuator. A Physical* 177:0 (2012), pp. 10–15.

Choi, S. & Chae, J.: Optimal biofilm formation and power generation in a micro-sized microbial fuel-cell (MFC). *Sensor. Actuator. A Physical* 195 (2013), pp. 206–212.

Choi, S., Lee, H.S., Yang, Y., Rittmann, B.E. & Chae, J.: A μL-scale micromachined microbial fuel-cell having high power density. *Lab on a Chip* 11:6 (2011), pp. 1110–1117.

Davila, D., Esquivel, J.P., Sabate, N. & Mas, J. Silicon-based microfabricated microbial fuel-cell toxicity sensor. *Biosens. Bioelectron.* 26:5 (2011), pp. 2426–2430.

Fan, Y., Hu, H. & Liu, H.: Enhanced Coulombic efficiency and power density of air-cathode microbial fuel-cells with an improved cell configuration. *J. Power Sources* 171:2 (2007), pp. 348–354.

Inoue, S.: Structural optimization of contact electrodes in microbial fuel-cells for current density enhancements. *Sensor. Actuator. A Physical* 177:7 (2012), pp. 30–36.

Liu, H., Cheng, S., Huang, L. & Logan, B.E.: Scale-up of membrane-free single-chamber microbial fuel-cells. *J. Power Sources* 179:1 (2008), pp. 274–279.

Logan, B.: *Microbial fuel-cells*. John Wiley & Sons, Inc., Hoboken, NJ, 2008.

Logan, B.E.: Peer reviewed: extracting hydrogen and electricity from renewable resources. *Environ. Sci. Technol.* 38:9 (2004), pp. 160A–167A.

Logan, B.E. & Rabaey, K.: Conversion of wastes into bioelectricity and chemicals by using microbial electrochemical technologies. *Science* 337:6095 (2012), pp. 686–690.

Malvankar, N.S., Vargas, M., Nevin, K.P., Franks, A.E., Leang, C., Kim, B.C. & Lovley, D.R.: Tunable metallic-like conductivity in microbial nanowire networks. *Nature Nanotechnol.* 6:9 (2011), pp. 573–579.

Marsili, E., Baron, D.B., Shikhare, I.D., Coursolle, D., Gralnick, J.A. & Bond, B.R.: Shewanella secretes flavins that mediate extracellular electron transfer. *PNAS* 105:10 (2008), pp. 3968–3973.

McCarty, P.L., Bae, J. & Kim J.: Domestic wastewater treatment as a net energy producer – Can this be achieved? *Environ. Sci. Technol.* 45:17 (2011), pp. 7100–7106.

Mink, J.E., Rojas, J.P., Logan. B.E. & Hussain, M.M.: Vertically grown multiwalled carbon nanotube anode and nickel silicide integrated high performance microsized (1.25 μL) microbial fuel-cell. *Nano Lett.* 12:2 (2012), pp. 791–795.

Parra, E. & Lin, L.: Microbial fuel-cell based on electrode-exoelectrogenic bacteria interface. Paper read at Micro Electro Mechanical Systems, 2009. MEMS 2009. *IEEE 22nd International Conference*, 2009.

Perlack, R.D, Wright, L.L., Turhollow, A.F., Graham, R.L., Stokes, B.J. & Erbach, D.C.: Biomass as feedstock for a bioenergy and bioproducts industry: the technical feasibility of a billion-ton annual supply. DTIC Document, 2005.

Pikul, J.H., Zhang, H., Cho, J. Braun, P. & King, W.P.: High-power lithium ion microbatteries from interdigitated three-dimensional bicontinuous nanoporous electrodes. *Nature Commun.* 4 (2013), 1732.

Pocaznoi, D., Erable, B., Delia, M. & Bergel, A.: Ultra microelectrodes increase the current density provided by electroactive biofilms by improving their electron transport ability. *Energy Environ. Sci.* 5:1 (2012), pp. 5287–5296.

Qian, F., Baum, M., Gu, Q. & Morse, D.E.: A 1.5 $\mu$L microbial fuel-cell for on-chip bioelectricity generation. *Lab on a Chip* 9:21 (2009), pp. 3076–3081.

Qian, F., He, Z., Thelen, M.P. & Li, Y.: A microfluidic microbial fuel-cell fabricated by soft lithography. *Bioresource Technol.* 102:10 (2011), pp. 5836–5840.

Rabaey, K., Boon, M., Siciliano, S.D., Verhaege, M. & Verstraete, W.: Biofuel-cells select for microbial consortia that self-mediate electron transfer. *Appl. Environ. Microbiol.* 70 (2004), pp. 5373–5382.

Ren, H. & Chae, J.: Scaling effect on MEMS based microbial fuel-cells toward a carbon-neutral miniaturized power source. In: *IEEE Solid-State Sensors and Actuators Workshop*. Hilton Head, SC, 2012.

Ren, H., Lee, H. & Chae, J.: Miniaturizing microbial fuel-cells for potential portable power sources: promises and challenges. *Microfluid. Nanofluid.* 13:3 (2012), pp. 353–381.

Ren, H., Rangaswami, S. Lee, H. & Chae, J.: A micro-scale microbial fuel cell (MFC) having ultramicro-electrode (UME) anode. Paper read at Micro Electro Mechanical Systems (MEMS), *2013 IEEE 26th International Conference*, 2013.

Ren, H., Pyo, S., Lee, J., Park, T., Gittleson, F.S., Leung, F.C., Kim, J., Taylor, A.D. & Chae, J.: A high power density miniaturized microbial fuel cell having carbon nanotube anodes. J. Power Sources 273 (2015), pp. 823–830.

Ren, H., Torres, C.I., Parameswaran, P., Rittmann, B.E. & Chae, J.: Improved current and power density with a micro-scale microbial fuel cell due to a small characteristic length. Biosens Bioelectron. 61 (2014), pp. 587–592.

Richter, H., McCarthy, K., Nevin, K.P., Johnson, J.P., Rotello, V.M. & Lovley, D.R.: Electricity generation by *Geobacter sulfurreducens* attached to gold electrodes. *Langmuir* 24:8 (2008), pp. 4376–4379.

Siu, C.P.B. & Chiao, M.: A microfabricated PDMS microbial fuel-cell. *J. Microelectromech. Syst.* 17:6 (2008), pp. 1329–1341.

Snider, R.M., Strycharz-Glaven, S.M., Tsoi, S.D., Erickson, J.S. & Tender, L.M.: Long-range electron transport in *Geobacter sulfurreducens* biofilms is redox gradient-driven. *PNAS* 109:38 (2012), pp. 15,467–15,472.

Sorrel S., Speirs, J., Bentley, R., Brandt, A. & Miller, R. Global oil depletion: an assessment of the evidence for a near-term peak in global oil production. A report produced by the Technology and Policy Assessment function of the UK Energy Research Centre, 2009.

Sorrell, S., Speirs, J., Bentley, R., Brandt, A. & Miller, M.: Global oil depletion: a review of the evidence. *Energy Policy* 38:9 (2010), pp. 5290–5295.

Xie, X., Yu, G.H., Liu, N., Bao, Z.N., Criddle, C.S. & Cui, Y.: Graphene-sponges as high-performance low-cost anodes for microbial fuel-cells. *Energy Environ. Sci.* 5:5 (2012), pp. 6862–6866.

# CHAPTER 9

## Microbial fuel cells: the microbes and materials

Keaton L. Lesnik & Hong Liu

### 9.1 INTRODUCTION

In contrast to traditional chemical fuel cells, microbial fuel cells (MFCs) rely on biologically driven chemical reactions to produce electricity, the ultimate fuel source of which is biologically degradable organic matter. It was over a century ago that M.C. Potter first noted that the liberation of electrical energy can be tied to the microbial oxidation of organic compounds (Potter, 1911), but it has not been until recently that engineers have managed to harness this ability as an energy source. Harnessing this power has allowed the direct generation of electricity from an immense range of degradable organic waste that 3.8 to 3.9 billion years of evolutionary time has imbued bacteria with the ability to oxidize.

In an MFC, this ability to oxidize organic compounds is used to catalyze the direct generation of electricity from those compounds. Currently, difficulties in obtaining high power densities have limited practical applications of MFCs for energy production (Lovley and Nevin, 2011). However, incorporation of MFCs energy harvesting abilities into a wastewater treatment infrastructure continues to spur research on scaling up MFC technology. Other interesting practical applications of MFCs do not emphasize power production, and instead MFCs are used as a way to enhance bioremediation efforts by providing an electrode as an electron acceptor (Zhang, T. et al., 2010) or as a way to monitor microbial respiration in anaerobic environments (Williams et al., 2010).

In order for envisioned applications of MFCs to become commercially viable, current limitations in MFC technology must be overcome. Understanding how microbes transfer electrons extracellularly and interact both with electrodes and other microbes along with developing efficient low-cost materials will lead to future breakthroughs in MFC technology. In this chapter, we will focus on highlighting these areas of research that will continue to push this technology forward in what is a rapidly developing field.

### 9.2 HOW MICROBIAL FUEL CELLS WORK

Electron transfer as a means to generate energy is universal to all life. One of the most common energy harvesting processes is respiration. During respiration, organic compounds are oxidized and the resulting electrons are transferred through various cellular respiratory enzymes and eventually to a terminal electron acceptor. This terminal electron acceptor can be oxygen (aerobic respiration), pyruvate/pyruvate derivatives (fermentation), or various inorganic compounds (anaerobic respiration). Many bacteria capable of anaerobic respiration have the ability to reduce insoluble minerals such as Fe(III) oxides. To effectively reduce these insoluble minerals bacteria must be able to transfer electrons outside of the cell; this is known as extracellular electron transfer (EET). EET is also required to transfer electrons to MFC electrodes. Bacteria capable of transferring electrons extracellularly are known as exoelectrogens and are the engines of the MFC (Logan, 2009).

The first step in the generation of electricity from MFCs is the exoelectrogen oxidizing an organic substrate in the anode chamber (Fig. 9.1). This reaction produces electrons as well as protons and carbon dioxide. While the electrons travel to the anode and through the circuit to the cathode, the protons diffuse through a chamber separator (optional) to the cathode. At the

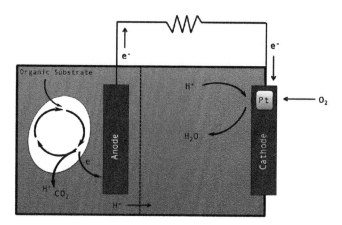

Figure 9.1.   Schematic of a microbial fuel cell. Bacteria oxidize organic compounds, electrons travel through microbial respiratory enzymes generating ATP for the cell. Electrons are then transferred extracellularly to the anode where they travel through the circuit to the cathode. At the cathode, electrons combine with protons generated by microbial respiration and ambient oxygen at the platinum catalyst to generate water.

cathode, the protons and electrons combine with the terminal electron acceptor. Oxygen is the most commonly utilized terminal electron acceptor in MFCs but others such as ferricyanide may be used (Rabaey *et al.*, 2004). If oxygen is employed as the terminal electron acceptor a catalyst such as platinum is typically utilized to increase the reaction rate of the oxygen reduction reaction. Anode, cathode, and separator material choices are of key importance and will be discussed at greater length in this chapter.

## 9.3   UNDERSTANDING EXOELECTROGENS

Understanding how bacteria interact with each other and electrodes is critical to the future development of MFC technology. Microbe-electrode interactions are important as different materials elicit varying transcriptional responses (Xu *et al.*, 2012). In addition to understanding microbe-electrode interfaces, elucidating microbe-microbe interactions, such as those in mixed consortium biofilms that result in higher currents than single species biofilms, have the potential to uncover mechanisms that could be used to optimize MFC systems.

### 9.3.1   *Origins of microbe-electrode interactions*

Many bacteria are capable of extracellular electron transfer (EET), meaning they have the ability to exchange electrons with insoluble minerals and other naturally occurring electron acceptors outside of their cells (Bond, 2010; Logan, 2009). EET has allowed many bacteria to interact electronically with electrodes as well even though electrodes have not been part of the natural environments in which they evolved (Lovley, 2008). Understanding how microbes evolved this ability will help better understand the microbe-electrode interaction.

Microorganisms inherently transfer electrons from a lower potential electron donor to a high potential electron acceptor as a part of their respiratory metabolism. In many anaerobic settings, microbes can use insoluble minerals as an electron acceptor by taking advantage of EET. However, just because a bacterium can transfer electrons to an insoluble mineral does not mean it can interact electronically with an electrode. *Pelobacter* is an example of a microorganism capable of growing on Fe(III) but not on electrodes (Richter *et al.*, 2007). Insoluble minerals and electrodes are actually quite different. Electrodes can supply electron-accepting abilities indefinitely while the electron-accepting abilities of minerals are exhausted quickly. Though most *Geobacter* species are capable of using both Fe(III) and electrodes to support their growth effectively there are significant transcriptional differences associated with growth using the differing electron acceptors (Childers *et al.*, 2002; Nevin *et al.*, 2009; Reguera *et al.*, 2006). When *Geobacter* species are grown on Fe(III) they express flagella and are planktonic, yet when grown on electrodes thick biofilms are formed and electron transfer rates are greatly increased (Childers *et al.*, 2002; Nevin *et al.*, 2009; Reguera *et al.*, 2006). These studies support the concept that there may be another evolutionary origin outside of the direct reduction of insoluble minerals that has conferred bacteria with the ability to interact with electrodes.

One theory on how bacteria may have developed this ability is through association with a geobattery (Lovley, 2012). Geobatteries are formed when graphite deposits are able to transmit electrons between anaerobic and oxic zones in the benthic subsurface (Bigalke and Grabner, 1997), creating an environment similar to that of an MFC. Alternatively, it could be that the ability to interact with an electrode is a product of bacteria evolving their ability to transfer electrons between microorganisms. This ability is known as interspecies electron transfer (IET). Mechanisms of IET that have been clearly identified involve electron transfer through hydrogen and other redox active molecules (Watanabe *et al.*, 2009) in addition to direct contact in cell aggregates (Summers *et al.*, 2010). Recent research suggests IET can also occur via electrical currents mediated by conductive nano-particles (Kato *et al.*, 2012). Similar nano-particles have also been shown to be capable of shuttling electrons to electrodes in electrochemical chambers (Kato *et al.*, 2010). It is clear that bacteria have evolved to interact electronically with a variety of conductive materials. Further understanding the origins and mechanisms behind these interactions is critical.

### 9.3.2  *Extracellular electron transfer (EET) mechanisms*

EET mechanisms are nearly as diverse as the bacteria that perform them. Some bacteria transfer electrons to electrodes efficiently on their own while others need help in the form of exogenous redox molecules. Some of the most proficient exoelectrogens likely employ some combination of the mechanisms discussed here (Fig. 9.2). Understanding these mechanisms and how they interplay is integral to future engineering efforts of MFCs along with other bioelectrochemical systems.

#### 9.3.2.1  *Redox shuttles/mediators*
The use of redox shuttles to mediate EET is perhaps the most widespread EET mechanism. Many bacteria otherwise incapable of interacting with electrodes can interact with them via the addition of exogenous mediators. EET mediators are typically small, soluble, highly reactive redox couples capable of performing multiple cycles of electron transfer between a cell and an electrode (Rabaey, 2010). The reactivity and soluble nature of these compounds significantly reduce overpotentials that may develop due to the distance between the redox centers of the microbial enzymes responsible for transferring the electrons and the electrodes themselves (Lojou *et al.*, 2005). The redox centers of many of these enzymes are located inside cells with thick cell walls making direct transfer problematic; in this case the use of redox shuttles is the only way to exchange electrons with insoluble electron acceptors. These same redox centers that exchange electrons with mediators may also be involved in direct transfer mechanisms diffusing the idea of dedicated pathways for each mechanism (Coursolle *et al.*, 2010).

Though artificial mediators such as resazurin, methylviologen, neutral red and thionine may be used for select purposes including immobilization onto electrodes, it is not practical to consider them as part of an efficient design as they can be expensive and possibly toxic (Lovley, 2006). Mediators in general do have their advantages though. One of the advantages is that mediators are non-specific. The mediators produced by a specific microorganism can be used by a wide range of other species. The same is true of natural mediators, many different species can effectively use mediators found naturally in the environment for their electron transfer needs. Some of these

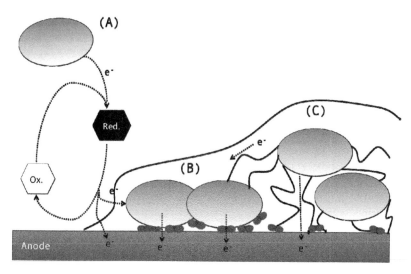

Figure 9.2.   Mechanisms of EET: (a) Electron transfer through use of redox mediator. Mediator is reduced by
bacterium and subsequently oxidized by electrode. Use of mediator to transfer electrons to other
bacteria may be possible. (b) Electron transfer through outer membrane c-type cytochromes.
Electrons are transferred through outer membrane proteins in direct contact with electrode. (c)
Electron transfer through use of microbial nano-wires. Conductive pili confer conductivity to
the biofilm allowing electron transfer to other bacteria as well as electrode; transfer of electrons
from nano-wires to electrodes likely catalyzed by c-type cytochromes dispersed in the biofilm.

natural mediators include humic acids and a variety of sulfur compounds such as cysteine (Rabaey,
2010).

One class of microbially produced redox mediators that have been identified are phenazines.
Multiple phenazines produced by *Pseudomonas aeruginosa* including pyocyanin, enable power
production by *P. aeruginosa* along with other bacteria in a mixed-consortium MFC (Rabaey *et al.*,
2004). *Geobacter* and *Shewanella* have become model organisms for understanding EET. Unlike
*Geobacter, Shewanella* does not produce thick conductive biofilms and is planktonic in MFCs
suggesting that excreted mediators may be responsible for *Shewanella's* EET ability (Lanthier
*et al.*, 2008). Flavins were later identified as mediators and following electrochemical studies
have indicated that they play a primary role in *Shewanella's* ability to reduce electrodes (Baron
*et al.*, 2009; Canstein *et al.*, 2008; Marsili *et al.*, 2008).

### 9.3.2.2   c-type cytochromes

Redox active proteins located on the outer membrane of some cells, such as c-type cytochromes
and iron-sulfur proteins, have the ability to transfer electrons directly to electrodes (Rabaey,
2010). This mechanism has been demonstrated for several different species of bacteria, but most
in-depth studies of direct contact mechanisms have focused on *Geobacter sulfurreducens* due to
the availability of a genome sequence and presence of over 100 different types of cytochromes
(Methe *et al.*, 2003).

Several of these outer membrane cytochromes have been purified and characterized. Of these,
OmcZ has been identified as playing a pivotal role in biofilms producing high currents. OmcZ
contains eight heme groups with redox potentials between $-420$ to $-60$ mV. Deletion of the
OmcZ gene significantly reduced current production in biofilms and increased resistance of
electron transfer from the biofilm to electrodes was observed. In contrast, deletion of other
c-type cytochromes did not affect current production significantly (Nevin *et al.*, 2009). This lack of
redundancy may mean that different cytochromes are associated with different electron acceptors
with OmcZ being the one that interacts electronically with solid insoluble electron acceptors

like electrodes. Engineering around mechanisms associated with OmcZ and other proteins with similar functions in other microorganisms would appear to be critical yet it is still unknown what exact properties confer electrode interacting abilities to certain c-type cytochromes. One reason why certain c-type cytochromes may be able to interact compared to others is it's physical position in the biofilm. Cytochromes such as OmcZ appear to be positioned along the interface between the electrode and the biofilm allowing them to be in contact with the electrode (Busalmen *et al.*, 2008; Dumas *et al.*, 2008; Inoue *et al.*, 2011). Other outer membrane proteins such as OmcS are associated with conductive pili that confer conductivity to the biofilm allowing long-range electron transport through thicker biofilms (Lovley, 2011).

### 9.3.2.3 *Conductive pili*
Electron flux would be limited by microbial respiration rates if only cells that were in contact with the electrode could transfer electrons. Similarly, if redox shuttles were the only way to exchange electrons over a long range (that is, distances many times larger than a single cell) transfer rates would be equal to the slow diffusion of these shuttles through a biofilm or solution. Observed electron flux in many MFC setups is significantly greater than these two mechanisms would predict. This is due to effective long-range electron transport through conductive pili known as microbial nano-wires (Reguera *et al.*, 2005). Microbial nano-wires, such as those produced by *G. metallireducens* and *G. sulfurreducens*, are microbially produced pili filaments that measure 3–5 nm in diameter and extrude between 10–30 μm from the cell (Lovley, 2012). These microbial nano-wire filaments are likely to interact and exchange electrons directly with known cytochromes such as OmcS that may be dispersed throughout the biofilm (Lovley, 2011). In *G. sulfurreducens,* the expression of a PilA subunit, which comprises microbial nano-wires was directly related to the conductivity of the biofilm (Malvankar *et al.*, 2011; Reguera *et al.*, 2006) and high current output (Yi *et al.*, 2009). While there is still considerable debate surrounding the nature of the conductivity (Bond *et al.*, 2012), current research indicates that these filaments allow for rapid electron transfer by conferring a metallic-like conductivity to the extracellular matrix of the biofilm, a first for a biological protein (Malvankar *et al.*, 2012). The features of these filaments that permit this phenomenon have not yet been characterized due in part to difficulties in crystallization.

### 9.3.3 *Interactions and implications*

Maximum power densities in biofilms with diverse communities of bacteria are generally higher than pure cultures, but when the bacteria responsible for impressive current outputs in mixed biofilms are isolated they produce less power than pure cultures (Logan, 2009). Reconstituting biofilms (Zhang *et al.*, 2012) and adding supernate of other bacteria (Pham *et al.*, 2008) can lead to power increases, yet the interspecies interactions in these complex communities of MFC anode biofilms are currently not well understood. Insight into how bacteria interact with electrodes is also important for optimizing bioelectrochemical systems.

## 9.4 ANODE MATERIALS AND MODIFICATIONS

There are clear differences between chemical and microbial fuel cell anodes. The most obvious difference is that anodes of MFCs must be able to support the growth of biological organisms. MFC anodes must also be highly conductive in order to efficiently collect electrons produced by bacteria as small increases in material resistance can have a significant impact on maximum power outputs. Other considerations when selecting an anode material include the expense of the material and the ability for it to be manufactured on a large scale.

Metals such as platinum (Schroder *et al.*, 2003), gold (Richter *et al.*, 2008), titanium (Ter Heijne *et al.*, 2008), stainless steel (Dumas *et al.*, 2007) and copper (Kargi and Eker, 2007) have been investigated as MFC anodes due to their high conductivity, but the overall performance of these materials has been underwhelming. This could be due to them having less than the optimal

Table 9.1.   Comparison of anode materials for microbial fuel cells.

| Anode | Advantages | Disadvantages | Max. Power[1] $(mW\ m^{-2})/(W\ m^{-3})$ | Ref. |
|---|---|---|---|---|
| Graphite plates/rods | Tractability Cost | Surface area | 480/– | Bond and Lovley (2003) |
| Graphite brushes | Surface area porosity | Electrode spacing | 1430/– | Logan *et al.* (2007) |
| Graphite granular | Surface area porosity | Conductivity | –/90 | Rabaey *et al.* (2005) |
| Carbon RVC | Conductivity | Biofilm support | 0.2/– | Larrosa-Guerrero *et al.* (2010) |
| Carbon cloth | Surface area conductivity electrode spacing | – | 4320/2080 | Fan *et al.* (2012) |

[1]Conclusions drawn from comparing maximum power between materials may not be valid due to large variance in MFC designs used.

surface properties for biofilm establishment and lower specific surface areas. The use of metals such as platinum and gold in large-scale applications would also be restricted due to their high cost while the practicality of metals such as copper are potentially corroded by MFC anodic communities. Carbon-based materials are currently the best option for most MFC setups.

### 9.4.1   *Carbon-based anode materials*

Carbonaceous anodes typically have larger surface areas, better biocompatibility, and increased chemical stability compared to metal anodes while maintaining high conductivity (Table 9.1). These characteristics make carbon-based anodes the most popular choice for use in MFCs. Many early MFC studies experimented with the use of solid graphite anodes, often as part of a sediment MFC setup where the anode is embedded in marine sediment (Tender *et al.*, 2002). Although these types of anodes are inexpensive and easy to work with the limited surface area limits maximum power densities (Bond and Lovley, 2003). Brush anodes have the largest surface area and highest porosities of all anode materials and have generated high current densities (1430 mW m$^{-2}$) (Logan *et al.*, 2007), but brush anode designs must take into account increased internal resistance due to increased electrode spacing. Granular graphite and carbon have a similar electrode spacing design problem in addition to being unable to establish electrical connections throughout the surface area of the granules, but were still able to generate volumetric power densities of up to 90 W m$^{-3}$ due to large surface areas (Rabaey *et al.*, 2005). Using carbon-based fabrics such as carbon cloth (Fan *et al.*, 2012; Nam *et al.*, 2010), felt (Kim *et al.*, 2002; Park and Zeikus, 1999), mesh (Wang *et al.*, 2009), and paper (Liu *et al.*, 2005) allows for reduced electrode spacing due to their planar nature thereby significantly reducing internal resistance and increasing surface area to volume ratios. Carbon fabrics combine large surface areas with good conductivity and excellent durability making them the material of choice for most MFC anode designs. Of all the carbon fabrics, carbon cloth anodes have generated some of the highest recorded power densities (4320 mW m$^{-2}$, 2080 W m$^{-3}$) (Fan *et al.*, 2012). Other carbonaceous materials such as reticulated vitreous carbon (RVC) may also be used (He *et al.*, 2005). RVC is highly conductive, but is limited to low power densities due to the inability to support robust biofilm growth and the buildup of concentration polarizations (Larrosa-Guerrero *et al.*, 2010). All in all, carbon cloth has so far proven to be the best material choice for an anode in an MFC.

## 9.4.2   *Anode modifications*

In order to improve MFC performance many different anode modifications have been investigated. Most modifications aim to increase performance by altering surface properties thereby enhancing biocompatibility, conductivity, and activity of the anode. To achieve reduced acclimation times anodes have been treated with ammonia gas (Cheng and Logan, 2007) or had quartz particles attached (Johnson and Logan, 1996) in order to increase the positive surface charge and the associated electrostatic attraction of cells to the surface. Both these modifications resulted in slight increases in MFC performance while the vapor deposition of iron oxide only reduced acclimation without improvements in performance (Kim *et al.*, 2005).

Improving the kinetic activity associated with the transfer of electrons to the anode could potentially improve MFC performance. To these ends, several different composites have been constructed and thoroughly tested. One study reported increased performance of graphite $-Fe^{3+}$, $-Mn^{4+}$ and $-$neutral red composites (Park and Zeikus, 2003). A subsequent study observed an increase in kinetic activity of composites 1.5 to 2.2 times that of those associated with graphite controls. The composites tested combined graphite with one or two of a selection of minerals ($Mn^{2+}$, $Ni^{2+}$, $Fe_3O_4$) and mediators (anthraquinone-1,6-disulfonic acid (AQDS), 1,4-napthoquinone (NQ)) (Lowy *et al.*, 2006). Additional research on various other anode metal modifications has shown limited success (Logan, 2008).

Conductivity of the anode is important to MFC performance. One method that has been successfully used to increase conductivity and current densities of MFC is the application of the conductive polymer polyaniline (Schroder *et al.*, 2003). However, performance was only improved temporarily as polyaniline was shown to be unstable and susceptible to microbial degradation (Niessen *et al.*, 2004). While these findings would seemingly limit polyaniline's potential contribution to future MFC designs research has suggested that it may be possible to improve both the stability and performance of polyaniline by making composites combined with fluorine, carbon nanotubes (Qiao *et al.*, 2007) and titanium dioxide (Qiao *et al.*, 2008).

Anodes of MFCs have also been modified by the application of varying alkanethiol self-assembled monolayers (SAMs). Strong hydrogen bonding between the carboxylic acid head groups of the SAMs and cytochromes in the bacteria may have been the reason for the increased power densities observed with these anode modifications (Crittenden *et al.*, 2006). The strong hydrogen bonds are hypothesized to lead to increased electronic interaction by stabilizing electron carriers from the cell to the electrode surface. This proposed mechanism is not without precedent as humates found in soils adsorb to metal oxide surfaces and play a similar role in those environments (Williams *et al.*, 2010).

Several nano-materials have been researched for MFC anodes applications, yet research is still in the early stages of understanding how these materials interact with the microbes to result in increased (or decreased) current production. When glassy carbon anodes were decorated with carbon nanotubes (CNT) a current density of $9.70 \, \mu A \, cm^{-2}$ was recorded, nearly 82 times greater than the glassy carbon control (Qiao *et al.*, 2007). In another experiment, multi-wall CNTs and polyelectrolyte polyethyleneimine (PEI) were combined in a layer-by-layer composite onto carbon paper electrodes resulting in a 20% increase in power density (Sun *et al.*, 2010). Gold nano-particles have also been used to decorate graphite resulting in 20-fold increase in current densities. However, additional experiments with palladium nano-particles did not result in greatly increased current densities (Fan *et al.*, 2011). The cost of using gold and other precious metals is likely to limit its practicality in large-scale applications. With this thought in mind, the use of iron nano-particles was also explored. Researchers reported a 5.9-fold increase in current densities with iron nano-particle decorated anodes. In response to these nano-particle decorations, biofilms were observed to have increased in thickness, increased the concentration of flavins, and increased number of outer membrane cytochromes (Xu *et al.*, 2012). The how and why for such responses are unclear though further improvements in anode modifications depend greatly on an enhanced understanding of the molecular mechanisms behind microbe-electrode interactions. Multiple studies have also suggested that graphene modifications of electrodes can be an effective way of improving MFC performance. Crumpled graphene particles were shown to increase power densities from 0.3 to 3.3 W m$^{-3}$ compared to unmodified carbon cloth (Xiao et al., 2012). Stainless steel anodes modified with graphene exhibited power densities 18 times larger than their unmodified counterparts (Zhang et al., 2011). Graphene composite electrodes have also been explored with promising results (Yong et al., 2012).

The field of MFC anode modifications is still developing, especially pertaining to the use of nano-materials, yet the research reviewed in this chapter suggests that several of these modifications can substantially aid electron transfer and improve MFC performance. However, research also indicates that reactions occurring at the anode are not the rate-limiting step for many of the most productive MFC designs. Additionally, modifications to current designs, such as the use of nano-materials, may be excessively expensive for large-scale applications. Therefore, improvements in cathode materials and designs are more critical to overall performance at this time.

## 9.5   CATHODE MATERIALS AND CATALYSTS

The development of a high-performance, low-cost cathode material represents a significant engineering challenge. For most MFC designs it is the chemical reaction occurring at the cathode, which limits power generation and hinders large-scale applications (Fan et al., 2008). While simply increasing the surface area of the cathode relative to the anode in setups limited by the cathode reaction will result in power increases research is rightly focused on developing newer, cheaper cathode materials and designs.

Ferricyanide (Rabaey et al., 2004) and permanganate (You et al., 2006) have been used as terminal electron acceptors, but the most common and practical MFC designs use passive air cathodes. Part of the challenge in using these air cathodes is getting the protons, electrons, and oxygen to meet at the catalyst (typically Pt) and react. The catalyst must be in contact with a conductive surface and be exposed to both the solution containing the protons as well as the air containing the oxygen. The reaction mechanism proceeds by the adsorption of oxygen to the catalyst surface. Subsequently, electrons and protons are supplied resulting in desorption of water as illustrated in the equation below:

$$O_2 + 4H^+ + 4e^- \rightarrow 2H_2O \; (E'^0 = 0.816\,V)$$

### 9.5.1   *Cathode construction*

The materials used for MFC air cathode conductive layers are similar to the materials used for MFC anodes. Carbon cloth is also the most common design choice, but designs using stainless steel current collectors have been used (Zhang, X. et al., 2010). In most designs, the catalyst faces the solution side with several diffusion layers on the air-exposed side. When a particulate catalyst like platinum is used, a conductive binding agent is often applied to coat the surface of the cathode. Most often Nafion, a perfluorinated proton exchange resin, is used in this role. Ideally, Nafion would be able to reduce the amount of platinum required while also serving a role facilitating charge transfer at neutral pH. However, Nafion is an expensive compound and recent evidence suggests it may not be necessary. Proton transfer to the catalyst is mainly carried out by anions in the neutral environments of MFCs, and a cation exchange resin like Nafion may actually increase the polarization resistance of the cathode. Binders with sulfonate groups (Saito et al., 2010) and binders that are exceedingly hydrophobic such as polytetrafluoroethylene (PTFE) also performed poorly (Cheng et al., 2006b; Wang et al., 2010). Simply replacing high cost binders such as Nafion with water was actually shown to improve the performance, however long-term stability of these binders is questionable (Huang et al., 2010). Binder stability is critical as biofouling represents a significant challenge in these systems (Saito et al., 2011). PTFE may not make a great binder, but is the most commonly used compound for the diffusion layer and enables air transfer to the catalyst while preventing the solution from leaking out of the MFC. Polydimethylsiloxane (PDMS) has also been used in a mesh cathode and demonstrated increased coulumbic efficiencies but similar power densities compared to other designs (Zhang, X. *et al.*, 2010). Binding agents, diffusion layers, current collectors, as well as the biological factors affecting MFCs certainly contribute to the challenge of engineering a cathode for these systems.

### 9.5.2 *Catalysts*

As mentioned previously, ferricyanide or permanganate may be used as catholytes and have resulted in low overpotentials and high power densities, however they must be replaced making them an undesirable option for many designs (Rabaey *et al.*, 2004). Ambient oxygen is therefore typically used as the terminal electron acceptor. A catalyst is not always required for the oxygen reduction reaction to proceed, but without catalysts reaction rates are less than one tenth of the most effectively catalyzed ones (Logan, 2008). Finding an oxygen reduction catalyst to function in the neutral pH conditions of MFCs is a difficult task. Platinum has been recognized as the most effective metal catalyst for this use and is a common part of MFC cathode designs. Platinum has a low overpotential for the oxygen reduction reaction conferring exceptional catalytic ability but is also extremely expensive. The high cost of platinum drives the search for ways to reduce the amount of Pt necessary in MFC cathodes as well as the search to identify new catalysts. Studies have indicated that it may be possible to reduce Pt concentrations from the $0.5-1.0 \, \mathrm{mg \, cm^{-2}}$ range commonly used, down to $0.1 \, \mathrm{mg \, cm^{-2}}$ without a significant drop in performance (Cheng *et al.*, 2006b).

Research has been conducted on many metals including Fe, Co, Mn and Pb in order to find an adequate replacement for Pt. Cobalt and iron have attracted the most attention. These two metals are typically combined with tetramethoxyphenylporphyrin (TMPP) or phthalocyanine (Pc) to form metal macrocyclic complexes. Pyrolyzed FePc and CoTMPP complexes have proven the most effective transition metal-based materials. FePc supported on Ketjen black carbon generated a power density of $634 \, \mathrm{mW \, m^{-2}}$ compared to $593 \, \mathrm{mW \, m^{-2}}$ produced by a Pt cathode in the same study (HaoYu, 2007). CoTMPP was also shown to produce currents comparable to Pt cathodes with improved performance at current densities above $0.6 \, \mathrm{mA \, cm^{-2}}$ (Cheng *et al.*, 2006b). Both pyr-FePc and CoTMPP catalysts are notably sensitive to proton and buffer concentrations. Lowering the phosphate buffer concentration in the MFC decreased the performance of the pyr-FePc catalyst by 40%, while a pH increase from 2.4 to 7 reduced the performance of the CoTMPP catalyst by 80% (Zhao *et al.*, 2006).

Iron chelated with ethylenediaminetetraacetic acid (FeEDTA) is another alternative to platinum. Pyrolyzed FeEDTA complexes can generate stable power densities of $1.122 \, \mathrm{W \, m^{-2}}$ compared to $1.166 \, \mathrm{W \, m^{-2}}$ generated by cathodes with a platinum concentration of $0.5 \, \mathrm{mg \, cm^{-2}}$ over the same time period (Aelterman *et al.*, 2009). A lead oxide ($PbO_2$) catalyst would be a good option but the potential of lead leaching affecting the microbial community in an MFC has tampered enthusiasm (Morris *et al.*, 2007). Other catalysts composed of $MnO_2$, Ni and Co were evaluated but were found to have maximum power densities less than 10% of those produced by Pt cathodes (Lefebvre *et al.*, 2009; Zhang *et al.*, 2009; Zhuang *et al.*, 2009).

Apart from metals, activated carbon may be able to provide comparable performance to a Pt cathode at a significantly lower cost. One design featured cold-pressed activated carbon with a PTFE binder and Ni mesh current collector as the cathode. The cathode design did not use carbon cloth or a catalyst typical of most MFC cathodes but produced a power density of $1.22 \, \mathrm{W \, m^{-2}}$ versus $1.06 \, \mathrm{W \, m^{-2}}$ generated by a common platinum/carbon cloth cathode (Zhang *et al.*, 2009). Studies on activated carbon fiber felt and $HNO_3$ treated carbon powder demonstrated similar performance at similarly low costs providing hope of identifying new low-cost high-performance catalysts in the future (Duteanu *et al.*, 2010).

### 9.5.3 *Cathode modifications*

Several different materials can be used to modify MFC cathodes and improve performance (Table 9.2). One class of these materials are nano-materials such as carbon nanotubes (CNT).

Table 9.2.   Maximum power outputs associated with alternate cathode materials.

| Exp. treatment | Exp. max. power [mW m$^{-2}$] | Control max. power[1] [mW m$^{-2}$] | Δ [%] | Ref. |
|---|---|---|---|---|
| Catalysts | | | | |
| Act. Carbon/Ni | 1220 | 1060 | +15.1 | Fu et al. (2011) |
| pyr-FePC/KJB | 634 | 593 | +6.9 | Zhang et al. (2009) |
| pyr-FeEDTA | 1122 | 1166 | −3.8 | HaoYu et al. (2007) |
| CoTMPP | 369 | 419 | −12 | Aelterman (2009) |
| Modifications | | | | |
| CNT | 329 | 151 | +117.8 | Cheng et al. (2006b) |
| PPy/AQS | 823 | 663 | +24.1 | Feng et al. (2010) |

[1]Control cathodes used Pt catalysts with concentrations between 0.5–1.0 mg cm$^{-2}$.

CNTs can be used to enhance Pt cathodes, improving catalytic activity and performance by increasing the catalyst surface area in the active layer of the cathode. Current densities have been improved by an order of magnitude higher than Pt cathodes by using a single-wall CNT sheet electrode infused with platinum nano-particles (Deng et al., 2010). A 2-fold increase of power density was observed using CNT-textile cathodes with only 19.3% of the Pt typically used (Sanchez, 2010). A a similar increase in power densities (329 vs. 151 mW m$^{-2}$) was seen when CNT mat cathodes were used (Xie et al., 2011). In addition to nano-materials, conductive polymers can be also used to increase performance of the oxygen reduction reaction in MFCs. Polypyrrole/anthraquinone-2-sulfonate (PPy/AQS) and Prussian blue/polyaniline (PB/PANI) modified cathodes also demonstrated good performance on par with a ferricyanide cathode (Feng et al., 2011).

### 9.5.4   Biocathodes

Much as bacteria catalyze the anode reaction in an MFC, bacteria can also catalyze the terminal reduction reaction at the cathode (Rabaey et al., 2003). Biocathodes are both low-cost and sustainable making them an attractive research area. However, biocathodes have not made much of an impact in the MFC field yet as power densities to date have been quite low. These low power densities appear to be mainly a product of oxygen mass transfer limitations associated with diffusion through a biofilm (Clauwaert et al., 2007). Measured current densities for biocathodes were significantly lower than their Pt counterparts but may have other advantages such as being able to reduce several different oxidized contaminates, such as Cr(IV) and Cr(OH)$_3$ (Clauwaert et al., 2007; Ter Heijne, 2010).

### 9.6   MEMBRANES/SEPARATORS

Unlike chemical fuel cells, fuel crossover is not an issue in MFCs. In fact, the primary roles of a separator in an MFC is to physically separate the anode and cathode thereby preventing electrical short circuits and oxygen diffusion to the anode where it could inhibit EET. In some designs a separator may not even be necessary. In membrane and separator-free air cathode MFC designs a biofilm generally develops on the cathode and can successfully prevent the diffusion of oxygen to the anode. Proton transport is also uninhibited, cost and complexity is decreased along with internal resistance resulting in higher power densities (Watson et al., 2011). Oxygen crossover may become an issue at electrode spacings less than 1–2 cm, portending a need for some sort of separator (Jiang and Li, 2009). The closer the electrodes can be placed without contact/allowing diffusion of oxygen, the lower the internal resistance of the system will be.

If membranes are to be used in MFCs, they must be able to transport protons efficiently from the anode to the cathode chambers. There are many different types of membranes including cation exchange membranes (CEM), anion exchange membranes (AEM), bipolar membranes (BPM), microfiltration membranes (MFM), and ultrafiltration membranes (UFM). Selection of a membrane is primarily dependent on the choice of charge carrying species in the system. When ferricyanide is used as the terminal electron acceptor, a CEM must be used to prevent diffusion of toxic ferricyanide into the anode chamber where it would greatly impact the microbial community. The CEM would allow protons to be transferred to the cathode chamber, ideally maintaining a pH balance near neutral in the anode chamber. Nafion membranes are a popular choice for use as a CEM in MFC research. Nafion preferentially allows for the transfer of protons over other ions. Unfortunately, in most MFC setups other cations, such as $Na^+$, $K^+$, $Ca^{2+}$ and $Mg^{2+}$, are present in much higher concentrations than protons, meaning charges will be transferred but a pH imbalance will be generated (Tandukar *et al.*, 2009). Power output will be significantly decreased as pH increases in the anode chamber and decreases in the cathode chamber. Nafion can also very expensive, making it impractical in large-scale designs. Other types of membranes should be considered for MFCs not using ferricyanide as a catholyte.

AEMs generally perform better than their CEM counterparts with recent research showing a 27% increase in power densities associated with the switch from CEM to AEM (Pham, 2004). AEMs in MFC systems are generally used in conjunction with phosphate buffers. The use of these buffers makes it possible to balance both pH and charge by the transfer of phosphate anions (Rozendal *et al.*, 2006). At a concentration of about 0.1 M, proton-carrying anion concentrations will be much higher than proton concentrations themselves leading to more efficient proton transfer. Power densities may also be increased by doubling buffer concentrations to 0.2 M (Rozendal *et al.*, 2006).

In addition to CEMs and AEMs, a bipolar membrane may also be used. A bipolar membrane sandwiches a catalytic intermediate layer, or junction layer, in-between a CEM and an AEM. When a current is passed through the system water molecules disassociate into proton and hydroxide ions. This then gives bipolar membranes much greater selectivity in which ions are transported leading to less severe pH gradients. However, splitting water molecules comes at a cost and polarization potential losses are more pronounced than in systems using CEMs or AEMs. These losses were not able to make up for the increased ability to selectively transport ions and lower voltages were recorded in a bioelectrochemical system running on wastewater (Rozendal *et al.*, 2006).

The use of filtration membranes is another attractive option due to the low cost and ability to reduce overall internal resistance of the system as most ions can freely pass through relatively large pores (Rozendal *et al.*, 2008; Sun *et al.*, 2009a). A disadvantage of these membranes is that oxygen can also pass through relatively easily. A smaller pore size or thicker membrane will prevent oxygen crossover, but it also inhibits proton transport resulting in decreased power (Sun *et al.*, 2009b; Zhang, X. *et al.*, 2010).

Porous fabrics can successfully be used as separators to prevent short circuits and oxygen diffusion while allowing efficient mass transfer of ions. Ideally, these fabrics are inexpensive, durable, non-conductive, and permeable to ion transport but not oxygen. In one study, the Coulombic efficiency of the MFC was doubled from 35% to 71% by including a separator (Cheng *et al.*, 2006a). This non-conductive layer also allows for electrode spacing to be reduced to less than 1 mm, greatly reducing the internal resistance (Fan *et al.*, 2008). Using this design generation of power densities of up to 2770 mW m$^{-2}$/1550 W m$^{-3}$ were recorded with room for optimization (Fan *et al.*, 2008). Overall, given the low cost and favorable trade-off between oxygen permeability and internal resistance, porous fabric separators make a good alternative to many membranes.

## 9.7   SUMMARY

Microbial fuel cells rely on bioelectrochemical reactions driven by microbial metabolism to pro-duce electricity while degrading organic substrates. Pure culture MFC studies with bacteria such as *Geobacter* and *Shewanella* have informed researchers on the mechanisms behind extracellular electron transfer (EET), but it is mixed consortium communities that typically produce the most power and have the most metabolic flexibility to handle a variety of substrates and environmental conditions.

Carbon-based anodes are currently the best choice for MFC anodes due to their high surface area and conductivity along with their chemical and biological compatibility. Modifications to MFC anodes including the incorporation of nano-materials are being explored to improve MFC performance further. However, it is the reaction occurring at the cathode that is rate-limiting. Platinum has been used as an effective oxygen reduction catalyst at the cathode but the inclusion of platinum along with other high cost materials such as Nafion into large-scale cathode designs is prohibitively expensive, especially for wastewater treatment. The search is on for cheaper, highly effective alternative catalysts and materials with some promising results demonstrated using inexpensive activated carbon. Breakthroughs in this area will greatly increase the commercial feasibility of MFC technology. The use of inexpensive separators that minimize distance between the anode and cathode while allowing the flow of ions has also proven to be an ideal design component that offers a cost advantage over more expensive membranes.

## 9.8   OUTLOOK

Demand for energy and clean water is on the rise with little sense that this demand will ebb anytime in the near future. For the past century, fossil fuels have been capable of satiating our energy needs, but it is clear now that our current energy trajectory is unsustainable. New solutions must be developed. There will be no one technology that will solve our energy needs, the solution lies in harnessing energy wherever it can be found. One of the most overlooked sources of energy is waste.

Wastewater is rich in organic matter and therefore energy rich, but converting the energy held in the chemical bonds of these organic compounds into a usable fuel has proven difficult. While it has been known for over a century that bacteria can generate electrical currents it has not been until recently that this phenomenon has been applied to generate energy from wastewater through use of microbial fuel cells (Liu *et al.*, 2004; Pahm, 2004). Since then MFCs have garnered much research interest due to their ability to generate electricity while concurrently treating wastewater. This interest has led to enormous improvements in MFC performance over the last decade (Fan *et al.*, 2008), though maximum power densities of MFCs are currently significantly less than chemical fuel cells ($1\,\mathrm{W\,m^{-2}}/1\,\mathrm{kW\,m^{-3}}$). However, continued advances in performance as well as decreases in material cost are expected to make MFCs a commercially viable technology. These advances will likely come from new materials some of which are likely to be the product of micro- and nano-engineering. With these breakthroughs, the commercial applications of MFC technology will expand greatly and perhaps one piece of tomorrow's energy solution will be realized.

## ACKNOWLEDGEMENTS

The authors acknowledge support from the U.S. National Science Foundation (CBET 0955124).

## REFERENCES

Aelterman, P., Versichele, M., Genettello, E., Verbeken, K. & Verstraete, W.: Microbial fuel-cells operated with iron-chelated air cathodes. *Electrochim. Acta* 54 (2009), pp. 5754–5760.

Baron, D., LaBelle, E., Coursolle, D., Gralnick, J.A. & Bond, D.R.: Electrochemical measurement of electron transfer kinetics by *Shewanella oneidensis* MR-1. *J. Biol. Chem.* 284 (2009), pp. 28,865–28,873.

Bigalke, J. & Grabner, E.W.: The geobattery model: a contribution to large scale electrochemistry. *Electrochim. Acta* 42 (1997), pp. 3443–3452.

Bond, D.R.: Electrodes as electron acceptors, and the bacteria who love them. In: L.L. Barton, M. Mandl & A. Loy (eds): *Geomicrobiology: Molecular and environmental perspective.* Springer, pp. 385–399, 2010.

Bond, D.R. & Lovley, D.R.: Electricity production by *Geobacter sulfurreducens* attached to electrodes. *Appl. Environ. Microbiol.* 69 (2003), pp. 1548–1555.

Bond, D.R., Strycharz-Glaven, S.M., Tender, L.M. & Torres, C.I.: On electron transport through geobacter biofilms. *ChemSusChem* 5 (2012), pp. 1099–1105.

Busalmen, J.P., Esteve-Núñez, A., Berná, A. & Feliu, J.M.: C-type cytochromes wire electricity-producing bacteria to electrodes. *Angew. Chemie* 120 (2008), pp. 4952–4955.

Canstein, H. von, Ogawa, J., Shimizu, S. & Lloyd, J. R.: Secretion of flavins by *Shewanella* species and their role in extracellular electron transfer. *Appl. Environ. Microbiol.* 74 (2008), pp. 615–623.

Cheng, S. & Logan, B.E.: Ammonia treatment of carbon cloth anodes to enhance power generation of microbial fuel-cells. *Electrochem. Commun.* 9 (2007), pp. 492–496.

Cheng, S., Liu, H. & Logan, B.E.: Increased power generation in a continuous flow MFC with advective flow through the porous anode and reduced electrode spacing. *Environ. Sci. Technol.* 40 (2006a), pp. 2426–2432.

Cheng, S., Liu, H. & Logan, B.E.: Power densities using different cathode catalysts (Pt and CoTMPP) and polymer binders (Nafion and PTFE) in single chamber microbial fuel-cells. *Environ. Sci. Technol.* 40 (2006b), pp. 364–369.

Childers, S.E., Ciufo, S. & Lovley, D.R.: *Geobacter metallireducens* accesses insoluble Fe(III) oxide by chemotaxis. *Nature* 416 (2002), pp. 767–769.

Clauwaert, P., Rabaey, K., Aelterman, P., De Schamphelaire, L, Pham, T.H., Boeckx, P., Boon, Ni. & Verstraete, W.: Biological denitrification in microbial fuel-cells. *Environ. Sci. Technol.* 41 (2007), pp. 3354–3360.

Coursolle, D., Baron, D.B., Bond, D.R. & Gralnick, J.A.: The Mtr respiratory pathway is essential for reducing flavins and electrodes in *Shewanella oneidensis*. *J. Bacteriol.* 192 (2010), pp. 467–474.

Crittenden, S.R., Sund, C.J. & Sumner, J.J.: Mediating electron transfer from bacteria to a gold electrode via a self-assembled monolayer. *Langmuir* 22 (2006), pp. 9473–9476.

Deng, Q., Li, X., Zuo, J., Ling, A. & Logan, B.E.: Power generation using an activated carbon fiber felt cathode in an upflow microbial fuel-cell. *J. Power Sources* 195 (2010), pp. 1130–1135.

Dumas, C.,Mollica, A., Feron, D., Basseguy, R., Etcheverry, L. & Bergel, A.: Marine microbial fuel-cell: Use of stainless steel electrodes as anode and cathode materials. *Electrochim. Acta* 53 (2007), pp. 468–473.

Dumas, C., Basséguy, R. & Bergel, A.: Electrochemical activity of *Geobacter sulfurreducens* biofilms on stainless steel anodes. *Electrochim. Acta* 53 (2008), pp. 5235–5241.

Duteanu, N., Erable, B., Senthil Kumar, S.M., Ghangrekar, M.M. & Scott, K.: Effect of chemically modified Vulcan XC-72R on the performance of air-breathing cathode in a single-chamber microbial fuel-cell. *Bioresource Technol.* 101 (2010), pp. 5250–5255.

Fan, Y., Sharbrough, E. & Liu, H.: Quantification of the internal resistance distribution of microbial muel-cells. *Environ. Sci. Technol.* 42 (2008), pp. 8101–8107.

Fan, Y., Xu, S., Schaller, R., Jiao, J., Chaplen, F. & Liu H.: Nanoparticle decorated anodes for enhanced current generation in microbial electrochemical cells. *Biosens. Bioelectron.* 26 (2011), pp. 1908–1912.

Fan, Y., Han, S.-K. & Liu, H.: Improved performance of CEA microbial fuel-cells with increased reactor size. *Energy Environ. Sci.* 5 (2012), pp. 8273–8280.

Feng, C., Li, F., Liu, H., Lang, X. & Fan, S.: A dual-chamber microbial fuel-cell with conductive film-modified anode and cathode and its application for the neutral electro-Fenton process. *Electrochim. Acta* 55 (2010), pp. 2048–2054.

Feng, C., Wan, Q., Lv, Z., Yue, X., Chen, Y., & Wei, C.: One-step fabrication of membraneless microbial fuel-cell cathode by electropolymerization of polypyrrole onto stainless steel mesh. *Biosens. Bioelectron.* 26 (2011), pp. 3953–3957.

Fu, L., You, S., Zhang, G., Yang, F., Fang, X., & Gong, Z.: PB/PANI-modified electrode used as a novel oxygen reduction cathode in microbial fuel-cell. *Biosens. Bioelectron.* 26 (2011), pp. 1975–1979.

HaoYu, E., Cheng, S., Scott, K. & Logan, B.: Microbial fuel-cell performance with non-Pt cathode catalysts. *J. Power Sources* 171 (2007), pp. 275–281.

He, Z., Minteer, S.D. & Angenent, L.T.: Electricity generation from artificial wastewater using an upflow microbial fuel-cell. *Environ. Sci. Technol.* 39 (2005), pp. 5262–5267.

Huang, Y., He, Z. & Mansfeld, F.: Performance of microbial fuel-cells with and without Nafion solution as cathode binding agent. *Bioelectrochemistry* 79 (2010), pp. 261–264.

Inoue, K., Leang, C., Franks, A.E., Woodard, T.L., Nevin, K.P., & Lovley, D.R.: Specific localization of the c-type cytochrome OmcZ at the anode surface in current-producing biofilms of *Geobacter sulfurreducens*. *Environ. Microbiol. Rep.* 3 (2011), pp. 211–217.

Jiang, D. & Li, B.: Novel electrode materials to enhance the bacterial adhesion and increase the power generation in microbial fuel-cells (MFCs). *Water Sci. Technol.* 59 (2009), pp. 557–563.

Johnson, W.P. & Logan, B.E.: Enhanced transport of bacteria in porous media by sediment-phase and aqueous-phase natural organic matter. *Water Res.* 30 (1996), pp. 923–931.

Kargi, F. & Eker, S.: Electricity generation with simultaneous wastewater treatment by a microbial fuel-cell (MFC) with Cu and Cu–Au electrodes. *J. Chem. Technol. Biotechnol.* 82 (2007), pp. 658–662.

Kato, S., Nakamura, R., Kai, F., Watanabe, K. & Hashimoto, K.: Respiratory interactions of soil bacteria with (semi)conductive iron-oxide minerals. *Environ. Microbiol.* 12 (2010), pp. 3114–3123.

Kato, S., Hashimoto, K. & Watanabe, K.: Microbial interspecies electron transfer via electric currents through conductive minerals. *PNAS* 109 (2012), pp. 10,042–10,046.

Kim, H.J., Park, H.S., Hyun, M.S., Chang, I.S., Kim, M. & Kim, B.H.: A mediator-less microbial fuel-cell using a metal reducing bacterium, *Shewanella putrefaciens*. *Enzyme Microb. Tech.* 30 (2002), pp. 145–152.

Kim, J.R., Min, B. & Logan, B.E.: Evaluation of procedures to acclimate a microbial fuel-cell for electricity production. *Appl. Microbiol. Biotechnol.* 68 (2005), pp. 23–30.

Lanthier, M., Gregory, K.B. & Lovley, D.R.: Growth with high planktonic biomass in *Shewanella oneidensis* fuel-cells. *FEMS Microbiol. Lett.* 278 (2008), pp. 29–35.

Larrosa-Guerrero, A., Scott, K., Katuri, K., Godinez, C., Head, I. & Curtis, T.: Open circuit versus closed circuit enrichment of anodic biofilms in MFC: effect on performance and anodic communities. *Appl. Microbiol. Biotechnol.* 87 (2010), pp. 1699–1713.

Lefebvre, O., Ooi, W.K., Tang, Z., Abdullaha-Al-Mamun, M., Chua, D.H.C. & Ng, H.: Optimization of a Pt-free cathode suitable for practical applications of microbial fuel-cells. *Bioresour. Technol.* 100 (2009), pp. 4907–4910.

Liu, H., Ramnarayanan, R. & Logan, B.E.: Production of electricity during wastewater treatment using a single chamber microbial fuel-cell. *Environ. Sci. Technol.* 38 (2004), pp. 2281–2285.

Liu, H., Cheng, S. & Logan, B.E.: Production of electricity from acetate or butyrate using a single-chamber microbial fuel-cell. *Environ. Sci. Technol.* 39 (2005), pp. 658–662.

Logan, B.E.: *Microbial fuel-cells*. John Wiley & Sons, 2008.

Logan, B.E.: Exoelectrogenic bacteria that power microbial fuel-cells. *Nature Rev. Microbiol.* 7 (2009), pp. 375–381.

Logan, B., Cheng, S., Watson, V. & Estadt, G.: Graphite fiber brush anodes for increased power production in air-cathode microbial fuel-cells. *Environ. Sci. Technol.* 41 (2007), pp. 3341–3346.

Lojou, E., Gludici-Orticoni, M.T. & Bianco, P.: Direct electrochemistry and enzymatic activity of bacterial polyhemic cytochrome c(3) incorporated in clay films. *J. Electroanal. Chem.* 579 (2005), pp. 199–213

Lovley, D.R.: Bug juice: harvesting electricity with microorganisms. *Nature Rev. Microbiol.* 4:7 (2006), pp. 4497–4508.

Lovley, D.R.: The microbe electric: conversion of organic matter to electricity. *Curr. Opin. Biotechnol.* 19 (2008), pp. 564–571.

Lovley, D.R.: Live wires: direct extracellular electron exchange for bioenergy and the bioremediation of energy-related contamination. *Energy Environ. Sci.* 4 (2011), pp. 4896–4906.

Lovley, D.R.: Electromicrobiology. *Ann. Rev. Microbiol.* 66 (2012), pp. 391–409.

Lovley, D.R. & Nevin, K.P.: A shift in the current: new applications and concepts for microbe-electrode electron exchange. *Curr. Opin. Biotechnol.* 22 (2011), pp. 441–448.

Lowy, D.A., Tender, L.M., Zeikus, J.G., Park, D.H. & Lovley, D.R.: Harvesting energy from the marine sediment–water interface II. *Biosens. Bioelectron.* 21 (2006), pp. 2058–2063.

Malvankar, N.S. & Lovley, D.R.: Microbial nanowires: A new paradigm for biological electron transfer and bioelectronics. *ChemSusChem* 5 (2012), pp. 1039–1046.

Malvankar, N.S., Vargas, M., Nevin, K.P., Franks, A.E., Leang, C., Kim, B., Inoue, K., Mester, T., Covalla, S., Johnson, J.P., Rotello, V.M., Tuominen, M.T. & Lovley, D.R.: Tunable metallic-like conductivity in microbial nanowire networks. *Nature Nanotechnol.* 6 (2011), pp. 573–579.

Marsili, E. Baron, D.B., Shikhare, I.D., Coursole, D., Gralnick, J.A. & Bond, D.R.: *Shewanella* secretes flavins that mediate extracellular electron transfer. *PNAS* 105 (2008), pp. 3968–3973.

Methé, B.A., Nelson, K.E., Eisen, J.A., Paulsen, I.T., Nelson, W., Heidelberg, J.F., Wu, D., Wu, M., Ward, N. & Beanan, M.J.: Genome of *Geobacter sulfurreducens*: Metal reduction in subsurface environments. *Science* 302 (2003), pp. 1967–1969.

Morris, J.M., Jin, S., Wang, J., Zhu, C. & Urynowicz, M.A.: Lead dioxide as an alternative catalyst to platinum in microbial fuel-cells. *Electrochem. Commun.* 9 (2007), pp. 1730–1734.

Nam, J.-Y., Kim, H.-W. & Shin, H.-S.: Ammonia inhibition of electricity generation in single-chambered microbial fuel-cells. *J. Power Sources* 195 (2010), pp. 6428–6433.

Nevin, K.P., Kim, B., Glaven, R.H., Johnson, J.P., Woodard, T.L., Methe, B.A., DiDonato, R.J., Covalla, S.F., Franks, A.E., Liu, A. & Lovley, D.R.: Anode biofilm transcriptomics reveals outer surface components essential for high density current production in *Geobacter sulfurreducens* fuel-cells. *PLoS ONE* 4 (2008), e5628.

Niessen, J., Schröder, U., Rosenbaum, M. & Scholz, F.: Fluorinated polyanilines as superior materials for electrocatalytic anodes in bacterial fuel-cells. *Electrochem. Commun.* 6 (2004), pp. 571–575.

Park, D.H. & Zeikus, J.G.: Utilization of electrically reduced neutral red by *Actinobacillus succinogenes*: Physiological function of neutral red in membrane-driven fumarate reduction and energy conservation. *J. Bacteriol.* 181 (1999), pp. 2403–2410.

Park, D.H. & Zeikus, J.G.: Improved fuel-cell and electrode designs for producing electricity from microbial degradation. *Biotechnol. Bioeng.* 81 (2003), pp. 348–355.

Pham, T., Boon, N., Aelterman, P., Clauwaert, P., De Schamphelaire, L., Vanhaecke, L., De Maeyer, K., Hofte, M., Verstraete, W. & Rabaey, K.: Metabolites produced by *Pseudomonas* sp. enable a gram-positive bacterium to achieve extracellular electron transfer. *Appl. Microbiol. Biotechnol.* 77 (2008), pp. 1119–1129.

Pham, T.H.: Improvement of cathode reaction of a mediatorless microbial fuel-cell. *J. Microbiol. Biotechnol.* 14 (2004), pp. 324–329.

Potter, M.C.: Electrical effects accompanying the decomposition of organic compounds. *Proceedings of the Royal Society of London.* Series B, Containing Papers of a Biological Character 84, 1911, pp. 260–276.

Qiao, Y., Li, C.M., Bao, S.-J. & Bao, Q.-L.: Carbon nanotube/polyaniline composite as anode material for microbial fuel-cells. *J. Power Sources* 170 (2007), pp. 79–84.

Qiao, Y., Bao, S., Li, C.M., Cui, X., Lu, Z. & Guo, J.: Nanostructured polyaniline/titanium dioxide composite anode for microbial fuel-cells. *ACS Nano* 2 (2008), pp. 113–119.

Rabaey, K.: *Bioelectrochemical systems: From extracellular electron transfer to biotechnological application.* IWA Publishing, 2010.

Rabaey, K., Lissens, G., Siciliano, S.D. & Verstraete, W.: A microbial fuel-cell capable of converting glucose to electricity at high rate and efficiency. *Biotechnol. Lett.* 25 (2003), pp. 1531–1535.

Rabaey, K., Boon, N., Siciliano, S.D., Verhaege, M. & Verstraete, W.: Biofuel-cells select for microbial consortia that self-mediate electron transfer. *Appl. Environ. Microbiol.* 70 (2004), pp. 5373–5382.

Rabaey, K., Clauwaert, P., Aelterman, P. & Verstraete, W.: Tubular microbial fuel-cells for efficient electricity generation. *Environ. Sci. Technol.* 39 (2005), pp. 8077–8082.

Reguera, G., McCarthy, K.D., Mehta, T., Nicoll, J.S., Tuominen, M.T. & Lovley, D.R.: Extracellular electron transfer via microbial nanowires. *Nature* 435 (2005), pp. 1098–1101.

Reguera, G., Nevin, K.P., Nicoll, J.S., Covall, S.F., Woodard, T.L. & Lovley, D.R.: Biofilm and nanowire production leads to increased current in *Geobacter sulfurreducens* fuel-cells. *Appl. Environ. Microbiol.* 72 (2006), pp. 7345–7348.

Richter, H., Lanthier, M., Nevin, K.P. & Lovley, D.R.: Lack of electricity production by *Pelobacter carbinolicus* indicates that the capacity for Fe(III) oxide reduction does not necessarily confer electron transfer ability to fuel-cell anodes. *Appl. Environ. Microbiol.* 73 (2007), pp. 5347–5353.

Richter, H., McCarthy, K., Nevin, K.P., Johnson, J.P., Rotello, V.M. & Lovley, D.R.: Electricity generation by *Geobacter sulfurreducens* attached to gold electrodes. *Langmuir* 24 (2008), pp. 4376–4379.

Rozendal, R.A., Hamelers, H.V. M. & Buisman, C.J.N.: Effects of membrane cation transport on pH and microbial fuel-cell performance. *Environ. Sci. Technol.* 40 (2006), pp. 5206–5211.

Rozendal, R.A., Sleutels, T.H.J.A., Hamelers, H.V.M. & Buisman, C.J.N.: Effect of the type of ion exchange membrane on performance, ion transport, and pH in biocatalyzed electrolysis of wastewater. *Water Sci. Technol.* 57 (2008), pp. 1757–1762.

Saito, T., Merrill, M.D., Watson, V.J., Logan, B.E. & Hickner, M.A.: Investigation of ionic polymer cathode binders for microbial fuel-cells. *Electrochim. Acta* 55 (2010), pp. 3398–3403.

Saito, T., Roberts, T.H., Long, T.E., Logan, B.E. & Hickner, M.A.: Neutral hydrophilic cathode catalyst binders for microbial fuel-cells. *Energy Environ. Sci.* 4 (2011), pp. 928–934.

Sanchez, D.V.P., Huynh, P., Kozlov, M.E., Baughman, R.H., Vidic, R.D. & Yun, M.: Carbon nanotube/ platinum (Pt) sheet as an improved cathode for microbial fuel-cells. *Energy Fuels* 24 (2010), pp. 5897–5902.

Schröder, U., Nießen, J. & Scholz, F.A.: Generation of microbial fuel-cells with current outputs boosted by more than one order of magnitude. *Angew. Chemie* Int. Ed. 42 (2003), pp. 2880–2883.

Summers, Z.M., Fogarty, H.E., Leang, C., Franks, A.E., Malvankar, N.S. & Lovley, D.R.: Direct exchange of electrons within aggregates of an evolved syntrophic coculture of anaerobic bacteria. *Science* 330 (2010), pp. 1413–1415.

Sun, J., Hu, Y., Bi, Z. & Cao, Y.: Improved performance of air-cathode single-chamber microbial fuel-cell for wastewater treatment using microfiltration membranes and multiple sludge inoculation. *J. Power Sources* 187 (2009a), pp. 471–479.

Sun, J., Hu, Y.-Y., Bi, Z. & Cao, Y.-Q.: Simultaneous decolorization of azo dye and bioelectricity generation using a microfiltration membrane air-cathode single-chamber microbial fuel-cell. *Bioresour. Technol.* 100 (2009b), pp. 3185–3192.

Sun, J.-J., Zhao, H.-Z., Yang, Q.-Z., Song, J. & Xue, A.: A novel layer-by-layer self-assembled carbon nanotube-based anode: Preparation, characterization, and application in microbial fuel-cell. *Electrochim. Acta* 55 (2010), pp. 3041–3047.

Tandukar, M., Huber, S.J., Onodera, T. & Pavlostathis, S.G.: Biological chromium(VI) reduction in the cathode of a microbial fuel-cell. *Environ. Sci. Technol.* 43 (2009), pp. 8159–8165.

Tender, L.M., Reimers, C.E., Stecher, H.A., Holmes, D.E., Bond, D.R., Lowy, D.A., Pilobello, K., Fertig, S.J. & Lovley, D.R.: Harnessing microbially generated power on the seafloor. *Nature Biotechnol.* 20 (2002), pp. 821–825.

Ter Heijne, A., Hamelers, H.V.M., Saakes, M. & Buisman, C.J.N.: Performance of non-porous graphite and titanium-based anodes in microbial fuel-cells. *Electrochim. Acta* 53 (2008), pp. 5697–5703.

Ter Heijne, A., Strik, D.P.B.T.B., Hamelers, H.V.M. & Buisman, C.J.N.: Cathode potential and mass transfer determine performance of oxygen reducing biocathodes in microbial fuel-cells. *Environ. Sci. Technol.* 44 (2010), pp. 7151–7156.

Wang, X., Cheng, S., Feng, Y., Merrill, M., Saito, T. & Logan, B.E.: Use of carbon mesh anodes and the effect of different pretreatment methods on power production in microbial fuel-cells. *Environ. Sci. Technol.* 43 (2009), pp. 6870–6874.

Wang, X., Feng, Y., Liu, J., Shi, X., Lee, H., Li, N. & Ren, N.: Power generation using adjustable Nafion/PTFE mixed binders in air-cathode microbial fuel-cells. *Biosens. Bioelectron.* 26 (2010), pp. 946–948.

Watanabe, K., Manefield, M., Lee, M. & Kouzuma, A.: Electron shuttles in biotechnology. *Curr. Opin. Biotechnol.* 20 (2009), pp. 633–641.

Watson, V.J., Saito, T., Hickner, M.A. & Logan, B.E.: Polymer coatings as separator layers for microbial fuel-cell cathodes. *J. Power Sources* 196 (2011), pp. 3009–3014.

Williams, K.H., Nevin, K.P., Franks, A., Englert, A., Long, P.E. & Lovley, D.R.: Electrode-based approach for monitoring in situ microbial activity during subsurface bioremediation. *Environ. Sci. Technol.* 44 (2010), pp. 47–54.

Xiao, L., Damien, J., Luo, J., Jang, H.D., Huang, J. & He, Z.: Crumpled graphene particles for microbial fuel-cell electrodes. *J. Power Sources* 208 (2012), pp. 187–192.

Xie, X., Pasta, M., Hu, L., Yang, Y., McDonough, J., Cha, J., Criddle, C.S. & Cui, Y.: Nano-structured textiles as high-performance aqueous cathodes for microbial fuel-cells. *Energy Environ. Sci.* 4 (2011), pp. 1293–1297.

Xu, S., Liu, H., Fan, Y., Schaller, R., Jiao, J. & Chaplen, F.: Enhanced performance and mechanism study of microbial electrolysis cells using Fe nanoparticle-decorated anodes. *Appl. Microbiol. Biotechnol.* 93 (2012), pp. 871–880.

Yi, H., Nevin, K.P., Kim, B., Franks, A.E., Kilmes, A., Tender, L.M. & Lovley, D.R.: Selection of a variant of *Geobacter sulfurreducens* with enhanced capacity for current production in microbial fuel-cells. *Biosens. Bioelectron.* 24 (2009), pp. 3498–3503.

Yong, Y.-C., Dong, X.-C., Chan-Park, M. B., Song, H. & Chen, P.: Macroporous and monolithic anode based on polyaniline hybridized three-dimensional graphene for high-performance microbial fuel-cells. *ACS Nano* 6 (2012), pp. 2394–2400.

You, S., Zhao, Q., Zhang, J., Jiang, J. & Zhao, S.: A microbial fuel-cell using permanganate as the cathodic electron acceptor. *J. Power Sources* 162 (2006), pp. 1409–1415.

Zhang, F., Cheng, S., Pant, D., Bogaert, G.V. & Logan, B.E.: Power generation using an activated carbon and metal mesh cathode in a microbial fuel-cell. *Electrochem. Commun.* 11 (2009), pp. 2177–2179.

Zhang, F., Saito, T., Cheng, S., Hickner, M.A. & Logan, B.E.: Microbial fuel-cell cathodes with poly(dimethylsiloxane) diffusion layers constructed around stainless steel mesh current collectors. *Environ. Sci. Technol.* 44 (2010), pp. 1490–1495.

Zhang, J., Zhang, E., Scott, K. & Burgess, J.G.: Enhanced electricity production by use of reconstituted artificial consortia of estuarine bacteria grown as biofilms. *Environ. Sci. Technol.* 46 (2012), pp. 2984–2992.

Zhang, L., Liu, C., Zhuang, L., Li, W., Zhou, S. & Zhang, J.: Manganese dioxide as an alternative cathodic catalyst to platinum in microbial fuel-cells. *Biosens. Bioelectron.* 24 (2009), pp. 2825–2829.

Zhang, T., Gannon, S.M., Nevin, K.P., Franks, A.E. & Lovley, D.R.: Stimulating the anaerobic degradation of aromatic hydrocarbons in contaminated sediments by providing an electrode as the electron acceptor. *Environ. Microbiol.* 12 (2010), pp. 1011–1020.

Zhang, X., Cheng, S., Huang, X. & Logan, B.E.: The use of nylon and glass fiber filter separators with different pore sizes in air-cathode single-chamber microbial fuel-cells. *Energy Environ. Sci.* 3 (2010), pp. 659–664.

Zhang, Y., Mo, G., Li, X., Zhang, W., Zhang, J., Ye, J., Huang, X. & Yu, C.: A graphene modified anode to improve the performance of microbial fuel-cells. *J. Power Sources* 196 (2011), pp. 5402–5407.

Zhao, F., Harnisch, F., Schroder, U., Scholz, F., Bogdanoff, P. & Herrmann, I.: Challenges and constraints of using oxygen cathodes in microbial fuel-cells. *Environ. Sci. Technol.* 40 (2006), pp. 5193–5199.

Zhuang, L., Zhou, S., Wang, Y., Liu, C. & Geng, S.: Membrane-less cloth cathode assembly (CCA) for scalable microbial fuel-cells. *Biosens. Bioelectron.* 24 (2009), pp. 3652–3656.

# CHAPTER 10

## Modeling and analysis of miniaturized packed-bed reactors for mobile devices powered by fuel cells

Srinivas Palanki & Nicholas D. Sylvester

## 10.1  INTRODUCTION

Currently, there is a lot of interest in developing power sources in the 10–100 W range for portable power applications (Kundu *et al.*, 2007; Service, 2002). One promising approach involves the use of a reformer that produces hydrogen from a hydrocarbon source, coupled with a fuel cell stack that utilizes the hydrogen to produce power. Such a device can be used to power radios, computers, electronic displays, and small unmanned air vehicles (National Research Council, 2004; Office of the Secretary of Defense, 2005).

There are a variety of commonly available hydrocarbons such as methane, methanol, propane, butane, gasoline, and diesel that can be used for reforming reactants to produce hydrogen. For instance, Pattekar and Kothare (2005) fabricated a radial flow micro packed-bed reactor via deep reactive ion etching that utilizes methanol to generate sufficient hydrogen for a 20 W power application. Shah and Besser (2008) developed an integrated silicon micro reactor based methanol steam reformer that produces sufficient hydrogen for 0.38 W of power. Mu *et al.* (2007) fabricated a miniature reformer that utilizes methanol to produce sufficient hydrogen to generate 100 W of power in a fuel cell stack. While there is considerable literature on fabrication of micro-reformers that demonstrates the feasibility of utilizing methanol to produce hydrogen for micro and small-scale applications, there are few papers in modeling and analysis of these reactors, which are necessary for optimizing performance.

In this chapter, the design equations of a packed-bed reactor in a tubular geometry that can be used to produce hydrogen are presented. Fundamental principles of reaction engineering, fluid mechanics, and heat transfer are utilized to develop this model. It is shown how the hydrogen generated from this reactor can be used to design a fuel cell stack to produce power. Two examples are presented in the last section to show the applicability of these design equations.

## 10.2  REACTOR AND FUEL CELL MODELING

A schematic of the reactor system under consideration is shown in Figure 10.1. The reactor is packed with catalyst particles, and reactants enter the reactor and produce hydrogen in the presence of the catalyst particles. There is a jacket around the reactor so that additional heat can be provided to the reactor. The resulting hydrogen gas mixture is sent to a fuel cell stack where an electrochemical reaction occurs to generate power. The fuel cell stack is composed of several identical hydrogen fuel cells in series.

### 10.2.1  *Design equations of the reactor*

In the packed-bed reactor, the reactant gases react in the presence of the catalyst particles. This is a multi-step process involving adsorption of the reactants onto the catalyst surface, reaction on the active sites, and then desorption of the product molecules. The slowest step (called the 'rate-limiting step') determines the reaction rate for this process and is determined experimentally.

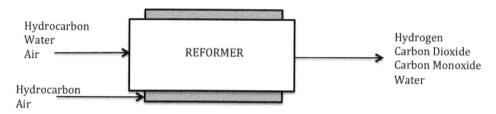

Figure 10.1.   Schematic of packed-bed reactor system.

The steady-state model equations for each species in the reactor are given by the following mole-balance equations (Fogler, 2006):

$$\frac{dF_i}{dt} = r_i A_c \tag{10.1}$$

where $F_i$ is the molar flow rates of species $i$, $r_i$ is the reaction rate with respect to species $i$, and $z$ is the axial dimension of the reactor. The term, $A_c$, represents the area of cross section of the reactor.

As the reactant and product gases flow through the packed-bed, there are two competing effects that can change the pressure. There is a drop in pressure due to resistance to fluid movement through the packing; however, when the reactions involved result in a total increase in the number of moles, this results in an increase in pressure. The overall pressure drop is the net sum of these two effects. The pressure drop in the pack-bed reactor is modeled via the Ergun equation as follows (Fogler, 2006):

$$\frac{dP}{dz} = -\frac{G}{\rho D_p}\left(\frac{1-\phi}{\phi^3}\right)\left[\frac{150(1-\phi)\mu}{D_p} + 1.75G\right] \tag{10.2}$$

where $P$ is the reactor pressure, $\phi$ is the void fraction, $D_p$ is the diameter of the catalyst particle in the reformer, $\mu$ is the viscosity of the gas mixture, $\rho$ is the gas mixture density, and $G$ is the superficial mass velocity. It is observed that the Ergun equation is not a direct function of the reaction rates. However, the mixture density and viscosity may be significantly affected by the reaction rate because these two quantities change as the composition of the mixture changes along the length of the reactor. The Ergun equation requires the computation of the gas mixture density, $\rho$, as well as the gas mixture viscosity, $\mu$, as a function of reactor length. The mixture density is typically estimated by computing the molar average density of the reacting gas mixture; however, due to the presence of hydrogen in the mixture, a more accurate method, such as Wilke's method (Poling et al., 2001), is utilized to estimate the gas mixture viscosity.

The reactions to form hydrogen are typically endothermic reactions and hence it is generally necessary to provide energy to the reactor from an external source. A steady-state energy balance on the reformer leads to the following equation (Fogler, 2006):

$$\frac{dT}{dt} = \frac{\left(Q + \sum r_i \Delta H_{r_i}\right)}{\sum F_j C_{pj}} A_c \tag{10.3}$$

where $T$ is the reformer temperature, $\Delta H_{r_i}$ is the heat of reaction for reaction $r_i$, $Q$ is the overall heat provided to the reactor, and $C_{pj}$ is the specific heat of species $j$.

### 10.2.2   Design equations for the fuel cell stack

The current generated by the fuel cell stack is directly proportional to the hydrogen consumption rate. The voltage generated by a cell depends on the current density in the cell, thermodynamic parameters (temperature and partial pressures), and the cell materials and design. For a given cell design and thermodynamic state, it is necessary to determine the voltage *versus* current density. This relationship, called the polarization curve of the fuel cell, is generally determined experimentally. The relation between the hydrogen flow rate, the current, and the number of cells, is given by Larminie and Dicks (2000):

$$I = \frac{2 F \varepsilon F_{H_2}}{n} \qquad (10.4)$$

where $I$ is the current, $F$ is Faraday's constant, $F_{H_2}$ is the molar flow rate of hydrogen entering the fuel cell stack, $\varepsilon$ is an efficiency factor, and $n$ is the number of cells in the fuel cell stack. The desired power generated by the fuel cell stack is computed from:

$$P = Vin \qquad (10.5)$$

where $P$ is the power and $V$ is the voltage.

## 10.3   APPLICATIONS

In this section, we consider two different applications: (i) a methanol-based reformer and fuel cell system, and (ii) an ammonia-based reactor and fuel cell system. Design issues for both systems are considered for generating 100 W of power.

### 10.3.1   *Methanol-based system*

The following reactions take place in the reformer:

$$CH_3OH + H_2O \rightarrow CO_2 + 3H_2 \qquad (10.6)$$

$$CH_3OH \rightarrow CO + 2H_2 \qquad (10.7)$$

$$CO + H_2O \rightarrow CO_2 + H_2 \qquad (10.8)$$

$$CH_3OH + 1.5O_2 \rightarrow CO_2 + 2H_2O \qquad (10.9)$$

Kinetic expressions for reactions 10.6–10.9 were developed by Peppley *et al.* (1999a; 1999b) and corrected in Peppley (2006), and are shown below:

$$r_1 = \frac{k_R K^*_{CH_3O(1)} \left( \frac{P_{CH_3OH}}{P_{H_2}^{0.5}} \right) \left( 1 - \frac{P_{H_2}^3 P_{CO_2}}{K_R P_{CH_3OH} P_{H_2O}} \right) C^T_{S_1} C^T_{S_{1a}} S_C \rho_b}{\left( 1 + K^*_{CH_3O(1)} \left( \frac{P_{CH_3OH}}{P_{H_2}^{0.5}} \right) + K^*_{HCOO(1)} P_{CO_2} P_{H_2}^{0.5} + K^*_{OH(1)} \left( \frac{P_{H_2O}}{P_{H_2}^{0.5}} \right) \right) \left( 1 + K^{0.5}_{H(1a)} P^{0.5}_{H_2} \right)} \qquad (10.10)$$

$$r_2 = \frac{k_D K^*_{CH_3O(2)} \left( \frac{P_{CH_3OH}}{P_{H_2}^{0.5}} \right) \left( 1 - \frac{P_{H_2}^2 P_{CO}}{K_D P_{CH_3OH}} \right) C^T_{S_2} C^T_{S_{2a}} S_C \rho_b}{\left( 1 + K^*_{CH_3O(2)} \left( \frac{P_{CH_3OH}}{P_{H_2}^{0.5}} \right) + K^*_{OH(2)} \left( \frac{P_{H_2O}}{P_{H_2}^{0.5}} \right) \right) \left( 1 + K^{0.5}_{H(2a)} P^{0.5}_{H_2} \right)} \qquad (10.11)$$

$$r_3 = \frac{k_W K^*_{OH(1)} \left( \frac{P_{CO} P_{H_2O}}{P_{H_2}^{0.5}} \right) \left( 1 - \frac{P_{H_2} P_{CO_2}}{K_W P_{CO} P_{H_2O}} \right) C^T_{S_1} C^T_{S_1} S_C \rho_b}{\left( 1 + K^*_{CH_3O(1)} \left( \frac{P_{CH_3OH}}{P_{H_2}^{0.5}} \right) + K^*_{HCOO(1)} P_{H_2}^{0.5} + K^*_{OH(1)} \left( \frac{P_{H_2O}}{P_{H_2}^{0.5}} \right) \right)^2} \qquad (10.12)$$

$$r_4 = K_{ox} \frac{P_{CH_3OH}^{0.18} P_{O_2}^{0.18}}{P_{H_2O}^{0.14}} \qquad (10.13)$$

where $r_1, r_2, r_3,$ and $r_4$ are rates of formation of carbon oxides in reactions 10.6 to 10.9 respectively, and $P_i$ is the partial pressure of species $i$.

When the kinetic rate expressions represented by equations (10.10)–(10.13) are substituted in Equation (10.10.1), the following equations are obtained:

$$\frac{dF_{CH_3OH}}{dz} = (-r_1 - r_2)A_c \tag{10.14}$$

$$\frac{dF_{H_2O}}{dz} = (-r_1 - r_3)A_c \tag{10.15}$$

$$\frac{dF_{CO_2}}{dz} = (r_1 + r_3) A_c \tag{10.16}$$

$$\frac{dF_{H_2}}{dz} = (3r_1 + 2r_2 + r_3) A_c \tag{10.17}$$

$$\frac{dF_{CO}}{dz} = (r_2 - r_3) A_c \tag{10.18}$$

where $F_{CH_3OH}$, $F_{H_2O}$, $F_{CO_2}$, $F_{H_2}$, and $F_{CO}$ are the molar flow rates of methanol, steam, carbon dioxide, hydrogen, and carbon monoxide respectively.

The pressure balance is given by Equation (10.2). In the energy balance, the external heat input term is $Q = Ua\Delta T$, where $U$ is the overall heat transfer coefficient between the jacket and the reactor, $a$ is the ratio of the heat transfer area and the reactor volume, and $\Delta T$ is the temperature difference between the jacket and the reactor at a length $z$. The overall coefficient, $U$, is constructed from the individual coefficients and the resistance of the tube wall. The overall heat transfer coefficient is calculated by the following equation (McCabe *et al.*, 2001):

$$U = \frac{1}{\frac{1}{h_i}\frac{D_o}{D_i} + \frac{x_w}{k_m}\left(\frac{D_o}{\frac{D_o - D_i}{\ln(D_o/D_i)}}\right) + \frac{1}{h_o}} \tag{10.19}$$

where $D_o$ is the outer diameter of the tubular reactor, $D_i$ is the inner diameter of the tubular reactor, $h_o$ is the individual heat transfer coefficient at the annulus side of the reactor, $h_i$ is the individual heat transfer coefficient for the packed-bed reactor side, $x_w$ is the reactor wall thickness, and $k_m$ is the thermal conductivity of the reactor wall. The Sieder-Tate equation is used to estimate the individual heat transfer coefficient on the annulus side (McCabe, 2001):

$$h_o = J_H \frac{k_{mi}}{D_o} \left(\frac{C_p \mu_m}{k_{mi}}\right)^{\frac{1}{3}} \tag{10.20}$$

where $J_H$ is the Colburn $j$ factor, $C_p$ is the average specific heat, $k_{mi}$ is the average thermal conductivity of the gas mixture in the reactor, and $\mu_m$ is the mixture viscosity. To predict the rate of heat transfer on the reactor side for different particle and tube sizes, gas flow rates, and gas properties, the coefficient $h_i$ is used to account for the resistance in the region very near the wall, and for the resistance in the rest of the packed-bed. This coefficient is estimated from the following empirical equation, which was determined by subtracting the calculated bed resistance from the measured overall resistance (Tosun, 2007):

$$h_i = (0.4Re_p^{0.5} + 0.2Re_p^{0.66})Pr^{0.4}\frac{1 - \phi}{\phi}\frac{k_{mi}}{D_p} \tag{10.21}$$

where $Re_p$ is the Reynolds number for the packed-bed reactor, $Pr$ is the Prandtl number, $\phi$ is the void fraction, and $k_{mi}$ is the average thermal conductivity of the gas mixture.

The design equations described by Equations (10.1)–(10.3) were integrated numerically. The kinetic parameters are given in Table 10.1, the process parameters are given in Table 10.2, the specific heats are given in Table 10.3, and the heats of reaction are given in Table 10.4. The steam reforming reactions are endothermic and it is necessary to provide additional heat to keep

Table 10.1.   Kinetic parameters for methanol-based system.

| Parameter | Value | |
|---|---|---|
| $k_R$ | $7.4 \times 10^{14} e^{\left(-\frac{102800}{RT}\right)}$ | $m^2\,mol^{-1}\,s^{-1}$ |
| $k_D$ | $3.8 \times 10^{20} e^{\left(-\frac{170000}{RT}\right)}$ | $m^2\,mol^{-1}\,s^{-1}$ |
| $k_W$ | $5.9 \times 10^{13} e^{\left(-\frac{87600}{RT}\right)}$ | $m^2\,mol^{-1}\,s^{-1}$ |
| $K_R$ | $10^{(1.4142\times10^{-13}T^5 - 4.2864\times10^{-10}T^4 + 5.3993\times10^{-7}T^3 - 3.6385\times10^{-4}T^2 + 1.4096\times10^{-1}T - 2.0258\times10^1) - 2}$ | $MPa^2$ |
| $K_D$ | $10^{(2.9463\times10^{-13}T^5 - 8.8919\times10^{-10}T^4 + 11.1130\times10^{-7}T^3 - 7.4160\times10^{-4}T^2 + 2.7969\times10^{-1}T - 4.4944\times10^1) - 1}$ | $MPa$ |
| $K_W$ | $10^{(-1.4936\times10^{-13}T^5 + 4.5026\times10^{-10}T^4 - 5.6216\times10^{-7}T^3 + 3.7206\times10^{-4}T^2 - 1.3726\times10^{-1}T + 2.4537\times10^1)}$ | |
| $K^*_{CH_3O(1)}$ | $3.16e^{\left(\frac{41.8}{R} - \frac{-20000}{RT}\right)}$ | $MPa^{-0.5}$ |
| $K^*_{CH_3O(2)}$ | $10e^{\left(\frac{30}{R} - \frac{-20000}{RT}\right)}$ | $MPa^{-1}$ |
| $K^*_{HCOO(1)}$ | $31.62e^{\left(\frac{-179200}{RT}\right)}$ | $MPa^{-1.5}$ |
| $K^*_{OH(1)}$ | $31.62e^{\left(\frac{44.5}{R} - \frac{-20000}{RT}\right)}$ | $MPa^{-1.5}$ |
| $K^*_{OH(2)}$ | $10e^{\left(\frac{30}{R} - \frac{-20000}{RT}\right)}$ | $MPa^{-1}$ |
| $K_{H(1a)}$ | $10e^{\left(\frac{-100.8}{R} - \frac{-50000}{RT}\right)}$ | $MPa^{-1}$ |
| $K_{H(2a)}$ | $10e^{\left(\frac{-46.2}{R} - \frac{-50000}{RT}\right)}$ | $MPa^{-1}$ |
| $C^T_{S_1}$ | $7.5 \times 10^{-6}$ | $mol\,m^{-2}$ |
| $C^T_{S_{1a}}$ | $1.5 \times 10^{-5}$ | $mol\,m^{-2}$ |
| $C^T_{S_2}$ | $7.5 \times 10^{-6}$ | $mol\,m^{-2}$ |
| $C^T_{S_{2a}}$ | $1.5 \times 10^{-5}$ | $mol\,m^{-2}$ |

Table 10.2.   Process parameters for methanol-based system.

| Process parameter | Value | |
|---|---|---|
| Catalyst particle diameter, $D_p$ | 0.005 | m |
| Catalyst density, $\rho_b$ | 1300 | $kg\,m^{-3}$ |
| Specific surface area, $S_C$ | 102000 | $m^2\,kg^{-1}$ |
| Void fraction, $\phi$ | 0.38 | |
| Reactor length, $h$ | 0.07 | m |
| Reactor inner diameter, $D_i$ | 0.035 | m |
| Reactor outer diameter, $D_o$ | 0.042 | m |
| Reactor wall thickness, $x_w$ | 0.0035 | m |
| Reactor wall thermal conductivity, $k_m$ | 16.258 | $W\,m^{-1}K^{-1}$ |
| Average thermal conductivity of gases in reactor, $k_{mi}$ | 0.071 | $W\,m^{-1}K^{-1}$ |

the reformer in a temperature range where it is possible to get close to 100% conversion. The required heat is provided via combustion of methanol in the jacket. To design the heat transfer jacket, the inner pipe diameter, annulus diameter, inlet pressure, inlet reactor temperature, and inlet jacket temperature are used to calculate the individual heat transfer coefficients for the inside and outside of the pipe (Equations (10.20) and (10.21)). The overall coefficient is constructed from the individual coefficients and the resistance of the tube wall using Equation (10.19). The overall heat transfer coefficient is a function of temperature and flow rate. For $T_i = 773\,K$, $T_o = 1600\,K$ and a methanol flow rate of $4 \times 10^{-4}\,mol\,s^{-1}$, the overall heat transfer coefficient was found to be $9.4\,J\,m^{-2}\,K^{-1}\,s^{-1}$. A variation of 10% in the flow rate and temperatures resulted in a change of less than 1% in the value of the heat transfer coefficient. For this reason, a constant value of $9.4\,J\,m^{-2}\,K^{-1}\,s^{-1}$ was used in the simulations.

Table 10.3.  Specific heat of gases for methanol-based system.

| Species | Specific heat [J mol$^{-1}$K$^{-1}$] |
|---|---|
| $H_2$ | $C_{p,H_2} = a_1 + b_1 \left(\dfrac{T}{100}\right) + c_1 \left(\dfrac{T}{100}\right)^2 + d_1 \left(\dfrac{T}{1000}\right)^3 + e_1 \left(\dfrac{1000}{T}\right)^2$ |
| $H_2O$ | $C_{p,H_2O} = a_2 + b_2 \left(\dfrac{T}{100}\right) + c_2 \left(\dfrac{T}{100}\right)^2 + d_2 \left(\dfrac{T}{1000}\right)^3 + e_2 \left(\dfrac{1000}{T}\right)^2$ |
| $CO_2$ | $C_{p,CO_2} = a_3 + b_3 \left(\dfrac{T}{100}\right) + c_3 \left(\dfrac{T}{100}\right)^2 + d_3 \left(\dfrac{T}{1000}\right)^3 + e_3 \left(\dfrac{1000}{T}\right)^2$ |
| $CO$ | $C_{p,CO} = a_4 + b_4 \left(\dfrac{T}{100}\right) + c_4 \left(\dfrac{T}{100}\right)^2 + d_4 \left(\dfrac{T}{1000}\right)^3 + e_4 \left(\dfrac{1000}{T}\right)^2$ |
| $O_2$ | $C_{p,O_2} = a_5 + b_5 \left(\dfrac{T}{100}\right) + c_5 \left(\dfrac{T}{100}\right)^2 + d_5 \left(\dfrac{T}{1000}\right)^3 + e_5 \left(\dfrac{1000}{T}\right)^2$ |
| $CH_3OH$ | $C_{p,CH_3OH} = 63.4$ |

where

| | | | | |
|---|---|---|---|---|
| $a_1 = 33.066178$ | $a_2 = 30.92000$ | $a_3 = 24.99735$ | $a_4 = 25.56759$ | $a_5 = 29.65900$ |
| $b_1 = -11.363417$ | $b_2 = 6.832514$ | $b_3 = 55.18696$ | $b_4 = 6.096130$ | $b_5 = 6.137261$ |
| $c_1 = 11.432816$ | $c_2 = 6.7934356$ | $c_3 = -33.69137$ | $c_4 = 4.054656$ | $c_5 = -1.186521$ |
| $d_1 = -2.772874$ | $d_2 = -2.534480$ | $d_3 = 7.948387$ | $d_4 = -2.671301$ | $d_5 = 0.095780$ |
| $e_1 = -0.158558$ | $e_2 = 0.0821398$ | $e_3 = -0.136638$ | $e_4 = 0.131021$ | $e_5 = -0.219663$ |

Table 10.4.  Heats of reaction for methanol-based system.

| | | |
|---|---|---|
| $\Delta H_{r_1}$ | $4.95 \times 10^4 + (C_{p,CO_2} + 3C_{p,H_2} - C_{p,CH_3OH} - C_{p,H_2O})(T - 298)$ | J mol$^{-1}$ |
| $\Delta H_{r_2}$ | $9.07 \times 10^4 + (C_{p,CO} + 2C_{p,H_2} - C_{p,CH_3OH})(T - 298)$ | J mol$^{-1}$ |
| $\Delta H_{r_3}$ | $-4.12 \times 10^4 + (C_{p,CO_2} + C_{p,H_2} - C_{p,H_2O} - C_{p,CO})(T - 298)$ | J mol$^{-1}$ |
| $\Delta H_{r_4}$ | $-6.752 \times 10^4 + (2C_{p,H_2O} + C_{p,CO_2} - C_{p,CH_3OH} - 1.5C_{p,O_2})(T - 298)$ | J mol$^{-1}$ |

Mu et al. (2007) fabricated a mini-reformer that has the same dimensions as the reformer considered in this simulation for autothermal reforming of methanol to produce hydrogen, and their conversion results were compared with simulation results. Simulations were conducted at various process conditions to determine a set of conditions where at least $8 \times 10^{-4}$ mol s$^{-1}$ of hydrogen is produced. A value of $0.5 \times 10^{-3}$ was used for the effectiveness factor in the simulations. It will be shown later that this flow rate is sufficient to produce 100 W of power. Figure 10.2 shows the comparison between the experimental results of Mu et al. (2007) and the simulation results of the model developed here, when the steam to methanol ratio is varied from 1.0 to 1.4 at an inlet temperature of 773 K and inlet pressure of 400 kPa. It is observed that there is excellent agreement between theoretical predictions and experimental results. Figure 10.3 shows the effect of changing the inlet temperature between 773 K and 833 K on the hydrogen flow rate out of the reformer at an inlet pressure of 300 kPa and a steam to methanol ratio of 1.2. It is observed that as the inlet temperature is increased, the outlet flow rate of hydrogen coming out of the reactor increases, due to higher reforming rates. While increasing temperature increases the amount of hydrogen produced, this comes at a cost of increased heat duty.

A suitable fuel cell stack was developed to verify that a hydrogen flow rate of $8 \times 10^{-4}$ mol s$^{-1}$ could, indeed, produce 100 W of power. The procedure developed by Alagharu et al. (2010) was used to design a fuel cell stack. The parameters of the stack are given in Table 10.5. The polarization curve shown in Figure 10.4 was utilized to model the relation between voltage and power density, which was developed from experimental data at 60°C by Chang et al. (2002).

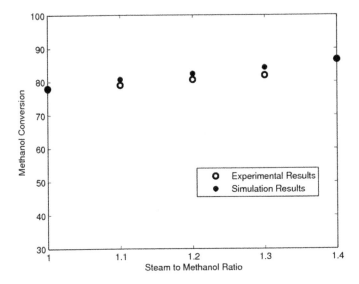

Figure 10.2.   Comparison between experimental results and theoretical predictions for methanol conversion.

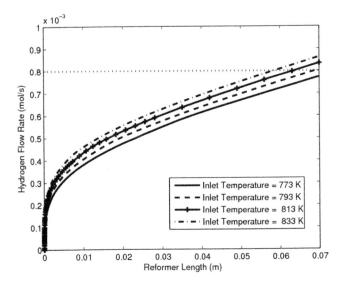

Figure 10.3.   Effect of varying inlet temperature on hydrogen flow rate.

It is observed that a stack of 25 cells results in a current of 5.86 A and a voltage of 0.68 V. Using Equation (10.5), a total power of 100 W is produced. The above simulations were done for a fixed flow rate of methanol to produce sufficient hydrogen for generating 100 W of power. Due to the polarization effect in the fuel cell, changing the flow rate of methanol has a non-linear effect on the power generated, since an increase in the hydrogen flow rate results in an increase in the current density, but a *decrease* in voltage. Temperature also affects the polarization curve characteristics significantly. The effects of flow rate and temperature on the total power produced in a fuel cell are discussed in more detail in Kolavennu *et al.* (2006).

Table 10.5.   Fuel cell parameters for methanol-based system.

| | | |
|---|---|---|
| Number of cells, $n$ | 25 | |
| Area of cross section, $A$ | 20.25 | $cm^2$ |
| Efficiency factor, $\varepsilon$ | 0.95 | |
| Faraday's constant, $F$ | 96,487 | $C\,mol^{-1}$ |
| Inlet flow rate of hydrogen, $F_{H_2}$ | $8 \times 10^{-4}$ | $mol\,s^{-1}$ |

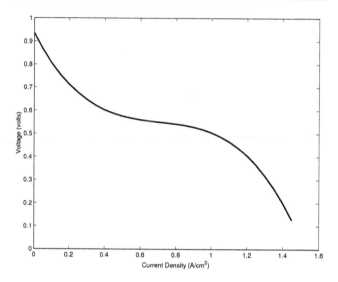

Figure 10.4.   Polarization curve.

### 10.3.2  Ammonia-based system

Ammonia decomposition to produce hydrogen can be represented by the following equation (Chellappa *et al.*, 2002):

$$NH_3 \rightarrow 1.5H_2 + 0.5N_2 \tag{10.22}$$

This reaction is carried out at a temperature range of 793–853 K and a pressure range of 0.1–0.2 MPa, using a Ni-Pt catalyst. This reaction is endothermic with a heat of reaction of 46 kJ per mole of ammonia. In the temperature range stated above, the reaction is irreversible and the reaction rate is represented as follows (Chellappa *et al.*, 2002):

$$r_{NH_3} = k_0 e^{\left(-\frac{E}{RT}\right)} P_{NH_3} \tag{10.23}$$

where $r_{NH_3}$ is the reaction rate for decomposition of ammonia, $k_0$ is the frequency factor, $E$ is the activation energy, $R$ is the universal gas constant, $T$ is the temperature, and $P_{NH_3}$ is the partial pressure of ammonia.

When the kinetic rate expressions are substituted in Equation (10.1), the following equations are obtained:

$$\frac{dF_{NH_3}}{dz} = -r_{NH_3} A_c \tag{10.24}$$

$$\frac{dF_{H_2}}{dz} = 1.5 r_{NH_3} A_c \tag{10.25}$$

$$\frac{dF_{N_2}}{dz} = 0.5 r_{NH_3} A_c \tag{10.26}$$

Table 10.6.  Kinetic and reactor parameters for ammonia-based system.

| Parameter | Value | |
|---|---|---|
| Frequency factor, $k_0$ | $1.33 \times 10^{11}$ | $\mathrm{mol\,m^{-3}\,s^{-1}\,Pa^{-1}}$ |
| Activation energy, E | $1.9 \times 10^5$ | $\mathrm{J\,mol^{-1}}$ |
| Catalyst particle diameter, $D_p$ | 0.00035 | m |
| Catalyst density, $\rho_b$ | 2000 | $\mathrm{kg\,m^{-3}}$ |
| Void fraction, $\phi$ | 0.30 | |
| Reactor length, $h$ | 0.31 | m |
| Reactor inner diameter, $D_i$ | 0.05 | m |
| Specific heat of ammonia, $C_{p,\mathrm{NH_3}}$ | $19.99 + 49.77T - 15.37T^2 + 1.92T^3 + \dfrac{0.18}{T^2}$ | $\mathrm{kJ\,kmol^{-1}\,K^{-1}}$ |
| Specific heat of hydrogen, $C_{p,\mathrm{H_2}}$ | $26.09 + 8.21T - 1.97T^2 + 0.1592T^3 + \dfrac{0.04}{T^2}$ | $\mathrm{kJ\,kmol^{-1}\,K^{-1}}$ |
| Specific heat of nitrogen, $C_{p,\mathrm{N_2}}$ | $33.06 - 11.36T + 11.43T^2 - 2.77T^3 + \dfrac{0.15}{T^2}$ | $\mathrm{kJ\,kmol^{-1}\,K^{-1}}$ |

where $F_{\mathrm{NH_3}}$, $F_{\mathrm{H_2}}$, and $F_{\mathrm{N_2}}$ are the molar flow rates of ammonia, hydrogen, and nitrogen respectively.

The pressure balance is given by Equation (10.2) and the energy balance is given by Equation (10.3). In this system, it is assumed that part of the power generated by the fuel cell is used to electrically heat the reactor and thereby provide the heat input $Q$. Table 10.6 lists the kinetic parameters and reactor parameters used in simulation studies.

The mathematical model developed above provides a relation between the ammonia flow rate into the reactor and the hydrogen flow rate out of the reactor. The reactor temperature changes as a function of the reactor length depending on the heat flux provided to the reactor, which in turn, affects conversion. There is a constraint that a 100% conversion of ammonia is required to avoid poisoning the PEM fuel cell catalyst. Increasing the flow rate of ammonia increases the heat required to achieve complete conversion, which in turn requires more power to be generated by the fuel cell stack. Thus, it is necessary to choose the flow rate of ammonia into the reactor, the heat flux to the reactor, and the size of the fuel cell stack to achieve optimal performance. These design parameters were set on a trial-and-error basis by running a large number of simulations. The design equations described by Equations (10.1)–(10.3) were integrated numerically. The flow rate of ammonia entering the reactor was set at $9 \times 10^{-4}$ mol s$^{-1}$. Simulations were conducted which assumed that there is a constant heat flux of 70 kJ m$^{-3}$s$^{-1}$ to the reactor via electrical heating. This is equivalent to 42 W of power for the reactor, and so sufficient hydrogen has to be generated to produce 142 W of power in the fuel cell, out of which 42 W is used for heating, and the remaining 100 W is available for external use. Figure 10.5 shows the plot of flow rates of ammonia, hydrogen, and nitrogen as a function of reactor volume when the inlet temperature is 793 K and the inlet pressure is 0.2 MPa. The flow rate of hydrogen coming out of the reactor is $1.35 \times 10^{-3}$ mol s$^{-1}$. Figure 10.6 shows the corresponding temperature profile, and it is observed that while the temperature initially decreases due to the endothermic reaction, it starts to increase due to the supply of heat to the reactor, thereby increasing the reaction rate. Figure 10.7 shows the corresponding pressure profile as a function of the reactor volume. It is observed that the pressure drop is small (about 8%). This indicates that the pressure drop due to flow through the packed-bed is approximately balanced by the pressure increase, due to the increase in total moles as a result of the decomposition of ammonia to hydrogen.

The polarization curve shown in Figure 10.4 is utilized to model the relation between voltage and power density. Using the above equations, an iterative calculation was performed to determine the number of cells that would produce 142 W of power. The fuel cell stack parameters

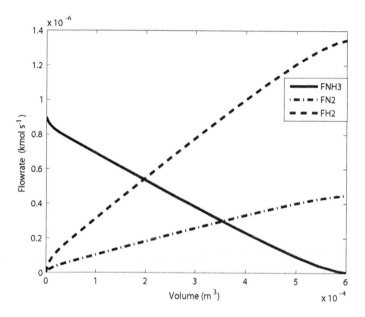

Figure 10.5.   Flow rates of ammonia, hydrogen, and nitrogen as a function of reactor volume.

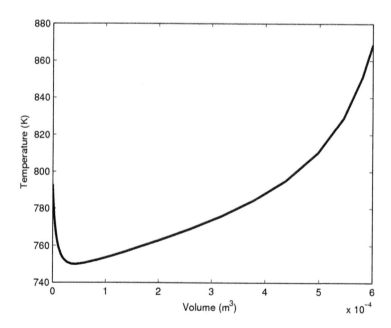

Figure 10.6.   Temperature profile as a function of reactor volume.

are shown in Table 10.7. It was determined that using a stack with 20 cells resulted in a total current of 13 A and a total voltage of 11.92 volts, resulting in a total power of 142 W. The heat requirement for the reactor was previously determined to be 42 W. Thus, a net power of 100 W is generated.

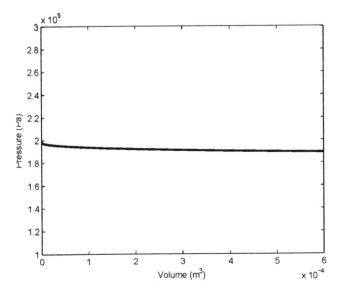

Figure 10.7.    Pressure profile as a function of reactor volume.

Table 10.7.    Fuel cell stack parameters for ammonia-based system.

| | | |
|---|---|---|
| Number of cells, $n$ | 20 | |
| Area of cross section, $A$ | 20.25 | $cm^2$ |
| Efficiency factor, $\varepsilon$ | 0.85 | |
| Faraday's constant, $F$ | 96,487 | $C\,mol^{-1}$ |

## 10.4   CONCLUSIONS

In this paper, a general methodology is presented for steady-state modeling of a reactor and fuel cell stack systems. These design equations were tested on two different systems: (i) a methanol-based system and (ii) an ammonia-based system. The simulation results of the methanol-based system were compared with the experimental results of Mu *et al.* (2007), and were found to be in excellent agreement. Furthermore, it was shown that this design methodology could be utilized for designing the operating conditions of a reactor that will provide the necessary hydrogen flow rate for a fuel cell stack to produce 100 W of power.

## REFERENCES

Alagharu, V., Palanki, S. & West, K.N.: Analysis of ammonia decomposition reactor to generate hydrogen for fuel cell applications. *J. Power Sources* 195 (2010), pp. 829–833.

Chang, H., Kim, J.R., Cho, J.H., Kim, H.K. & Choi, K.H.: Materials and processes for small fuel cells. *Solid State Ionics* 148 (2002), pp. 601–606.

Chellappa, A.S., Fischer, C.M. & Thomson, W.J.: Ammonia decomposition kinetics over Ni-Pt/Al$_2$O$_3$ for PEM fuel cell applications. *Appl. Catal. A General* 227 (2002), pp. 231–240.

Fogler, H.S.: *Elements of chemical reactor engineering*. Prentice-Hall, Upper Saddle River, 2006.

Kolavennu, P., Telotte, J.C. & Palanki, S.: Design of a fuel cell power system for automotive applications. *Int. J. Chem. Reactor Eng.* 4 (2006), Article A19.

Kundu, A., Jang, J.H., Gil, J.H., Jung, C.R., Lee, H.R., Kim, S.H., Ku, B. & Oh, Y.S.: Micro-fuel cells-current development and applications. *J. Power Sources* 170 (2007), pp. 67–78.

Larminie, J. & Dicks, A.: *Fuel cell systems*. Wiley, New York, 2000.

McCabe, W.L., Smith, J.C. & Harriot, P.: *Unit operations of chemical engineering*. 6th ed., McGraw-Hill, NY, 2001.

Mu, X., Pan, L., Liu, N., Zhang, C., Li, S., Sun, G. & Wang, S.: Autothermal reforming of methanol in a mini-reactor for a miniature fuel cell. *Int. J. Hydrogen Energy* 32 (2007), pp. 3327–3334.

National Research Council: Meeting the energy needs of future warriors. http://www.nap.edu/catalog/11065.html, 2004.

Office of the Secretary of Defense: Unmanned Aircraft Systems Roadmap 2005–2030, U.S. DOD, 2005.

Pattekar A.V. & Kothare, M.V.: A radial microfluidic fuel processor. *J. Power Sources* 147 (2005), pp. 116–127.

Peppley, B.A.: Personal Communication. 2006, www.southalabama.edu/engineering/chemical/faculty/palanki/research/res/peppley06.pdf.

Peppley, B.A., Amphlett, J.C., Kearns, L.M. & Mann, R.F.: Methanol-steam reforming on $Cu\backslash ZnO\backslash Al_2O_3$ catalysts. Part 1. The reaction network. *Appl. Catal.* A 179 (1999a), pp. 21–29.

Peppley, B.A., Amphlett, J.C., Kearns, L.M. & Mann R.F. Methanol-steam reforming on $Cu\backslash ZnO\backslash Al_2O_3$ catalysts. Part 2. A comprehensive kinetic model. *Appl. Catal.* A 179 (1999b), pp. 31–49.

Poling, B.E., Prausnitz J.M. & O'Connell, J.P.: *The properties of gases and liquids*. 5th ed., McGraw-Hill, 2001.

Service, R.F.: Shrinking fuel cells promise power in your pocket. *Science* 296 (2002), pp. 1222–1224.

Shah K. & Besser, R.S.: Understanding thermal integration issues and heat loss pathways in a planar microscale fuel processor: Demonstration of an integrated silicon microreactor-based methanol steam reformer. *Chem. Eng. J.* 135S (2008), pp. S46–S56.

Tosun, I.: *Modeling in transport phenomena*. 2nd ed., Elsevier, UK, 2007.

# CHAPTER 11

## Photocatalytic fuel cells

Michael K.H. Leung, Bin Wang, Li Li & Yiyi She

## 11.1  INTRODUCTION

Photocatalytic fuel cell (PFC) is a synergistic integration between two emerging technologies, namely, photocatalysis (PC) and fuel cell (FC) technologies. The PFC activity depends highly on the nano-structure (morphology and dimensions) of the photocatalyst and the micro-scale design of the fuel cell. Recent studies have demonstrated the effectiveness of the new PFC approach in solving environmental and energy problems. A solar PFC can decompose organic compounds in wastewater for wastewater treatment and simultaneously produce electricity or renewable hydrogen fuel. The dual benefits are particularly valuable for remote areas where energy sources are scarce for the residents. PFC is also applicable to food waste, agricultural residues, sludge, industrial organic waste, etc. The potential of a PFC is enormous. In this chapter, we present the photocatalytic electrochemical mechanisms of PFC and its development.

## 11.2  PFC CONCEPT

### 11.2.1  *Fuel cell*

A fuel cell converts a fuel into electricity by electrochemical reactions with no combustion. There are various types of fuel cells: alkaline fuel cells (Shen *et al.*, 1999), proton exchange membrane (PEM) fuel cells (Cheng *et al.*, 2007; Zhao and Xu, 2009), solid oxide fuel cells (SOFC) (Ni *et al.*, 2009), microbial fuel cells (Logan *et al.*, 2006), etc. For the PEM electrochemical energy conversion, at the anode, with the effect of an electro-catalyst, the hydrogen molecules are decomposed to protons and electrons. Then, the protons migrate to the cathode through the proton exchange membrane, while the electrons migrate to the cathode through an external circuit to produce electricity. At the cathode, oxygen molecules are reduced and combined with electrons and protons to form water.

### 11.2.2  *Photocatalysis*

When a photocatalyst is irradiated by its activating spectrum, electron-hole pairs will be produced. They can be effectively applied to perform various green functions, such as detoxification, disinfection, hydrogen fuel production etc. (Leung *et al.*, 2010; Ni *et al.*, 2007). Among many candidate photocatalysts ($TiO_2$, ZnO, ZnS, $WO_3$, $SrTiO_3$), $TiO_2$ is most popular because it is abundant, chemically stable, and relatively non-toxic. The band gap of $TiO_2$ (3.2 eV) can be overcome by UV radiation (Asahi *et al.*, 2001). As a fraction of the solar spectrum is UV, $TiO_2$ is regarded as a solar energy harvesting material and its solar utilization has been extensively investigated (Schloegl, 2008).

The reaction rate of photocatalysis is commonly slow because of the phenomenon of electron-hole recombination. The limited activating UV spectrum in sunlight for $TiO_2$ also explains why solar photocatalysis has a low photonic efficiency. A number of strategies have been developed to modify photocatalysts for visible-light response, such as using metal (Lin *et al.*, 2010), metal oxide (Peng *et al.*, 2011), and metal sulfide (Li *et al.*, 2010) as modifiers.

## 11.2.3  *Photocatalytic fuel cell*

Photocatalysis and fuel cell electrochemistry can be integrated to accomplish the concept of the photocatalytic fuel cell (PFC). A PFC is composed of two major coupling parts: a photocatalytic (PC) reactor and a fuel cell (FC). The integration will yield a synergy due to the current doubling effect (Lianos, 2011). The photocatalyst and electro-catalyst used in a PFC can mutually enhance each other. A solar PFC can be practically applied for wastewater treatment and simultaneous production of electricity and/or hydrogen. At the anode side, pollutant molecules can be electro-chemically and photocatalytically oxidized by electro-catalysts and photocatalysts, respectively, releasing electrons to flow through the external circuit to the cathode. At the cathode side, several electrochemical reactions may occur, e.g. reducing protons to produce hydrogen. The double environmental benefits, namely, waste treatment and simultaneous energy production, make the PFC an attractive environmental-friendly technology. The details of the PFC architecture and synergistic mechanisms are provided in the following sections.

## 11.3  PFC ARCHITECTURE AND MECHANISMS

### 11.3.1  *Cell configurations*

A photocatalytic fuel cell with a single compartment structure basically involves two electrodes immersed in an aqueous electrolyte. The schematic of a photocatalytic fuel cell constructed with a single compartment is illustrated in Figure 11.1. The photoanode is deposited with a thin layer of photocatalyst. When the photoanode is exposed to activating light, photo-induced electrons ($e^-$) and holes ($h^+$) will be produced at the photocatalytic sites. The electrons can move through an electric circuit to the cathode side. The holes can oxidize the $OH^-$ to produce hydroxyl free radicals or photodegradable organic matter in the solution, thereby forming $CO_2$, electrons and protons. Moreover, the organic matter can be oxidized by generated OH radicals if the electrolyte is filled with a basic medium. It should be noted that in order to avoid light irradiance loss, a quartz window is installed in the cell for light exposure to the photoanode. At the counter electrode side, such as at the Pt cathode, the reaction varies depending on the availability of oxygen. If no oxygen is present, the reduction reaction involved is hydrogen production. If oxygen is available in the solution, the reaction is oxygen reduction to produce water. Therefore, one of the practical applications of a PFC is to convert organic matter including pollutants, into electricity and/or hydrogen fuel.

A PFC can also be constructed in a two-compartment cell configuration (Fig. 11.2). A membrane or glass frit is used to separate the anodic and cathodic compartments. A proton exchange membrane, such as a Nafion membrane, is usually adopted to separate the two compartments. In addition, two different types of electrolytes can be used in the anodic and cathodic compartments. The advantage of using two different electrolytes over a single electrolyte is that a chemical bias can be formed in the photoelectrochemical cell. The photo-induced holes have to be consumed in the anodic compartment to avoid electron-hole recombination. Therefore, filling the anodic compartment with an alkaline solution is preferred. In addition, electrons generated from photocatalytic reactions or oxidation of organic matter are expected to react with proton ions and/or oxygen. Thus, filling the cathodic compartment with an acidic solution is preferred.

### 11.3.2  *Bifunctional photoanode*

The anode is a critical component which determines the overall PFC performance. Both PC and FC reactions take place simultaneously at the anode. The photoanode is usually made by coating a thin layer of photocatalysts (e.g. $TiO_2$, ZnO) onto an electrical conducting substrate.

#### 11.3.2.1  *Photocatalyst*
Many metal oxides and metal sulfides have been developed to be photocatalysts. Owing to its strong chemical stability, non-toxicity, and low cost, nano-crystalline titanium dioxide ($TiO_2$) is

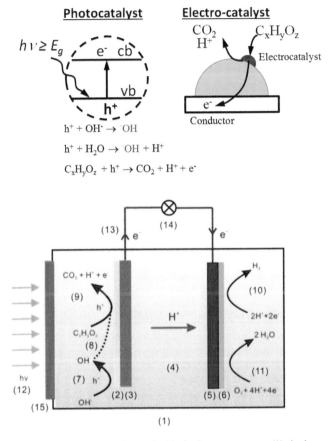

**Photocatalyst**

**Electro-catalyst**

$hv \geq E_g$     e⁻  cb

vb

h⁺

$CO_2$
$H^+$     $C_xH_yO_z$

Electrocatalyst

e⁻

Conductor

$h^+ + OH^- \rightarrow \ OH$

$h^+ + H_2O \rightarrow \ OH + H^+$

$C_xH_yO_z + h^+ \rightarrow CO_2 + H^+ + e^-$

(13)  e⁻   (14)   e⁻

$CO_2 + H^+ + e$

(9)

h⁺

$C_xH_yO_z$
(8)

OH

(7)  h⁺

OH⁻

(2)(3)

hv
(12)

(15)

H⁺

(4)

$H_2$

(10)

$2H^+ + 2e$

$2 H_2O$

(11)

(5) (6)   $O_2 + 4H^+ + 4e$

(1)

Figure 11.1.   Schematic of photocatalytic fuel cell with single compartment: (1) single compartment photo-electrochemical cell; (2) photocatalyst thin film; (3) anode substrate; (4) aqueous electrolyte; (5) cathode support; (6) cathode catalyst; (7) oxidization reaction of OH⁻ by holes in basic media; (8) oxidization of organic matter by OH radicals; (9) oxidization reaction of organic matter in neutral media; (10) proton reduction to produce hydrogen in anaerobic condition; (11) oxygen reduction to generate water in aerobic condition; (12) light irradiation; (13) electric circuit; (14) loading; (15) quartz window.

considered as the most promising photocatalyst, which has shown a great potential in various photocatalytic applications (Murdoch *et al.*, 2011; Yu *et al.*, 2010). It is well known that there are three main types of TiO₂ structures: anatase, rutile, and brookite. Previous studies suggest that anatase is the most stable phase for TiO₂ nano-particles below 11 nm (Fujishima *et al.*, 2008; Zhang and Banfield, 2000). However, the commercially available TiO₂ of the anatase phase is usually smaller than 50 nm in size, and these particles have a band gap of 3.2 eV, corresponding to a UV wavelength of 385 nm. In contrast, the TiO₂ of the rutile phase has a band gap of 3.0 eV with an excitation wavelength that extends to the visible light region (Hurum *et al.*, 2003).

Recently, Degussa P25 TiO₂, a mixed phase photocatalyst, has received considerable interest in the application of photocatalytic fuel cells. The particle size and surface area of Degussa P25 TiO₂ nano-particles are 30 nm and 49 m² g⁻¹, respectively (Jafry *et al.*, 2011). In addition, the P25 powder is composed of 70% anatase phase and 30% rutile phase (Ohno *et al.*, 2001). Compared with pure phase TiO₂, the enhanced activity of mixed phase is due to the efficient electron-transfer from anatase to lower energy rutile electron trapping sites, leading to effective electron-hole separation (Bickley *et al.*, 1991).

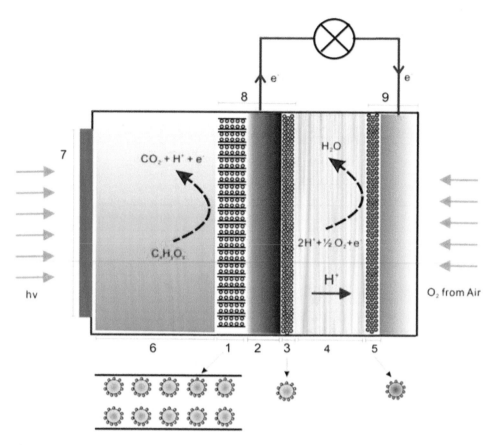

Figure 11.2.    Schematic of photocatalytic fuel cell with two-compartment photoelectrochemical cell: (1) photocatalyst layer: Pt-WO₃ decorated TiO₂ NTs; (2) carbon cloth; (3) electro-catalyst layer: Pt-WO₃; (4) polymer electrolyte membrane; (5) Pt/C electro-catalyst for oxygen reduction; (6) wastewater chamber; (7) light window; (8) bifunctional anode; (9) gas diffusion cathode.

A $TiO_2$ nanotube (TNT) is a promising alternative due to its high specific area, good electronic transport, and high mechanical strength. Moreover, there are several ways to modify TNTs either by band gap engineering (doping and semiconductor alloys) or by sensitization (organic dye sensitization and composite semiconductors) for activation under visible light for solar PFC applications (Asahi *et al.*, 2001; Baker and Kamat, 2009; Bingham and Daoud, 2011; Daskalaki *et al.*, 2010; Dong *et al.*, 2008; Khan *et al.*, 2002; Kukovecz *et al.*, 2005; Lee *et al.*, 2009a; 2009b; Murphy, 2008; Nakamura *et al.*, 2000; Ni *et al.*, 2007; Sun *et al.*, 2009; Valentin *et al.*, 2007; Wang *et al.*, 2013; Zhang *et al.*, 2008).

### 11.3.2.2    *Substrate materials*
The purpose of the substrate is to effectively transport the generated electrons at the anode to the external circuit, finally arriving at the cathode for the reduction reactions, such as oxygen reduction and proton reduction. The requirements for the suitable substrates include high stability in the used electrolyte and low electrical resistance. Substrates also provide necessary support to anchor photocatalyst materials and facilitate electron collection. In the reported literatures, transparent conducting glass, titanium substrate, and carbon materials are the major substrates for the deposition of photocatalyst layers.

In most cases, transparent conducting glasses, for example, FTO (fluorine-doped tin oxide) and ITO (indium-tin oxide), are preferred to be used as the substrate. The use of FTO and ITO glass in a photocatalytic fuel cell can be found in recent works (Antoniadou *et al.*, 2010; Nishikiori *et al.*, 2011). These materials are already commercially available from Sigma-Aldrich. The surface resistivity varies from 7–13 $\Omega$/sq. Conducting glass of the lower resistivity usually has the thicker conductive coating. In order to make non-planar reactors, plastic substrates with a coating of transparent conductive oxide (TCO) are also available. Although both sides of the glass or plastic material can be coated with transparent conductive oxide, only one side is needed for the deposition of photocatalysts. When the photoanode is exposed to the light irradiation from the front face, the generated electrons will be transported away by the conductive layer. The main advantage of transparent conducting glass is that light can be back irradiated to the photocatalysts by passing through the glass layer. It should be mentioned that in the back irradiation process, the glass layer can also adsorb UV light irradiation, which can reduce the irradiance to the photocatalyst layers. Therefore, it is important to choose the right substrate for the photoanode preparation.

Prior to the photocatalyst deposition, the transparent conducting glass is usually required to be cleaned in order to remove possible contaminants. A typical procedure of glass cleaning is described as follow: the glass is ultrasonically cleaned in distilled water, ethanol, and acetone, and then naturally dried in air. The clean surface will ensure the successful coating of photocatalyst and enhance the adhesion between the transparent conductive oxide layer and the photocatalyst layer.

A titanium substrate, particularly a Ti sheet, is one of the good candidates for the current collector due to its excellent resistance to corrosion, as well as its low electrical resistivity. Recently, the direct growth of $TiO_2$ nanotube arrays onto Ti substrates has received considerable attention due to the promising application in photocatalysis. Liu *et al.* (2011a; 2011b) recently reported the use of $TiO_2$ nanotube arrays fabricated from Ti sheets for photocatalytic fuel cell application. Since $TiO_2$ nanotube arrays can be directly fabricated by anodization in the fluoride containing electrolyte, they are regarded as one of the promising materials for harvesting solar energy with simultaneous wastewater treatment.

Similar to transparent conducting glasses, a titanium substrate has to be cleaned before the growth of nano-structured $TiO_2$ materials. A typical titanium pretreatment process can be found in published literatures (Liu *et al.*, 2011a). In brief, it is first cleaned with acetone followed by absolute ethanol. Then it is washed with distilled water. Afterwards, it is boiled in 10 wt.% oxalic acid for 30 min to remove the potential oxide layer. Finally, it is ultrasonically cleaned in water for 15 min followed by air-drying in the atmosphere.

Carbon materials, such as carbon cloth and carbon paper, are also good substrates for the deposition of photocatalysts, due to their low resistivity and cost. They are also widely used in the proton exchange membrane fuel cell (PEMFC), and are also commercially available at www.fuelcellstore.com. Carbon cloth/paper is supplied with an untreated surface or reinforced with PTFE. Since the photocatalytic reactions at the anode side involve three phases: liquid electrolyte, solid anode photocatalyst, and the produced gaseous $CO_2$, carbon materials with hydrophilic surfaces are preferred in the fabrication of photoanodes. Compared to the previous two substrates, carbon material can be used directly without any pretreatment.

### 11.3.2.3 *Catalyst deposition methods*

The approach of coating photocatalysts varies depending on the types of substrates used. Due to the wide application of $TiO_2$ in photocatalytic fuel cells, emphasis will be given to the deposition of $TiO_2$ on electrically conductive substrates.

For the deposition of powered photocatalysts onto transparent conducting glass, the approach generally includes doctor blade methods, spin coating, dip coating, sputtering, and screen printing. Antoniadou *et al.* (2010) reported the use of doctor blade techniques to coat $TiO_2$ onto glass. Generally, doctor blade and dip coating methods are useful strategies for laboratory investigations. The main disadvantage is that it is difficult to obtain $TiO_2$ thin film with uniform and controllable thickness. For mass production of photoanodes, an homogeneous coating of $TiO_2$ thin film can be achieved by a spin coater, sputter, and screen printer.

It should be noted that a $TiO_2$ slurry is usually made before spin coating or screen printing. Previous photocatalytic fuel cell studies showed that $TiO_2$ slurry can be easily prepared by using

Triton X-100 and acetylacetone (Kaneko *et al.*, 2009a), or polyethylene glycol (PEG, 2000) mixed with ethanol (Antoniadou *et al.*, 2008; 2010). It is also believed that the fabrication of $TiO_2$ photoanodes for dye sensitized solar cell (DSSC) application is also helpful for photocatalytic fuel cell studies. The details of paste fabrication from $TiO_2$ powder can be found in Ito *et al.* (2007). The key components in paste fabrication include acetic acid, terpineol, and ethyl cellulose. In order to improve the adhesion between $TiO_2$ and the FTO surface, sintering is applied to the thin film. This is because strong chemical bonding can be formed when hydroxides ($-OH$) on the surface of $TiO_2$ and the FTO surface react with each other by dehydration (Ito *et al.*, 2007). The involved reaction is expressed as follows:

$$M_1\text{-}OH + HO\text{-}M_2 \rightarrow M_1\text{-}O\text{-}M_2 + H_2O$$

where $M_1$ stands for a Ti atom on a surface of one particle and $M_2$ means a Ti atom of another particle or a Sn atom on the surface of the FTO glass. It should be noted that large particles can easily lead to the shrinkage of the $TiO_2$ thin film during sintering. The sintering temperature is in the range of 450–550°C (Antoniadou *et al.*, 2008; Kaneko *et al.*, 2009a).

For $TiO_2$ nanotube arrays, there is no need for the deposition of $TiO_2$ thin film onto Ti substrates. The fabrication of $TiO_2$ nanotube thin film is achieved by Ti anodization in fluoride containing electrolytes under constant voltage. Both water and organic solvents can be used in the Ti anodization. Typical fluorides used for electrolytes include $NH_4F$ (Shankar *et al.*, 2008) and HF (Mor *et al.*, 2006a). Non-aqueous organic polar electrolytes, e.g. N-methylformamide, dimethyl sulfoxide (DMSO), formamide (FA), and ethylene glycol (EG), are also helpful for the fabrication of longer $TiO_2$ nanotube arrays (Shankar *et al.*, 2007). The applied voltage is typically in the range of 20–60 V (Allam and Grimes, 2009; Prakasam *et al.*, 2007). The dimensions of $TiO_2$ nanotube arrays can be tuned by experimental conditions, such as applied voltage, electrolyte used, and anodization duration. After the anodization, the prepared $TiO_2$ nanotube array has to be cleaned with ethanol followed by water. More details of $TiO_2$ nanotube arrays can be found in recent review papers (Grimes, 2007; Mor *et al.*, 2006b; Rani *et al.*, 2010).

Recently, growth of $TiO_2$ nano-wire and nano-rod arrays on transparent glass (e.g. FTO) is successfully achieved by hydrothermal methods (Feng *et al.*, 2008; Liu and Aydil, 2009). The effective solar energy harvesting by $TiO_2$ nano-rods or nano-wire structures is also investigated in recent work (Huang *et al.*, 2011; Sun *et al.*, 2012). This growth technique will have promising applications in photocatalytic fuel cells, even though no such work is reported so far. The detachment of the photocatalyst layer from the substrate is the common problem for the photocatalytic applications. However, $TiO_2$ nano-wires and nano-rods can be directly grown onto FTO glass with strong adhesion. Previous study suggested that $SnO_2$ and $TiO_2$ could intermix mutually in their lattice, due to the ion radius difference between $Sn^{4+}$ (0.83 Å) and $Ti^{4+}$ (0.68 Å) (Li *et al.*, 2009). Recently, we found that $TiO_2$ nano-flowers can also be fabricated on FTO glass.

### 11.3.3  *Cathode*

Although there is a huge variety of choice for the anode materials in the photocatalytic fuel cell, the candidates for cathodes are very limited. So far, the most effective and widely used cathode materials are Pt-based electro-catalysts. Much research has directly used Pt wire or Pt foil as the cathode for proton reduction to produce hydrogen under anaerobic conditions. The involved reaction can be expressed as: $2H^+ + 2e^- \rightarrow H_2$.

To maximize the electrode surface area and lower the cost of the cathode, platinized Pt foil and platinized stainless steel mesh are also good choices for the cathode materials. Ueno *et al.* (2009) reported the electro-deposition of Pt black onto Pt foil ($1 \times 1$ cm) used as an oxygen reduction cathode. In addition, Kaneko, M. and co-workers reported the use of Pt-coated stainless steel (SUS) mesh ($4 \times 4$ cm, 100 mesh) as the cathode for oxygen reduction in UV light-activated photocatalytic fuel cells (Kaneko *et al.*, 2009b). Platinized substrates can be easily achieved by electrochemical reduction of the corresponding platinum precursors, such as $H_2PtCl_6$ and $K_2PtCl_6$.

However, due to the fact that the main focus in photocatalytic fuel cell research is the development of the photoanode, most of the current work uses the commercially available cathode electro-catalysts, such as Pt black. It is a fine, black powder of platinum, and can be sprayed or hot-pressed onto the membrane or carbon cloth/paper surface. The commercial Pt black is guaranteed to have a specific surface of 24.4 to 29.2 $m^2 g^{-1}$ (Mills, 2007). Therefore, the surface area for proton reduction is much higher than the geometrical area. Liu *et al.* (2011b) investigated the effect of cathode material on photocatalytic fuel cell performance, and they found that the Pt black coated Pt electrode showed much better performance than a pure Pt plate. This is due to the more active sites provided by Pt black for oxygen reduction.

Compared with proton reduction, oxygen reduction reactions (ORR) are preferred in the photocatalytic fuel cell to enhance the fuel cell performance. This is due to the higher redox potential of the oxygen reduction reaction than that of the proton reduction. In the aerobic environment, the oxygen can be reduced to water by reacting with protons and electrons. The involved reaction is expressed as: $O_2 + 2 H^+ + e^- \rightarrow H_2O$.

Due to the low solubility of oxygen in water, the mass transfer of oxygen at the cathode can severely affect the cell performance. Therefore, it is desirable to develop an air-breathing cathode for oxygen reduction. The air-breathing cathode is also called a gas diffusion electrode (GDE), which is widely used in fuel cell technology. The details of air-breathing cathode fabrication can be found in previous literature.

Although Pt is the best electro-catalyst for proton reduction or oxygen reduction reactions, the extremely high cost of this material hinders the wide application of this technology in fuel cells. Therefore, it is urgent for research to develop alternative electrode materials. Such work can be found in the study of the fuel cell or the microbial fuel cell. Recently, Kaneko *et al.* (2009a) reported the use of $MnO_2$ as a cathode material for oxygen reduction by mixing it with active carbon and Nafion solution. In addition, a Pt-free photocatalytic fuel cell was proposed by Iyatani *et al.* (2013). They used a carbon rod as the cathode material to facilitate the $I^{3-}/I^-$ redox reaction. The use of Pt-free catalysts will definitely reduce the cost and accelerate the commercialization of the photocatalytic fuel cell.

## 11.4   ELECTROCHEMICAL KINETICS

Various characterization techniques can be applied to measure electrochemical properties and thus evaluate the fuel cell system performance. Below, we describe several of the most widely used characterization techniques applicable for photocatalytic fuel cells.

### 11.4.1   *Current-voltage characteristics*

From thermodynamics, we can get the ideal voltage for a fuel cell. However, in practice, the actual cell voltage is less than the ideal thermodynamically predicted voltage due to irreversible losses, even when the open circuit voltage is measured. Normally, the losses can be broken up into three major types: activation losses, ohmic losses and concentration losses. The real cell voltage can thus be written as below:

$$V = E_{thermo} - \eta_{act} - \eta_{ohmic} - \eta_{conc} \tag{11.1}$$

where $V$ is the actual cell voltage, $E_{thermo}$ is the ideal thermodynamically predicted voltage, $\eta_{act}$ is the activation losses, $\eta_{ohmic}$ is the ohmic losses and $\eta_{conc}$ is the concentration losses.

High-performance fuel cells will exhibit less loss and thus a higher voltage for a given current load (O'Hayre, 2009). In the experiment, we use potentiostatic techniques to measure the current at an imposed voltage or galvanostatic techniques to measure the output voltage at an imposed current. By stepping the voltage or current demand, the current-voltage response, i.e. the *j-V* curve is attained, which gives an overall quantitative evaluation of the fuel cell performance and fuel cell power density. In theory, the three major losses mentioned above give a fuel cell *j-V* curve its characteristic shape but each has a different effect and contributes in its own way. Below, we will talk about the characteristics of the *j-V* curve in detail.

Table 11.1.  Summary of photocatalytic fuel cell designs.

| No. | Anode material | Anode substrate | Deposition method | Cathode material | Electrolyte | Cell architecture | Fuel | Light source | Reference |
|---|---|---|---|---|---|---|---|---|---|
| 1 | Degussa P25 $TiO_2$ | FTO | Doctor blade method | Pt/carbon black | 20% EtOH + 0.2 M NaOH; 20% glycerol + 0.2 M NaOH | Single- and two-compartment cell | Ethanol, glycerol etc. | 450 W Xe lamp | Antoniadou et al. (2010) |
| 2 | Degussa P25 $TiO_2$ | FTO | Paste casting | Pt wire | Anolyte: NaOH+EtOH Catholyte: $H_2SO_4$ | Two-compartment cell | Ethanol | 4 black light tubes (3.2 mW cm$^{-2}$) | Antoniadou et al. (2008) |
| 3 | Degussa P25 $TiO_2$ | FTO | Spin coating | Pt black/FTO; $MnO_2$/carbon paper | 0.5 M glycine + 0.1 M $Na_2SO_4$; 0.4 M urea | One-compartment cell | Glycine, urea, biomass, etc. | 500 W Xe lamp (100 mW cm$^{-2}$) | Kaneko et al. (2009a) |
| 4 | Degussa P25 $TiO_2$ | FTO | Spin coating | Pt foil | 10 M $NH_3$ + 0.1 M $KNO_3$ | One-compartment cell | Ammonia | 500 W Xe lamp (469 mW cm$^{-2}$) | Kaneko et al. (2005) |
| 5 | Degussa P25 $TiO_2$ | FTO | Screen printing | Pt black/carbon cloth | 20% EtOH + 1.0 M NaOH | Two-compartment cell | Ethanol | 4 black light fluorescent tubes (3.5 mW cm$^{-2}$) | Panagiotopoulou et al. (2010) |
| 6 | Degussa P25 $TiO_2$ | FTO | – | Pt/carbon cloth | 20% EtOH + 0.2 M NaOH | Two-compartment cell | Ethanol | 4 black light fluorescent tubes (3.5 mW cm$^{-2}$) | Antoniadou and Lianos (2010) |
| 7 | Degussa P25 $TiO_2$ | FTO | Spin coating | Pt plate | 10 M $NH_3$ + 0.1 M $Na_2SO_4$ | One-compartment cell | Ammonia | 500 W Xe lamp (503 mW cm$^{-2}$) | Kaneko et al. (2006) |
| 8 | Degussa P25 $TiO_2$ | FTO | Spin coating | Pt/FTO Pt black/ SUS mesh | 60 mM $NH_3$ solution; pig urine; food suspension | One-compartment cell | Biomass and biowaste | 500 W Xe lamp (100 mW cm$^{-2}$) | Kaneko et al. (2009b) |
| 9 | Degussa P25 $TiO_2$ | Carbon paper | Drop casting | Pt/carbon paper | 1.0 M formic acid + 0.1 M $H_2SO_4$ | MEA structured cell | Formic acid, methanol, formaldehyde, etc. | 1000 W Xe lamp; AM 1.5 filter (100 mW cm$^{-2}$) | Seger et al. (2012) |
| 10 | $TiO_2$ | ITO | RF magnetron sputtering deposition | Carbon rod; Pt mesh | Anolyte: $CH_3OH$ + $Na_2SO_4$; Catholyte: 0.5 M NaI + 0.025 M $I_2$ + $Na_2SO_4$ | Two-compartment cell | Biomass derivatives | AM 1.5 (100 mW cm$^{-2}$) | Iyatani et al. (2012) |
| 11 | Pt-$TiO_2$ (Degussa P25) | Carbon cloth | Dropping and baking | Pt/Ti mesh | 1.0 M MeOH + 0.5 M $H_2SO_4$ | One-compartment cell | Methanol | 100 W Hg lamp and three 4 W UV tubes | Lin et al. (2011) |
| 12 | CdS-$TiO_2$ (Degussa P25) | FTO | Paste casting | Pt/FTO | Anolyte: 1.0 M NaOH Catholyte: 1.0 M $H_2SO_4$ | Two-compartment cell | Ethanol, glycerol, ammonia, Triton X-100, etc. | 4 black light tubes (3.2 mW cm$^{-2}$) | Antoniadou and Lianos (2009) |
| 13 | Glucose-$TiO_2$ | ITO | Sol-gel dip coating | Pt | Iodine-based electrolyte ($I_2$/LiI) | One-compartment cell | Glucose | 150 W Xe lamp | Nishikiori et al. (2011) |
| 14 | Pt-Ru-$TiO_2$ (Degussa P25) | Carbon paper | Paint brush method | Pt/carbon cloth | Nafion membrane; 1.0 M MeOH | MEA structured cell | Methanol | 150 W Hg lamp (28 mW cm$^{-2}$) | Drew et al. (2005) |
| 15 | $TiO_2$ nanotube | Ti sheet | Anodization | Ti; Pt; Pt black/Pt | 0.05 M acetate + 0.1 M $Na_2SO_4$ | One-compartment cell | Acetic acid, glucose, urea, etc. | UV light (3.8 mW cm$^{-2}$) | Liu et al. (2011b) |
| 16 | $TiO_2$ nanotube (modified by CdS and $Cu_2O$) | Ti sheet | Anodization | Pt black/Pt | 0.05 M phenol + 0.1 M $Na_2SO_4$ | One-compartment cell | Organic compounds | UV and solar light | Liu et al. (2011a) |

### 11.4.1.1  *Ideal thermodynamically predicted voltage*

Thermodynamics provides theoretical limits for fuel cell performance. The ideal cell voltage under standard state conditions (room temperature, atmospheric pressure, unit activities of all species) is:

$$E^0 = -\frac{\Delta \hat{g}^0_{rxn}}{nF} \tag{11.2}$$

where $E^0$ is the standard state ideal voltage, $\Delta \hat{g}^0_{rxn}$ is the standard state free energy change for the reaction, $n$ is the number of moles of electrons transferred and $F$ is Faraday's constant.

Actually, however, fuel cells are frequently operated under non-standard state conditions that vary greatly from the standard state. After analyzing how the reversible cell voltage varies with temperature, pressure, and activity, a full expression considering all the influencing factors can be written as:

$$E = E^0 + \frac{\Delta \hat{s}}{nF}(T - T_0) - \frac{RT}{nF} \ln \frac{\prod a^{v_i}_{products}}{\prod a^{v_i}_{reactants}} \tag{11.3}$$

where $E$ is the non-standard state ideal voltage, $E^0$ is the standard state ideal voltage, $\Delta \hat{s}$ is equal to $\left(-\left(\frac{d(\Delta \hat{g})}{dT}\right)_P\right)$, $a_{reactants}$ and $a_{products}$ are the activities of reactants and products repectively, and $v^i$ is the stoichiometric coefficient of species i.

If we assume that $T = T_0$, Equation (11.3) will become:

$$E = E^0 - \frac{RT}{nF} \ln \frac{\prod a^{v_i}_{products}}{\prod a^{v_i}_{reactants}} \tag{11.4}$$

This is known as the Nernst equation which is the centerpiece of fuel cell thermodynamics. It is widely used in the calculation of the ideal cell voltage.

### 11.4.1.2  *Activation losses*

Activation losses are due to reaction kinetics and are associated with the initial dramatic voltage losses of a fuel cell $j$-$V$ curve. Next, we will discuss the origin of activation losses and ways to improve kinetic performances.

It is obvious that the rate of an electrochemical reaction is finite because an energy barrier impedes the conversion of reactants into products. Therefore the current produced by an electrochemical reaction is limited. The relation between the energy barrier and the current density is as follows:

$$j_1 = nFc^*_R \frac{kT}{h} e^{-\frac{\Delta G_1}{RT}} \tag{11.5}$$

where $j_1$ is the current density in the forward direction (reactants → products), $n$ is the number of moles of electrons transferred, $F$ is Faraday's constant, $c^*_R$ is the reactant surface concentration, $k$ is Boltzmann's constant, $h$ is Planck's constant, $\Delta G_1$ is the size of the energy barrier between the reactant and activated states, $R$ is the gas constant and $T$ is the temperature. The reverse current density is then given by:

$$j_2 = nFc^*_P \frac{kT}{h} e^{-\frac{\Delta G_2}{RT}} \tag{11.6}$$

where $c^*_P$ is the product surface concentration and $\Delta G_2$ is the activation barrier for the reverse reaction.

At thermodynamic equilibrium, the forward and reverse current densities are equal, which is shown as below:

$$j_1 = j_2 = j_0 \tag{11.7}$$

where $j_0$ is called the 'exchange current density' for the reaction.

We can manipulate the size of the activation barrier by varying the cell voltage, which is a distinguishing feature of the electrochemical reaction. The forward activation barrier is decreased by $\alpha nF_\eta$, while the reverse activation barrier is increased by $(1 - \alpha)nF_\eta$. Here $\eta$ is the voltage loss, and $\alpha$ is the transfer coefficient whose value is always between 0 and 1 depending on the symmetry of the activation barrier. Thus away from equilibrium, the new forward and reverse current densities can be written by starting from $j_0$ and taking into account the changes in the forward and reverse activation barriers:

$$j_1 = j_0^0 \frac{c_R^*}{c_R^{0*}} e^{(\alpha n F_\eta/(RT))} \tag{11.8}$$

$$j_2 = j_0^0 \frac{c_P^*}{c_P^{0*}} e^{-(1-\alpha)n F_\eta/(RT)} \tag{11.9}$$

where $c_R^*$ and $c_P^*$ are the actual surface concentrations of the rate limiting species in the reaction, $\eta$ is the voltage loss and $j_0^0$ represents the exchange current density measured at the reference reactant and product concentration values $c_R^{0*}$ and $c_P^{0*}$. The net current is then:

$$j = j_1 - j_2 = j_0^0 \left( \frac{c_R^*}{c_R^{0*}} e^{(\alpha n F_\eta/(RT))} - \frac{c_P^*}{c_P^{0*}} e^{-(1-\alpha)n F_\eta/(RT)} \right) \tag{11.10}$$

Equation (11.10) is known as the Butler-Volmer equation. Here $\eta$ is labeled as the activation loss and given the subscript act, as in $\eta_{act}$ which represents the voltage that is sacrificed to overcome the activation barrier and forces the reaction to completion. From the Butler-Volmer equation we know that current density increases exponentially with activation loss.

From the theoretical analysis above, we know that in order to keep the activation loss at a minimum, there are several things that can be done, including improving the reactant concentration and operational temperature to increase the exchange current density, employing a more effective catalyst to decrease the activation barrier and increasing the catalytic presence to provide more reaction sites. A photocatalytic fuel cell contains a photoanode that bears the semiconductor photocatalyst, and a cathode that bears the electro-catalyst. Choosing effective catalysts is undoubtedly, a straightforward way to minimize the activation loss.

### 11.4.1.3   *Ohmic losses*

Ohmic losses, which are prevalent in every electronic device, also exist in photocatalytic fuel cells. They are most apparent in the middle section of the fuel cell $j$-$V$ curve and occur because of resistance to charge transport. Below, we will give an analysis of the origin of ohmic losses and also ways to minimize these kinds of losses.

Charge transport is an important step in the electrochemical generation of electricity for it completes the circuit in an electrochemical system. Unfortunately, charge transport occurs at a cost of cell voltage as it is not a frictionless process. These types of losses obey Ohm's law and that is why they are so named. The relation between the ohmic loss and current density can be written as:

$$\eta_{ohmic} = jA_{\text{fuel cell}} R_{ohmic} \tag{11.11}$$

where $\eta_{ohmic}$ is the ohmic loss, $j$ is the current density, $A_{\text{fuel cell}}$ is the area and $R_{ohmic}$ is the fuel cell resistance that comes from different components of the device.

Recall the original definition of resistance:

$$R = \frac{L}{A\sigma} \tag{11.12}$$

where $R$ is resistance, $A$ is cross-sectional area, $L$ is thickness, and $\sigma$ is conductivity.

Combining Equation (11.12) with Equation (11.11), we can conclude that:

$$\eta_{ohmic} = j \sum_i \frac{L_i}{\sigma_i} \tag{11.13}$$

where $i$ refers to different components of the fuel cell that contribute to resistance such as the electrical interconnections, anode, cathode, catalyst layer, electrolyte, and so on. It is obvious that the ohmic loss is directly proportional to thickness and inversely proportional to conductivity. Minimizing ohmic losses focuses on decreasing thickness and increasing conductivity.

There are two major types of charged species: electrons and ions. In most fuel cells, ionic charge transport is far more difficult than electronic charge transport as ionic conductivity is generally 4–8 orders of magnitude lower than the electronic conductivity. Therefore, the ionic contribution to ohmic losses tends to be the dominating factor in fuel cell kinetics. From Equation (11.13), we also know that the ohmic loss is proportional to the electrolyte thickness. Hence, fuel cell electrolytes are designed to be as thin as possible in order to reduce the ohmic loss.

However, the thickness of the electrolyte cannot be reduced boundlessly because extremely thin electrolytes risk mechanical breakage, electrical shorting, fuel crossover and so on. Besides, the contact resistance which is associated with the interface between the electrolyte and the electrode is independent of electrolyte thickness. Usually, the practical limit on thickness is about 10–100 μm depending on the electrolyte (O'Hayre, 2009).

As for the design of photocatalytic fuel cells, researchers have constructed many different cell geometries to minimize the electrolyte thickness. A photoanode carrying a photocatalyst and a cathode carrying an electro-catalyst sandwich a Nafion membrane to make the distance between the two electrodes minimal (Bahnemann *et al.*, 2007). A system of two electrodes, where the photoanode is placed very close to the cathode made of a steel mesh deposited with an electro-catalyst, is submerged in the electrolyte to make a one-compartment cell (Mrowetz *et al.*, 2004; Konstantinou and Albanis, 2004). Finally, a cathode made of carbon cloth loaded with a Pt carbon black electro-catalyst is attached to a Nafion membrane to form a half membrane-electrode assembly (Yang and Dionysiou, 2004). All of these examples give us good detailed information in finding more effective cell geometry designs.

The ohmic loss is inversely proportional to conductivity so developing high-conductivity electrodes and electrolyte materials is critical. From the above examples, we know that ionic charge transport in the electrolyte accounts for most of the ohmic loss. However unfortunately, the development of satisfactory ionic conductors is still challenging because a good fuel cell electrolyte must have high ionic conductivity and stability at the same time. The three most widely used material classes for fuel cells are aqueous electrolytes, polymer electrolytes and ceramic electrolytes.

### 11.4.1.4 *Concentration losses*

Mass transport, which governs the process of supplying reactants and removing products, is another important fuel cell process besides the electrochemical reaction process and the charge transport process as stated above. Poor mass transport leads to significant fuel cell performance losses due to reactant depletion and product accumulation in the catalyst layer. This type of loss is called a 'concentration loss' and is most significant in the tail of the fuel cell *j-V* curve.

A fuel cell operating at a high current density consumes reactants and forms products quickly, and this leads to reactant depletion and product accumulation in the vicinity of the catalyst layer. Fuel cell performance is determined by the reactant and product concentrations within the catalyst layer and the reactant depletion and product accumulation affects fuel cell performance in two ways: by Nernstian losses and reaction losses. Due to the electrochemical reaction at the catalyst layer, we have:

$$c_R^* < c_R^0 \tag{11.14}$$

$$c_P^* > c_P^0 \tag{11.15}$$

where $c_R^*$, $c_P^*$ are the catalyst layer reactant and product concentrations respectively and $c_R^0$, $c_P^0$ are the bulk reactant and product concentrations respectively.

Recall the form of a Nernst equation written in Equation (11.4). Assuming that the species concentration is equal to the relative species activity, and considering a fuel cell with a single reactant species and a single products species, we can get the incremental voltage loss:

$$
\begin{aligned}
\eta_{\mathrm{conc}} &= E^0_{\mathrm{Nernst}} - E^*_{\mathrm{Nernst}} \\
&= \left( E^0 - \frac{RT}{nF} \ln \frac{c^0_P}{c^0_R} \right) - \left( E^0 - \frac{RT}{nF} \ln \frac{c^*_P}{c^*_R} \right) \\
&= \frac{RT}{nF} \ln \left( \frac{c^0_R c^*_P}{c^*_R c^0_P} \right)
\end{aligned}
\tag{11.16}
$$

By combining Equations (11.14)–(11.16), it is obvious that $\eta_{\mathrm{conc}} > 0$, so the reversible fuel cell voltage decreases.

Recall the Butler-Volmer equation (Equation (11.10)). Considering the high current density region where the concentration effects are most significant, we can neglect the second term in the Butler-Volmer equation and the simplified expression is as follows:

$$
j = j^0_0 \left( \frac{c^*_R}{c^{0*}_R} e^{(\alpha n F_\eta /(RT))} \right)
\tag{11.17}
$$

Written in terms of the activation loss, we have:

$$
\eta_{\mathrm{act}} = \frac{RT}{\alpha n F} \ln \frac{jc^{0*}_R}{j^0_0 c^*_R}
\tag{11.18}
$$

In the same way, we can get the incremental voltage loss:

$$
\begin{aligned}
\eta_{\mathrm{conc}} &= \eta^*_{\mathrm{act}} - \eta^0_{\mathrm{act}} \\
&= \left( \frac{RT}{\alpha n F} \ln \frac{jc^{0*}_R}{j^0_0 c^*_R} \right) - \left( \frac{RT}{\alpha n F} \ln \frac{jc^{0*}_R}{j^0_0 c^0_R} \right) \\
&= \frac{RT}{\alpha n F} \ln \frac{c^0_R}{c^*_R}
\end{aligned}
\tag{11.19}
$$

By combining Equations (11.4) and (11.19), it is obvious that $\eta_{\mathrm{conc}} > 0$, so the activation loss will increase. The combination of these two loss effects is what we refer to as the 'fuel cell's concentration loss'.

Bulk reactant concentration is not a constant within fuel cell flow channels. Instead, it decreases with distance along a fuel cell flow channel because reactants are consumed. From Equations (11.16) and (11.19), we know that maintaining a consistent, high bulk reactant concentration is often the best way to minimize the concentration losses in a fuel cell. In real fuel cells, intricate flow structures containing many small flow channels are often employed to minimize concentration losses because, compared with a single-chamber design, small flow channel designs can not only keep the reactants constantly flowing across the fuel cell, but can also provide more contact points across the surface of the electrode from which the current can be harvested.

If all the losses that we have looked at, activation, ohmic and concentration, are combined then the actual operational graph of a fuel cell, the $j$-$V$ curve, is produced. The current is usually expressed as current density $J$, i.e. the quotient of the current divided by the geometrical surface of the electrodes. Thus from the $j$-$V$ curve, we can get the short-circuit current density $J_{\mathrm{SC}}$, the open-circuit voltage $V_{\mathrm{OC}}$, and the fuel cell power density, i.e., the $J \times V$ product. All of these parameters are very important in the evaluation of the photocatalytic fuel cell performance.

The fill factor gives the extent of diversion between the actual maximum power density that can be produced by the cell, $(JV)_{\mathrm{max}}$ (Antoniadou and Lianos, 2009) Thus, the fill factor can be calculated by (Antoniadou and Lianos, 2010; Lianos, 2011):

$$
\mathrm{FF} = \frac{(JV)_{\mathrm{max}}}{J_{\mathrm{SC}} V_{\mathrm{OC}}}
\tag{11.20}
$$

The quality of a cell is directly related to its fill factor (Antoniadou and Lianos, 2009).

### 11.4.2 Efficiency of a photocatalytic fuel cell

To calculate the efficiency of a photocatalytic fuel cell, all related energies should be taken into account, including the energy of the exciting radiation, the effective output energy related to the fuel produced, the chemical energy liberated by the photodegradable substance(s), and the energy input by any applied bias. So far, there is no common consensus to standardize the efficiency calculation. Below, we introduce several definitions of efficiency which are generally employed for photocatalytic fuel cells.

#### 11.4.2.1 Pseudo-photovoltaic efficiency

The pseudo-photovoltaic efficiency $\eta$ can be calculated by plotting the $j$-$V$ curve under reverse bias. By definition (Kelly and Gibson, 2006; Nazeeruddin *et al.*, 1993), $\eta$ is the ratio of the actual maximum power produced by the cell to the incident light power:

$$\eta = \frac{(JV)_{\max}}{P} \tag{11.21}$$

#### 11.4.2.2 External quantum efficiency

External quantum efficiency (EQE) is the ratio of the number of charge carriers produced by the cell to the number of photons incident on the cell (Lianos, 2011). Incident photon to current conversion efficiency (IPCE) (Antoniadou *et al.*, 2010; Brillet *et al.*, 2010; Tode *et al.*, 2010; Varghese and Grimes, 2008; Yu *et al.*, 2010), is one way to express EQE, which is defined as:

$$IPCE = \frac{1240 \times J_{SC} \,(\mathrm{mA\ cm^{-2}})}{\lambda \,(\mathrm{nm}) \times P \,(\mathrm{mW\ cm^{-2}})} \tag{11.22}$$

where $J_{SC}$ is the short-circuit current density and $P$ is the incident radiation intensity at a given wavelength $\lambda$. IPCE measures the effectiveness of a cell in converting the incident photons of monochromatic radiation into an electric current. The analysis can be also used to evaluate the effectiveness of a photocatalyst or combined photocatalyst or sensitized catalyst for spectral response (Lianos, 2011).

#### 11.4.2.3 Internal quantum efficiency

Internal quantum efficiency (IQE) is defined as the ratio of the number of charge carriers produced by the cell to the actual number of photons absorbed by the cell. It is also a pure number varying between 0 and 1 but larger than EQE as not all incident photons are absorbed by the photocatalytic fuel cell. A way to express IQE is as follows (Kaneko *et al.*, 2009b):

$$\eta = \frac{N_m \times n}{N_{eff}} \tag{11.23}$$

where $N_m$ is the number of molecules decomposed in the unit of time, $n$ is the number of electrons involved in the decomposition of each molecule, and $N_{eff}$ is the number of effective photons involved in the photodecomposition per unit time.

#### 11.4.2.4 Current doubling effect

The reason for the unusually high efficiency of photocatalytic fuel cells may be explained by the current doubling effect. The photoanodic decomposition of organic substances causes intermediate radical formation (Lee *et al.*, 1984; Lianos, 2011; Seger and Kamat, 2009). More electrons will be promoted to the conduction band and thus the current is increased. The holes will be scavenged by the organic agent, especially at high pH, and by $OH^-$ (Nazeeruddin *et al.*, 1993; Yu *et al.*, 2010).

## 11.5   PFC APPLICATIONS

### 11.5.1   *Wastewater problems*

Wastewater is one of the major environmental problems in urban cities. The problem is worsening due to rapid urbanization in many parts of the world. Most wastewater treatment methods merely aim to safely dispose of waste substances. In fact, the organic compounds in wastewater, such as carbohydrates, fatty acids, and amino acids, store substantial chemical energy (Sharma *et al.*, 2008). It makes a lot of sense to attempt a recovery of the potential energy in the wastewater treatment process for a green waste management approach. The microbial fuel cell is a candidate technology for achieving wastewater treatment and simultaneous production of electricity. However, the current microbial fuel cell technologies reveal slow bioelectrochemical activity, complex bacterial cultivation, a long start-up time and stringent working condition requirements (Logan *et al.*, 2006).

### 11.5.2   *Practical micro-fluidic photocatalytic fuel cell (MPFC) applications*

MPFC, using photocatalysts and electro-catalysts, is a more promising approach due to the presence of a synergistic current doubling effect (Lianos, 2011). The MPFC innovation has a high potential for waste treatment and electricity production. In principle, MPFC can degrade organic compounds in municipal and industrial wastewater; and simultaneously recover chemical energy stored in wastewater.

## 11.6   CONCLUSION

Recent studies conducted by the authors and other investigators have demonstrated the potential of using photocatalytic fuel cell (PFC) to contribute to the field of energy and the environment. One of the desirable applications is to utilize solar PFCs to decompose wastewater pollutants and simultaneously convert the chemical energy stored in pollutant compounds into electricity or hydrogen fuel. For the present technological development of the PFC, there is still a lot of room for improvement in the bifunctional anode. Further research efforts on the fabrication of photocatalysts and electro-catalysts immobilized onto the anode, and their synergistic interactions will be fruitful.

## ACKNOWLEDGEMENTS

The authors acknowledge financial support from the CityU Ability R&D Energy Research Centre, Hong Kong.

## REFERENCES

Allam, N.K. & Grimes, C.A.: Effect of rapid infrared annealing on the photoelectrochemical properties of anodically fabricated $TiO_2$ nanotube arrays. *J. Phys. Chem.* C 113:19 (2009), pp. 7996–7999.

Antoniadou M. & Lianos, P.: Photoelectrochemical oxidation of organic substances over nanocrystalline titania: Optimization of the photoelectrochemical cell. *Catal. Today* 144:1–2 (2009), pp. 166–171.

Antoniadou, M. & Lianos, P.: Production of electricity by photoelectrochemical oxidation of ethanol in a photofuel cell. *Appl. Catal.* B *Environ.* 99:1 (2010), pp. 307–313.

Antoniadou, M., Bouras, P., Strataki, N. & Lianos, P.: Hydrogen and electricity generation by photoelectrochemical decomposition of ethanol over nanocrystalline titania. *Int. J. Hydrogen Energy* 33:19 (2008), pp. 5045–5051.

Antoniadou, M., Kondarides, D.I., Labou, D., Neophytides, S. & Lianos, P.: An efficient photoelectrochemical cell functioning in the presence of organic wastes. *Sol. Energy Mat. Sol.* C 94:3 (2010), pp. 592–597.

Asahi, R., Morikawa, T., Ohwaki, T., Aoki, K. & Taga, Y.: Visible-light photocatalysis in nitrogen-doped titanium oxides. *Science* 293:5528 (2001), pp. 269–271.

Bahnemann, W., Muneer, M. & Haque, M.M.: Titanium dioxide-mediated photocatalysed degradation of few selected organic pollutants in aqueous suspensions. *Catal. Today* 124:3–4 (2007), pp. 133–148.

Baker, D.R. & Kamat, P.V.: Photosensitization of $TiO_2$ nanostructures with CdS quantum dots: Particulate versus tubular support architectures. *Adv. Funct. Mater.* 19:5 (2009), pp. 805–811.

Bickley, R.I., Gonzalezcarreno, T., Lees, J.S., Palmisano, L. & Tilley, R.J.D.: A structure investigation of titanium-dioxide photocatalysts. *J. Solid State Chem.* 92:1 (1991), pp. 178–190.

Bingham, S. & Daoud, W.A.: Recent advances in making nano-sized $TiO_2$ visible-light active through rare-earth metal doping. *J. Mater. Chem.* 21 (2011), pp. 2041–2050.

Brillet, J., Cornuz, M., Formal, F.L., Yum, J.H., Gratzel, M. & Silvula, K.: Examining architectures of photoanode-photovoltaic tandem cells for solar water splitting. *J. Mater. Res.* 25:1 (2010), pp. 17–24.

Cheng, X., Shi, Z., Glass, N., Zhang, L., Zhang, J., Song, D., Liu, Z.S, Wang, H. & Shen, J.: A review of PEM hydrogen fuel-cell contamination: Impacts, mechanisms, and mitigation. *J. Power Sources* 165:2 (2007), pp. 739–756.

Daskalaki, V.M., Antoniadou, M., Puma, G.L., Kondarides, D.I. & Lianos, P.: Solar light-responsive Pt/CdS/$TiO_2$ photocatalysts for hydrogen production and simultaneous degradation of inorganic or organic sacrificial agents in wastewater. *Environ. Sci. Technol.* 44:19 (2010), pp. 7200–7205.

Dong, F., Zhao, W. & Wu, Z.: Characterization and photocatalytic activities of C, N and S co-doped $TiO_2$ with 1D nanostructure prepared by the nano-confinement effect. *Nanotechnology* 19:36 (2008), 365607.

Feng, X. J., Shankar, K., Varghese, O.K., Paulose, M., LaTempa, T.J. & Grimes, C.A.: Vertically aligned single crystal $TiO_2$ nanowire arrays grown directly on transparent conducting oxide coated glass: Synthesis details and applications. *Nano Lett.* 8:11 (2008), pp. 3781–3786.

Fujishima, A., Zhang, X.T. & Tryk, D.A.: $TiO_2$ photocatalysis and related surface phenomena. *Surface Sci. Rep.* 63:12 (2008), pp. 515–582.

Grimes, C.A.: Synthesis and application of highly ordered arrays of $TiO_2$ nanotubes. *J. Mater. Chem.* 17:15 (2007), pp. 1451–1457.

Huang, Q.L., Zhou, G., Fang, L., Hu, L.P. & Wang, Z.S.: $TiO_2$ nanorod arrays grown from a mixed acid medium for efficient dye-sensitized solar cells. *Energy Environ. Sci.* 4:6 (2011), pp. 2145–2151.

Hurum, D.C., Agrios, A.G., Gray, K.A., Rajh, T. & Thurnauer, M.C.: Explaining the enhanced photocatalytic activity of Degussa P25 mixed-phase $TiO_2$ using EPR. *J. Phys. Chem.* B 107:19 (2003), pp. 4545–4549.

Iyatani, K., Horiuchi, Y., Fukumoto, S., Takeuchi, M., Anpo, M. & Matsuoka, M.: Separate-type Pt-free photofuel cell based on a visible light-responsive $TiO_2$ photoanode: effect of hydrofluoric acid treatment of the photoanode. *Appl. Catal.* A *General* 458 (2013), pp. 162–168.

Ito, S., Chen, P., Comte, P., Nazeeruddin, M.K., Liska, P., Pechy, P. & Gratzel, M.: Fabrication of screen-printing pastes from $TiO_2$ powders for dye-sensitised solar cells. *Prog. Photovoltaics* 15:7 (2007), pp. 603–612.

Jafry, H.R., Liga, M.V., Li, Q.L. & Barron, A.R.: Simple route to enhanced photocatalytic activity of P25 titanium dioxide nanoparticles by silica addition. *Environ. Sci. Technol.* 45:4 (2011), pp. 1563–1568.

Kaneko, M., Ueno, H., Saito, R., Suzuki, S., Nemoto, J. & Fujii, Y.: Biophotochemical cell (BPCC) to photodecompose biomass and bio-related compounds by UV irradiation with simultaneous electrical power generation. *J. Photochem. Photobiol.* A *Chem.* 205:2–3 (2009a), pp. 168–172.

Kaneko, M., Ueno, H., Saito, R., Yamaguchi, S., Fujii, Y. & Nemoto, J.: UV light-activated decomposition/cleaning of concentrated biomass wastes involving also solid suspensions with remarkably high quantum efficiency. *Appl. Catal.* B *Environ* 91:1–2 (2009b), pp. 254–261.

Kelly, N.A. & Gibson, T.L.: Design and characterization of a robust photoelectrochemical device to generate hydrogen using solar water splitting. *Int. J. Hydrogen Energy* 31:12 (2006), pp. 1658–1673.

Khan, S.U.M., Al-Shahry, M. & Ingler Jr. W.B.: Efficient photochemical water splitting by a chemically modified n-$TiO_2$. *Science* 297:5590 (2002), pp. 2243–2245.

Konstantinou, I.K. & Albanis, T.A.: $TiO_2$-assisted photocatalytic degradation of azo dyes in aqueous solution: kinetic and mechanistic investigations – A review. *Appl. Catal.* B *Environ* 49:1 (2004), pp. 1–14.

Kukovecz, A., Hodos, M., Konya, Z. & Kiricsi, I.: Complex-assisted one-step synthesis of ion-exchangeable titanate nanotubes decorated with CdS nanoparticles. *Chem. Phys. Lett.* 411:4–6 (2005), pp. 445–449.

Lee, J., Kato, T., Fujishima, A. & Honda, K.: Photoelectrochemical oxidation of alcohols on polycrystalline zinc-oxide. *Bull. Chem. Soc. Jap.* 57:5 (1984), pp. 1179–1183.

Lee, W., Kang, S.H., Kim, J.Y., Kolekar, G.B., Sung, Y.E. & Han, S.H.: $TiO_2$ nanotubes with a ZnO thin energy barrier for improved current efficiency of CdSe quantum-dot-sensitized solar cells. *Nanotechnology* 20:33 (2009a), 335706.

Lee, Y.L. & Lo, Y.S.: Highly efficient quantum-dot-sensitized solar cell based on co-sensitization of CdS/CdSe. *Adv. Funct. Mater.* 19:4 (2009b), pp. 604–609.

Leung, D.Y.C., Fu, X.L., Wang, C.F., Ni, M., Leung, M.K.H., Wang, X.X. & Fu, X.Z.: Hydrogen production over titania-based photocatalysts. *ChemSusChem* 3:6 (2010), pp. 681–694.

Li, J.H., Wang, H., Bai, Y.S., Zhang, H., Zhang, Z.H. & Guo, L.: CdS quantum dots-sensitized $TiO_2$ nanorod array on transparent conductive glass photoelectrodes. *J. Phys. Chem.* C 114:39 (2010), pp. 16,451–16,455.

Li, P.Q., Zhao, G.H., Cui, X., Zhang, Y.G. & Tang, Y.T.: Constructing stake structured $TiO_2$-NTs/Sb-doped $SnO_2$ electrode simultaneously with high electrocatalytic and photocatalytic performance for complete mineralization of refractory aromatic acid. *J. Phys. Chem.* C 113:6 (2009), pp. 2375–2383.

Lianos, P.: Production of electricity and hydrogen by photocatalytic degradation of organic wastes in a photoelectrochemical cell: The concept of the photofuelcell: A review of a re-emerging research field. *J. Hazard. Mater.* 185:2–3 (2011), pp. 575–590.

Lin, C.J., Lai, Y.K., Zhuang, H.F., Xie, K.P., Gong, D.G., Tang, Y.X., Sun, L. & Chen, Z.: Fabrication of uniform Ag/$TiO_2$ nanotube array structures with enhanced photoelectrochemical performance. *New J. Chem.* 34:7 (2010), pp. 1335–1340.

Liu, B. & Aydil, E.S.: Growth of oriented single-crystalline rutile $TiO_2$ nanorods on transparent conducting substrates for dye-sensitized solar cells. *J. Amer. Chem. Soc.* 131:11 (2009), pp. 3985–3990.

Liu, Y.B., Li, J.H., Zhou, B.X., Chen, H.C., Wang, Z.S. & Cai, W.M.: A $TiO_2$-nanotube-array-based photocatalytic fuel-cell using refractory organic compounds as substrates for electricity generation. *Chem. Commun.* 47:37 (2011a), pp. 10,314–10,316.

Liu, Y.B., Li, J.H., Zhou, B.X., Li, X.J., Chen, H.C., Chen, Q.P., Wang, Z.S., Li, L., Wang, J. L. & Cai, W.M.: Efficient electricity production and simultaneously wastewater treatment via a high-performance photocatalytic fuel-cell. *Water Res.* 45:13 (2011b), pp. 3991–3998.

Logan, B.E., Hamelers, B., Rozendal, R.A., Schrorder, U., Keller, J., Freguia, S., Aelterman, P., Verstraete, W. & Rabaey, K.: Microbial fuel-cells: methodology and technology. *Environ. Sci. Technol.* 40:17 (2006), pp. 5181–5192.

Mills, A.: Porous platinum morphologies: Platinised, sponge and black. *Platinum Met. Rev.* 51:1 (2007), p. 52.

Mor, G.K., Shankar, K., Paulose, M., Varghese, O.K. & Grimes, C.A.: Use of highly-ordered $TiO_2$ nanotube arrays in dye-sensitized solar cells. *Nano Lett.* 6:2 (2006a), pp. 215–218.

Mor, G.K., Varghese, O.K., Paulose, M., Shankar, K. & Grimes, C.A.: A review on highly ordered, vertically oriented $TiO_2$ nanotube arrays: Fabrication, material properties, and solar energy applications. *Sol. Energy Mat. Sol. C.* 90:14 (2006b), pp. 2011–2075.

Mrowetz, M., Balcerski, W., Colussi, A.J. & Hoffmann, M.R.: Oxidative power of nitrogen-doped $TiO_2$ photocatalysts under visible illumination. *J. Phys. Chem.* B 108:45 (2004), pp. 17,269–17,273.

Murdoch, M., Waterhouse, G.I.N., Nadeem, M.A., Metson, J.B., Keane, M.A., Howe, R.F., Llorca, J. & Idriss, H.: The effect of gold loading and particle size on photocatalytic hydrogen production from ethanol over Au/$TiO_2$ nanoparticles. *Nature Chem.* 3:6 (2011), pp. 489–492.

Murphy, A.B.: Does carbon doping of $TiO_2$ allow water splitting in visible light? Comments on 'Nanotube enhanced photoresponse of carbon modified (CM)-n-$TiO_2$ for efficient water splitting'. *Sol. Energy Mat. Sol. C* 92:3 (2008), pp. 363–367.

Nakamura, I., Negishi, N., Kutsuna, S., Ihara, T. Sugihara, S. & Takeuchi, K.: Role of oxygen vacancy in the plasma-treated $TiO_2$ photocatalyst with visible light activity for NO removal. *J. Molec. Catal.* A *Chem.* 161:1–2 (2000), pp. 205–212.

Nazeeruddin, M.K., Kay, A., Rodicio, I., Humphry-Baker, R., Muller, E., Liska, P., Vlachopoulos, N. & Gratzel, M.: Conversion of light to electricity by cis-X2bis(2,2′-bipyridyl-4,4′-dicarboxylate)ruthenium(II) charge-transfer sensitizers (X = Cl$^-$, Br$^-$, I$^-$, CN$^-$, and SCN$^-$) on nanocrystalline titanium dioxide electrodes. *J. Amer. Chem. Soc.* 115:14 (1993), pp. 6382–6390.

Ni, M., Leung, M.K.H., Leung, D.Y.C. & Sumathy, K.: A review and recent developments in photocatalytic water-splitting using $TiO_2$ for hydrogen production. *Renew. Sust. Energy Rev.* 11:3 (2007), pp. 401–425.

Ni, M., Leung, M.K.H. & Leung, D.Y.C.: Ammonia fed solid oxide fuel-cells (SOFCs) for power generation – A review. *Int. J. Energy Res.* 33:11 (2009), pp. 943–959.

Nishikiori, H., Isomura, K., Uesugi, Y. & Fujii, T.: Photofuel-cells using glucose-doped titania. *Appl. Catal.* B *Environ.* 106:1–2 (2011), pp. 250–254.

O'Hayre, R.P.: *Fuel-cell fundamentals.* John Wiley & Sons, Hoboken, NJ, 2009.

Ohno, T., Sarukawa, K., Tokieda, K. & Matsumura, M.: Morphology of a $TiO_2$ photocatalyst (Degussa, P-25) consisting of anatase and rutile crystalline phases. *J. Catal.* 203:1 (2001), pp. 82–86.

Peng, F., Zhang, S.S., Zhang, S.Q., Zhang, H.M., Liu, H.W. & Zhao, H.J.: Electrodeposition of polyhedral $Cu_2O$ on $TiO_2$ nanotube arrays for enhancing visible light photocatalytic performance. *Electrochem. Commun.* 13:8 (2011), pp. 861–864.

Prakasam, H.E., Shankar, K., Paulose, M., Varghese, O.K. & Grimes, C.A.: A new benchmark for $TiO_2$ nanotube array growth by anodization. *J. Phys. Chem.* C 111:20 (2007), pp. 7235–7241.

Rani, S., Roy, S.C., Paulose, M., Varghese, O.K., Mor, G.K., Kim, S., Yoriya, S., LaTempa, T.J. & Grimes, C.A.: Synthesis and applications of electrochemically self-assembled titania nanotube arrays. *Phys. Chem. Chem. Phys.* 12:12 (2010), pp. 2780–2800.

Schloegl, R.: Energy – fuel for thought. *Nature Mater.* 7:10 (2008), pp. 772–774.

Seger, B. & Kamat, P.V.: Fuel-cell geared in reverse: Photocatalytic hydrogen production using a $TiO_2$/Nafion/Pt membrane assembly with no applied bias. *J. Phys. Chem.* C 113:43 (2009), pp. 18,946–18,952.

Shankar, K., Mor, G.K., Prakasam, H.E., Yoriya, S., Paulose, M., Varghese, O.K. & Grimes, C.A.: Highly-ordered $TiO_2$ nanotube arrays up to 220 μm in length: use in water photoelectrolysis and dye-sensitized solar cells. *Nanotechnology* 18:6 (2007), 065707.

Shankar, K., Bandara, J., Paulose, M., Wietasch, H., Varghese, O.K., Mor, G.K., LaTempa, T.J., Thelakkat, M. & Grimes, C.A.: Highly efficient solar cells using $TiO_2$ nanotube arrays sensitized with a donor-antenna dye. *Nano Lett.* 8:6 (2008), pp. 1654–1659.

Sharma, T., Reddy, A.L.M., Chandra, T.S. & Ramaprabhu, S.: Development of carbon nanotubes and nanofluids based microbial fuel-cell. *Int. J. Hydrogen Energy* 33:22 (2008), pp. 6749–6754.

Shen, Y., Kordesch, K. & Aronson, R.R.: A comparison between the polymer electrolyte membrane fuel-cells and the alkaline fuel-cells (AFC) with liquid electrolyte. *The NPC'99 Nagoya International Battery and Power Sources Conference*, September 1999, Nagoya, Japan, 1999.

Sun, L., Li, J., Wang, C.L., Li, S.F., Chen, H.B. & Lin, C.J.: An electrochemical strategy of doping $Fe^{3+}$ into $TiO_2$ nanotube array films for enhancement in photocatalytic activity. *Sol. Energy Mat. Sol.* C 93:10 (2009), pp. 1875–1880.

Sun, P.P., Zhang, X.T., Liu, X.P., Wang, L.L., Wang, C.H., Yang, J.K. & Liu, Y.C.: Growth of single-crystalline rutile $TiO_2$ nanowire array on titanate nanosheet film for dye-sensitized solar cells. *J. Mater. Chem.* 22:13 (2012), pp. 6389–6393.

Tode, R., Ebrahimi, A., Fukumoto, S., Iyatani, K., Takeuchi, M., Matsuoka, M., Lee, C. H., Jiang, C.S. & Anpo, M.: Photocatalytic decomposition of water on double-layered visible light-responsive $TiO_2$ thin films prepared by a magnetron sputtering deposition method. *Catal. Lett.* 135:1–2 (2010), pp. 10–15.

Ueno, H., Nemoto, J., Ohnuki, K., Horikawa, M., Hoshino, M. & Kaneko, M.: Photoelectrochemical reaction of biomass-related compounds in a biophotochemical cell comprising a nanoporous $TiO_2$ film photoanode and an $O_2$-reducing cathode. *J. Appl. Electrochem.* 39:10 (2009), pp. 1897–1905.

Valentin, C.D., Finazzi, E., Pacchioni, G., Selloni, A., Livraghi, S., Paganini, M.C. & Giamello, E.: N-doped $TiO_2$: Theory and experiment. *Chem. Phys.* 339:1–3 (2007), pp. 44–56.

Varghese, O.K. & Grimes, C.A.: Appropriate strategies for determining the photoconversion efficiency of water photo electrolysis cells: A review with examples using titania nanotube array photoanodes. *Sol. Energy Mat. Sol.* C. 92:4 (2008), pp. 374–384.

Wang, B., Leung, M.K.H., Lu, X.Y. & Chen, S.Y.: Synthesis and photocatalytic activity of boron and fluorine codoped $TiO_2$ nanosheets with reactive facets. *Appl. Energy* 112 (2013), pp. 1190–1197.

Yang, Q.L. & Dionysiou, D.D.: Photolytic degradation of chlorinated phenols in room temperature ionic liquids. *J. Photochem. Photobiol. A Chem.* 165:1–3 (2004), pp. 229–240.

Yu, C.L., Zhou, W.Q., Yang, K. & Rong, G.: Hydrothermal synthesis of hemisphere-like F-doped anatase $TiO_2$ with visible light photocatalytic activity. *J. Mater. Sci.* 45:21 (2010), pp. 5756–5761.

Yu, Y.H., Kamat, P.V. & Kuno, M.: A CdSe nanowire/quantum dot hybrid architecture for improving solar cell performance. *Adv. Funct. Mater.* 20:9 (2010), pp. 1464–1472.

Zhang, H.Z. & Banfield, J.F.: Understanding polymorphic phase transformation behavior during growth of nanocrystalline aggregates: Insights from $TiO_2$. *J. Phys. Chem.* B 104:15 (2010), pp. 3481–3487.

Zhang, S., Chen, Y., Yu, Y., Wu, H., Wang, S., Zhu, B., Huang, W. & Wu, S.: Synthesis, characterization of Cr-doped $TiO_2$ nanotubes with high photocatalytic activity. *J. Nanopart. Res.* 10:5 (2008), pp. 871–875.

Zhao, T.S. & Xu, C.L.: Fuel-cells – Direct alcohol fuel-cells: Direct methanol fuel-cell: overview performance and operational conditions. In: *Encyclopedia of electrochemical power sources*. Elsevier, 2009, pp. 381–389.

# CHAPTER 12

## Transport phenomena and reactions in micro-fluidic aluminum-air fuel cells

Huizhi Wang, Dennis Y.C. Leung, Kwong-Yu Chan, Jin Xuan & Hao Zhang

## 12.1  INTRODUCTION

The development of high density energy storage systems is of great scientific and technological importance. While in past years lithium-ion batteries (LIBs) have captured the most attention, even the highest energy density achievable for LIBs is insufficient to satisfy the requirements of full electric vehicles and modern electronic devices with highly integrated functions. Metal-air semi-fuel cells, which can theoretically offer 5–9-fold increases in energy density compared with LIBs at a lower cost, have recently received renewed interest (Cheng and Chen, 2012; Wang *et al.*, 2013; Zhong *et al.*, 2012; Zu and Li, 2011). Of different metal anodes, aluminum has the highest volumetric electrochemical equivalent ($8.04 \, \text{Ah cm}^{-3}$, compared with 2.06 for lithium, 5.85 for zinc, and 3.83 for magnesium), and therefore, it is particularly worth studying for portable or mobile applications (Li and Bjerrum, 2002). Aluminum-air cells usually operate with an electrolyte flowing in microchannels. The submillimeter to millimeter characteristic dimensions of the electrolyte channels can greatly enhance the cell performance by providing a reduced internal resistance and rapid species transport. However, to date, aluminum-air technologies have met very limited commercial success. The competing processes of surface passivation and self-corrosion that occur on aluminum during discharge, pose a great challenge in developing appropriate anode alloys and electrolytes (Abdel-Gaber *et al.*, 2008; Beck and Rüetschi, 2000; Ferrando, 2004; Tang *et al.*, 2004). An alternative strategy that can increase the fuel efficiency of aluminum is to incorporate the hydrogen production into an alkaline aluminum-air cell (Zhuk *et al.*, 2006). To some extent, this strategy loosens the restrictions on both anode alloys and electrolytes since it treats the gas-evolving parasitic corrosion of aluminum as a route for hydrogen production, instead of a worthless but detrimental reaction that needs to be suppressed. Meanwhile, however, it raises another critical issue of hydrogen management inside the cell, which requires a thorough understanding on the intricate physico-electrochemical processes associated with two-phase flow in the micro-fluidic cell channel.

Mathematical models capable of providing insights into working cells can be an important tool to help address the above issue. Although a few aluminum-air modeling studies have been carried out (Chan and Savinell, 1991; Savinell and Chase, 1988; Yang and Knickle, 2003; Yang *et al.*, 2006), none of them took full account of the coupled processes of two-phase hydrodynamics, mass transport, or electrochemical kinetics that govern the cell behavior. Neglecting the two-phase effects and mass transport, a two-dimensional Laplace's equation for the potential was solved in the papers of Savinell and Chase (1988) and Yang *et al.* (2006) to study the edge effects in an aluminum-air cell. More comprehensive models of aluminum-air cells were developed by Chan and Savinell (1991) and Yang and Knickle (2003) to include the effects of gas evolution and species transport. However, these models failed to solve the two-phase flow field within the cell. Instead, they simplified the two-phase problem as plug flow with uniform gas fraction across the cell gap, which was obtained by Tobias' solution (Tobias, 1959). Also, a thin boundary layer model that assumes linear distributions of potential and reactant concentration inside boundary layers, was invoked. In this study, we aim to present a mathematical model for aluminum-air cells to

275

incorporate all the mentioned governing processes so as to gain an improved understanding of these processes and associated cell behaviors. A finite-volume-based computational fluid dynamics (CFD) technique is employed to solve a set of conservation equations of mass, momentum, species, and charge, together with an electrochemical kinetic model. The present model outperforms previous models in its readiness to be extended to handle other complexities related to the practical operation of aluminum-air cells, such as non-isothermal conditions and wedge-shaped electrodes.

In what follows, the model formulations are firstly described in Section 12.2 and Section 12.3. In Section 12.4, modeling results including model validation through comparisons with previous experimental data in the literature, and analyses on distributions of hydrogen, velocity, species, current density, and potential in the cell are shown and discussed. Conclusions from the study are finally summarized in Section 12.5. It should be mentioned that all computations carried out in the study are under steady-state assumption, though the equations are derived in a time-dependent form.

## 12.2   MATHEMATICAL MODEL

### 12.2.1   *Problem description*

Consideration is given to an aluminum-air cell with a geometry shown in Figure 12.1. Two parallel plane electrodes (aluminum anode and fuel cell type air cathode) with a distance of $S$ are located between $y=0$ and $y=L$. Thus, $S$ is the cell channel gap and $L$ is the electrode length. As mentioned before, $S$ is normally in a millimeter length scale. An electrolytic solution of potassium hydroxide (KOH) is fed into the cell with a uniform initial velocity, $v_0$, from the inlet at the bottom. Both inlet and outlet boundaries are placed at a sufficient distance, denoted as $L'$, from the electrode region to avoid their influences on modeling results, though this does not belong to the case in reality (Chan and Savinell, 1991). The minimum value of $L'$ is estimated from Sparrow's formula (Pickett, 1979). The reactions taking place during the cell discharge are (Yang and Knickle, 2003):

(i) main reaction at the anode:

$$Al + 4OH^- \rightarrow Al(OH)_4^- + 3e^- \tag{12.1}$$

(ii) parasitic reaction at the anode:

$$Al + 3H_2O + OH^- \rightarrow 1.5H_2 + Al(OH)_4^- \tag{12.2}$$

(iii) reaction at the cathode:

$$O_2 + 2H_2O + 4e^- \rightarrow 4OH^- \tag{12.3}$$

Consequently, three ionic species need to be considered: $K^+$, $OH^-$, and $Al(OH)_4^-$. General assumptions applied to the following model development include: (i) laminar incompressible flow, (ii) spherical hydrogen bubbles with an identical size, and (iii) existence of isothermal conditions. Justification for the first assumption is given by the nature of small Reynolds numbers for both bubbles ($Re_g = 0.02$ based on the terminal velocity for a single bubble in a quiescent fluid), and the electrolyte flow ($Re \leq 460$) studied here, preventing the transition to turbulence. The second assumption is appropriate because of the negligible hydrostatic pressure variation along the electrode height compared to the atmospheric pressure, the high surface tension, and the impeded bubble coalescence, possibly caused by the electrostatic repulsive forces acting between bubbles (Dahlkild, 2001; Shah et al., 2010). The final isothermal restriction can be easily relieved by including heat sources and an energy equation.

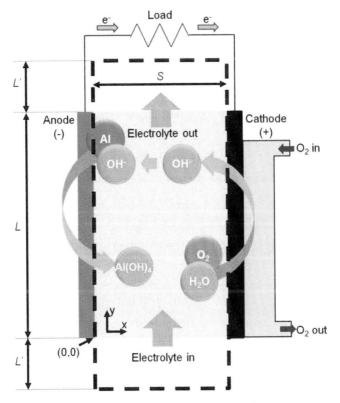

Figure 12.1.    Schematic of the aluminum-air cell (not to scale). Computational domain is denoted by a dashed-line box.

## 12.2.2   *Cell hydrodynamics*

The bubbly electrolyte is treated using the mixture model (Manninen and Taivassalo, 1996) which solves the continuity and momentum equations for the mixture, the volume fraction equation for the dispersed phase, and the algebraic approximation for the dispersed phase's momentum, as summarized below.

The mass and momentum conservations for the mixture of bubbles and electrolyte are, respectively:

$$\frac{\partial(\rho_{mix})}{\partial t} + \nabla(\rho_{mix} v_{mix}) = 0, \text{ and} \tag{12.4}$$

$$\frac{\partial(\rho_{mix} v_{mix})}{\partial t} + \nabla(\rho_{mix} v_{mix} v_{mix}) = -\nabla p + \rho_{mix} g + \nabla \tau_{mix} - \nabla \left[ \left( 1 - \frac{f_g \rho_g}{\rho_{mix}} \right) f_g \rho_g v_{lg} v_{lg} \right] \tag{12.5}$$

where the same pressure, $p$, is assumed for both phases as there is no bubble expansion during the process (Manninen and Taivassalo, 1996; Shah *et al.*, 2010). The last term of Equation (12.5) represents the momentum diffusion due to the phase slip. $v_{lg}$ denotes the slip velocity between the gas and the liquid phase, which has been formulated by Manninen and Taivassalo (1996) to be:

$$v_{lg} = \left[ \frac{4 d_g^2 \left( \rho_g - \rho_{mix} \right)}{3 C_D Re_g \mu_l} \right] \left[ g - \frac{\partial v_{mix}}{\partial t} - (v_{mix} \cdot \nabla) v_{mix} \right] \tag{12.6}$$

where $C_D$ is the drag coefficient determined from the Morsi and Alexander drag model (Philippe et al., 2005): $C_D = a_1 + a_2/\, Re_g + a_3/\, Re_g^2$ ($a_{1\sim3}$ are semi-empirical constants obtained with spherical particles). It should be noticed that Manninen and Taivassalo's formulation for the slip velocity did not include any non-drag forces responsible for the lateral drift of bubbles. Because of the uncertainty of the lateral dispersion mechanisms for electrochemically produced bubbles (Philippe et al., 2005), the present study ignores their effects, which may intensify the non-uniformity of the lateral gas distribution.

The transport equation for the gas volume fraction is derived by rearranging the continuity equation of gas:

$$\frac{\partial(f_g\rho_g)}{\partial t} + \nabla(f_g\rho_g v_{mix}) = -\nabla(f_g\rho_g v_{drift,g}) + S_g \tag{12.7}$$

where the drift velocity of the gas phase can be written in terms of the slip velocity:

$$v_{drift,g} = \left(1 - \frac{f_g\rho_g}{\rho_{mix}}\right) v_{lg} \tag{12.8}$$

and $S_g$ corresponds to the volumetric mass production rate of hydrogen from the parasitic reaction and can be estimated using Faraday's law as:

$$S_g = \frac{M_{H_2}}{2F}\left|\frac{j_p}{\Delta x}\right| \tag{12.9}$$

In the above equation, $j_p$ is area-based and thus divided by the width of the near-wall grid cell ($|\Delta x|$) to describe the volumetric electrochemical reaction (Ni, 2009).

As the mixture is fully saturated, the volume fraction of the electrolyte can be calculated from a solved gas fraction with the following relationship:

$$f_l = 1 - f_g \tag{12.10}$$

### 12.2.3   Charge conservation

Current density in the electrolytic solution, caused by the motion of charged particles, is expressed by (Newman and Thomas-Alyea, 2004):

$$j = F\sum_h z_h N_h \tag{12.11}$$

where the flux of each dissolved ionic species in the two-phase medium, $N$, is assumed to be in the form of the Nernst-Planck equation:

$$N_i = -D_i^{eff}\nabla C_i - \frac{z_i FD_i^{eff} C_i}{RT}\nabla\phi + v_l(1 - f_g)C_i \tag{12.12}$$

The first, second, and third terms on the right-hand side of Equation (12.12) respectively, represent the three governing transport mechanisms of species: diffusion, migration, and convection. Combining Equations (12.11) and (12.12) with the condition of electro-neutrality ($\Sigma_h z_h C_h = 0$) gives:

$$j = -F\sum_h z_h D_h^{eff}\nabla C_h - \kappa^{eff}\nabla\phi \tag{12.13}$$

The effective species diffusivity ($D_i^{eff}$) and electrical conductivity ($\kappa^{eff}$) are subject to a Bruggeman correction (Bruggeman, 1935) to account for the gas-induced reduction of these properties:

$$D_i^{eff} = D_{i,0}(1 - f_g)^{1.5} \tag{12.14}$$

and

$$\kappa^{eff} = \kappa_0(1 - f_g)^{1.5} \tag{12.15}$$

where $D_{i,0}$ and $\kappa_0$ respectively denote the gas-free values of diffusivity and conductivity. In dilute solutions, it is shown that:

$$\kappa_0 = \frac{F^2}{RT}\sum_h z_h^2 D_{h,0} C_h, \tag{12.16}$$

whereas for more concentrated solutions, the gas-free solution conductivity is suggested to be empirically evaluated (Gu *et al.*, 1997). Although the empirical conductivity is usually concentration-dependent, the constant $\kappa_0$ adopted in other aluminum-air modeling works (Chan and Savinell, 1991; Savinell and Chase, 1988; Yang and Knickle, 2003; Yang *et al.*, 2006) is kept for the present model.

The charge flux in the electrolyte has been related to the electrochemical kinetics occurring at the electrode/electrolyte interface (Gu *et al.*, 1997):

$$\nabla j = J \tag{12.17}$$

where $J$ is the volumetric transfer current resulting from the electrochemical reaction at the electrode/electrolyte interface and, consequently, non-zero only in the control volumes adjacent to the corresponding interface:

$$J = \begin{cases} \dfrac{j_m}{|\Delta x|} & \text{at the anode} \\ 0 & \text{in the electrolyte reservoir} \\ \dfrac{j_c}{|\Delta x|} & \text{at the cathode} \end{cases} \tag{12.18}$$

In the above, $j_m$ is the positive current leaving the anode and entering the electrolyte and $j_c$ is the negative current leaving the electrolyte and entering the cathode.

Substituting Equation (12.13) into Equation (12.17) yields the potential field in the electrolyte:

$$\nabla(-\kappa^{\text{eff}}\nabla\phi) = J + F\sum_h z_h \nabla\left(D_h^{\text{eff}}\nabla C_h\right) \tag{12.19}$$

### 12.2.4 *Ionic species transport*

Rearranging Equation (12.13) results in:

$$\nabla\phi = -\frac{j}{\kappa^{\text{eff}}} - \frac{F}{\kappa^{\text{eff}}}\sum_h z_h D_h^{\text{eff}}\nabla C_h \tag{12.20}$$

Inserting the above equation into the Nernst-Planck equation (Equation (12.12)) eliminates the potential term as follows:

$$N_i = -D_{i,0}(1 - f_g)^{1.5}\nabla C_i + \frac{t_i}{z_i F}j + \frac{t_i}{z_i}\sum_h z_h D_{h,0}(1 - f_g)^{1.5}\nabla C_h + v_l(1 - f_g)C_i \tag{12.21}$$

For simplicity, we postulate the ion transference number ($t_i$) to be a constant equal to that in the bulk.

Inserting Equation (12.21) into the material balance equation:

$$\frac{\partial[(1 - f_g)C_i]}{\partial t} = -\nabla N_i + S_i \tag{12.22}$$

gives the following conservation equation for each species:

$$\frac{\partial[(1 - f_g)C_i]}{\partial t} + \nabla\left[v_l(1 - f_g)C_i - D_{i,0}(1 - f_g)^{1.5}\nabla C_i\right]$$

$$= -\frac{t_i}{z_i}\sum_h z_h \nabla\left[D_{h,0}(1 - f_g)^{1.5}\nabla C_h\right] - \frac{t_i}{z_i F}J + S_i \tag{12.23}$$

The last term in Equation (12.23) stands for the volumetric species production or consumption rate from the electrochemical reactions, which is confined in the control volumes neighboring the electrode/electrolyte interface, and is equal to zero in most of the computational zone.

In the present model, Equation (12.23) is solved for two anions: $Al(OH)_4^-$ and $OH^-$, and the concentration of cation, $K^+$ is deduced with the electro-neutrality condition stated before from the known concentrations of $Al(OH)_4^-$ and $OH^-$ The production terms $(S_i)$ for $OH^-$ and $Al(OH)_4^-$, evaluated using Faraday's law, respectively are:

$$
S_{OH^-} = \begin{cases} -\dfrac{4j_m}{3F|\Delta x|} + \dfrac{j_p}{3F|\Delta x|} & \text{at the anode} \\[2ex] -\dfrac{j_c}{F|\Delta x|} & \text{at the cathode} \end{cases}
\tag{12.24}
$$

for the hydroxyl ion, and

$$
S_{Al(OH)_4^-} = \begin{cases} \dfrac{j_m}{3F|\Delta x|} - \dfrac{j_p}{3F|\Delta x|} & \text{at the anode} \\[2ex] 0 & \text{at the cathode} \end{cases}
\tag{12.25}
$$

for the aluminate ion.

### 12.2.5   Electrode kinetics

#### 12.2.5.1   Anode kinetics

Main and parasitic reaction kinetics at the anode are stated in a simple Bulter-Volmer form of:

$$
j_m = j_{m0} \left[ \left( \frac{C_{OH^-}^{SA}}{C_{OH^-}^{ref}} \right)^{\gamma_m} \exp\left( \frac{\alpha_m n_m \eta_m F}{RT} \right) - \left( \frac{C_{Al(OH)_4^-}^{SA}}{C_{Al(OH)_4^-}^{ref}} \right)^{\lambda_m} \exp\left( \frac{-\beta_m n_m \eta_m F}{RT} \right) \right]
\tag{12.26}
$$

for the main reaction, and

$$
j_p = j_{p0} \left[ \left( \frac{C_{OH^-}^{SA}}{C_{OH^-}^{ref}} \right)^{\gamma_p} \left( \frac{C_{H_2}^{SA}}{C_{H_2}^{ref}} \right)^{\lambda_p} \exp\left( \frac{\alpha_p n_p \eta_p F}{RT} \right) - \exp\left( \frac{-\beta_p n_p \eta_p F}{RT} \right) \right]
\tag{12.27}
$$

for the parasitic reaction. Equation (12.27) can be further reduced to Equation (12.28) because the first term on the right-hand side is negligible for most operating voltages of an aluminum-air cell (Chan and Savinell, 1991):

$$
j_p = j_{p0} \left[ -\exp\left( \frac{-\beta_p n_p \eta_p F}{RT} \right) \right]
\tag{12.28}
$$

In light of the low parasitic current density studied here, the surface coverage functions (Eigeldinger and Vogt, 200) reflecting the reduction of the active electrode area caused by the adhering gas bubbles are not introduced into our kinetic expressions. Typical values of parameters in the kinetic expressions in Equation (12.26) and Equation (12.28) have been determined for pure aluminum by Chan and Savinell (1991) through regression of the polarization data in the literature (Rudd, 1976), and which are summarized in Table 12.1 and adopted by the present model.

#### 12.2.5.2   Cathode kinetics

To facilitate the comparison between the present model and previous models, the kinetics data of 5% cobalt tetramethoxyphenyl porphyrin (CoTMPP) on steam-activated Sawinigan acetylene black as reported in the literature (Carbonio *et al.*, 1987) is used for the study. A three-constant

Table 12.1. Input parameters to numerical simulations.

| | Parameters |
|---|---|
| Substance physical properties | $D_{OH^-,0} = 5.26 \times 10^{-9} \text{ m}^2\text{s}^{-1}$, $D^-_{Al(OH)4,0} = 10^{-9} \text{ m}^2\text{s}^{-1}$, $\rho_l = 1150 \text{ kg m}^{-3}$, $\mu_l = 0.0008 \text{ kg (m s)}^{-1}$, $\rho_g = 0.08189 \text{ kg m}^{-3*}$, $\mu_g = 8.411 \times 10^{-6} \text{ kg (m s)}^{-1*}$, $d_g = 2.6 \times 10^{-5} \text{ m}$, $\kappa_0 = 80 \text{ S m}^{-1}$ |
| Anode kinetics: | |
| *Main* | $j_{m0} = 137.1 \text{ A m}^{-2}$, $\alpha_m = 0.07956$, $\beta_m = 0.9204$, $\gamma_m = 0.5$, $\lambda_m = 1$ $n_m = 1$ $E^{eq}_m = -2.4403 \text{ V}$ |
| *Parasitic* | $j_{p0} = 111.9 \text{ A m}^{-2}$, $\beta_p = 0.0591$ $n_p = 1 E^{eq}_p = -0.9508 \text{ V}$ |
| Cathode kinetics | $A = 3.898 \times 10^4 \text{ A m}^{-2}\text{V}^{-1}$, $B = 1.3904 \times 10^4 \text{A m}^{-2}$, $E^{eq}_c = 0.2857 \text{ V}$ |
| Operating conditions | $T = 333 \text{ K}$, $C^{ref}_{OH^-} = 5000 \text{ mol m}^{-3}$, $C^{-ref}_{Al(OH)4} = 500 \text{ mol m}^{-3}$ |

*From the database of FLUENT®CFD software.

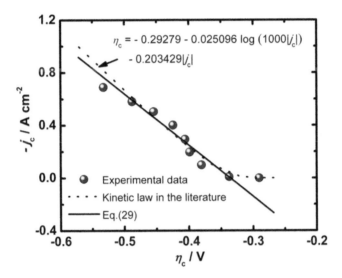

$$\eta_c = -0.29279 - 0.025096 \log (1000|j_c|) - 0.203429|j_c|$$

Figure 12.2. Cathode kinetics fitted by different equations. Symbols represent the measurement by Carbonio *et al.* (1987). The dotted line represents the fitting results using kinetic law by Chan and Savinell (1991).

empirical equation assuming the superposition of an exponential Tafel region and an ohmic control linear region was recommended to represent the kinetics for porous gas-diffusion electrodes as the polarization curve measured from such a type of electrode exhibited a steeper decline than the usual Tafel (or Butler-Volmer) kinetics (Chan and Savinell, 1991). This empirical equation, however, is a transcendental function, which causes difficulty in obtaining a solution. For simplicity, we assume a linear kinetic profile for the cathode:

$$j_c = A\eta_c + B \tag{12.29}$$

where constants $A$ and $B$ have been determined to be $3.898 \times 10^4 \text{ A m}^{-2} \text{ V}^{-1}$ and $1.3904 \times 10^4 \text{ A m}^{-2}$ for the CoTMPP cathode by Yang *et al.* (2006). Following a comparison between Equation (12.29) and the empirical kinetic-ohmic control representation together with the measuring data in Figure 12.2, it is concluded that the linear simplification in Equation (12.29) is sufficient to describe the CoTMPP cathode kinetics in a wide range of overpotential values except for the low polarization region, which is not involved in this study.

### 12.2.5.3   *Expression of overpotentials*

For an electrode reaction out of equilibrium, the galvanic potential deviates from the equilibrium potential by the overpotential:

$$\phi_s - \phi_l = E^{eq} + \eta \tag{12.30}$$

For the aluminum-air couple, applying Equation (12.30) along with the mixed potential theory yields the following correlation at the anode:

$$\phi_{s,a} - \phi_{l,a} = E_m^{eq} + \eta_m = E_p^{eq} + \eta_p \tag{12.31}$$

and

$$\phi_{s,c} - \phi_{l,c} = E_c^{eq} + \eta_c \tag{12.32}$$

at the cathode. In the above two equations, $\phi_s$ denotes the electrode potential and $\phi_l$ denotes the solution-phase potential adjacent to the electrode.

Expressions of three overpotentials involved in the model computations are obtained by combining Equations (12.31) and (12.32):

$$\text{Main reaction at the anode: } \eta_m = E_c^{eq} - E_m^{eq} - V_{cell} - \phi_{l,a} \tag{12.33}$$

$$\text{Reaction at the cathode: } \eta_c = -\phi_{l,c} \tag{12.34}$$

$$\text{Parasitic reaction at the anode: } \eta_p = E_c^{eq} - E_p^{eq} - V_{cell} - \phi_{l,a} \tag{12.35}$$

where the cell voltage $V_{cell} = \phi_{s,c} - \phi_{s,a}$, is prescribed before each computation.

### 12.2.6   *Boundary conditions*

Boundary conditions specified to complete the modeling equation system are:

Hydrodynamics:

$$v_l = v_0, f_g = 0 \text{ at the inlet} \tag{12.36}$$

$$p = p_{out} \text{ at the outlet} \tag{12.37}$$

Concentrations:

$$C_i = C_i^{ref} \text{ at the inlet} \tag{12.38}$$

$$\frac{\partial C_i}{\partial \overline{n}} = 0 \text{ at other boundaries} \tag{12.39}$$

Potential:

$$\frac{\partial \phi}{\partial \overline{n}} = 0 \text{ at all boundaries} \tag{12.40}$$

It should be noted that non-flux conditions are prescribed here for both concentrations and potential at the electrodes, as any production or consumption of species and charge has been included in source terms in the foregoing-derived governing equations.

## 12.3   NUMERICAL PROCEDURES

In summary, the present numerical problem consists of solutions to Equations (12.4), (12.5), (12.7), (12.10), (12.19), and (12.23) subject to boundary conditions. Equations (12.36)–(12.40) are for $(i + 5)$ unknowns: $v_{mix}$, $p$, $f_g$, $f_l$, $\phi$ and $C_i$. All governing equations in a convection–diffusion-source form (Wang, 2004), are discretized by the finite-volume method, and solved using the commercial CFD code, FLUENT 6.2. Charge and species conservations (Equations (12.19) and (12.23)) are solved as user-defined scalars implemented with a set of user-defined functions written in C language, and dynamically linked to the FLUENT source code. A non-uniform $24 \times 200$ grid scheme in the $x$, $y$ directions with mesh refinement near the electrodes, is selected for our studied cell design after a grid independence check.

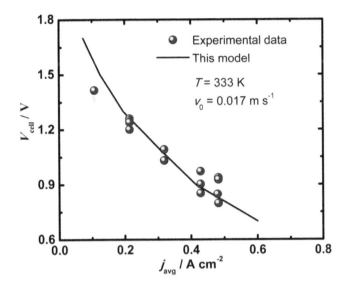

Figure 12.3. The model prediction compared with the AT400 cell performance with a pure aluminum anode (Chan and Savinell, 1991).

## 12.4 RESULTS AND DISCUSSION

### 12.4.1 *Model validation*

Figure 12.3 displays the model predicted $j$-$V$ characteristic curve compared with the measured one, for an AT400 aluminum-air cell (Chan and Savinell, 1991). The AT400 design of the aluminum-air cell was built by the Eltech Systems Corporation, Chardin, Ohio decades ago (Chan and Savinell, 1991), with the cell gap ($S$) of 0.0014 m and electrode height ($L$) of 0.13 m. An observed deviation of less than 12% of the experimental data in Figure 12.3 falls within the error limits, and thus validates the present model.

As no information of velocity and gas distribution was documented for aluminum-air cells, further examination of the model's predictability on the two-phase flow is carried out through comparisons with published experimental data (Boissonneau and Byrne, 2000; Riegel *et al.*, 1998) on gas-evolving electrolyzers. Employing laser Doppler velocimetry, Boissonneau and Byrne (2000) measured the velocity profile from the electrolysis of a $50 \, g \, L^{-1}$ sodium sulfate ($Na_2SO_4$) electrolyte in a vertical cell ($S = 0.003$ m and $L = 0.04$ m). Their results together with our model's predictions are shown in Figure 12.4a. A reasonable agreement between the two is found. The M-shaped velocity profiles with two peaks close to the electrodes against the Poiseuille flow, prior to entering the cell in Figure 12.4a are attributed to the gas accumulation in the upper part of the cell. Velocity near the cathode is observed to be higher than that near the anode, explained by the fact that the hydrogen production rate is twice the oxygen production rate during the water splitting process. In contrast to Boissonneau and Byrne (2000), Riegel *et al.* (1998) adopted a more indirect way, based on the correlation between the gas content and electrolyte conductivity, to get hydrogen distributions for an alkaline electrolyzer ($S = 0.008$ m and $L = 0.12$ m). Comparison of their measured gas fraction and the corresponding simulation results are made in Figure 12.4b. Our simulated profiles are found to diverge from the measured ones especially at a large current density, probably because of the ignorance of lateral forces in the model, and that their experimental turbulence flow regime ($Re = 15,870$) is beyond the scope of the present model. Nevertheless, a qualitative agreement is still ensured.

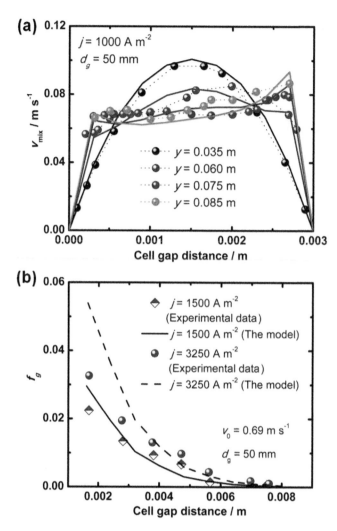

Figure 12.4.    Comparisons of the modeling results and experimental data in literature: (a) Mixture velocity profiles. Comparison between the model's predictions (solid line) and experimental data of Boissonneau and Byrne (2000) (symbol with dotted line). Oxygen is evolved at the anode ($x = 0$) and hydrogen at the cathode ($x = 0.003$). (b) Gas distributions. The simulation results compared with experimental data of Riegel $et\ al.$ (1998).

In general, the foregoing comparisons indicate that this model is reliable for further studies on cell properties in the following subsections. To save the computational time, a cell design of 0.002 m cell gap and 0.04 m electrode height is used for the simulations.

## 12.4.2    Hydrogen distribution

The presence of gas bubbles is known to significantly alter the heat and mass transfer, overpotential, limiting current density and ohmic resistance in an electrochemical device (Eigeldinger and Vogt, 2000). Determination of the parasitically produced hydrogen distribution in the aluminum-air cell is therefore of interest. Figure 12.5a illustrates the computed hydrogen volume fraction distribution under a cell voltage of 1.3 V. The prediction from the previous BLM model with

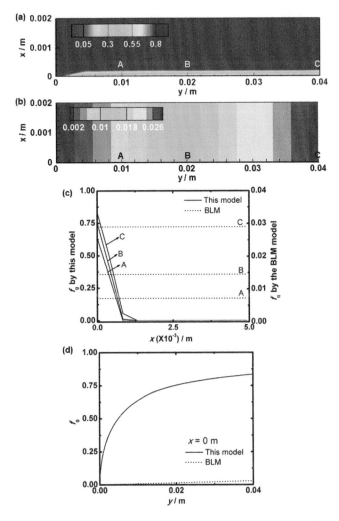

Figure 12.5. Hydrogen distribution in the aluminum-air cell. Comparison between the present model and the BLM model ($v_0 = 0.017\,\mathrm{m\,s^{-1}}$; concentration polarization ignored).

Tobias' solution to gas phase is also shown in Figure 12.5b for a comparison. To facilitate the comparison, concentration polarization is excluded at this stage. Figure 12.5a and 12.5b reveal a considerable discrepancy between the two models' solutions. Clearer views of gas distribution in the lateral direction and along the gas-evolving anode are given in Figure 12.5c and Figure 12.5d, respectively. The present CFD-predicted lateral hydrogen distribution, in which the gas volume fraction peaks at the anode and then vanishes sharply within a thin layer vicinity to the anode, against the unrealistic uniform one obtained by the BLM model, is shown in Figure 12.5c. Accordingly, in Figure 12.5d, a higher anode hydrogen concentration is computed from the present model. In view of the close connection between the effective conductivity and the local gas fraction implied by the Bruggeman formula (Equation (12.15)), it is necessary to identify how the discrepancy in gas distribution affects the current density evaluation. Figure 12.6 shows that the confined lateral hydrogen distribution estimated by the present model, leads to a steeper current density decline along the electrodes, compared with that predicted by the BLM model, which was attributed to the correspondingly lower conductivity. As the ignorance of the lateral drift of hydrogen bubbles in the present model somewhat overestimates the decaying behavior

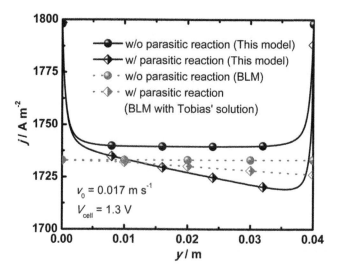

Figure 12.6.    Current density distribution along the electrodes. Comparison between the present model and the BLM model (concentration polarization ignored).

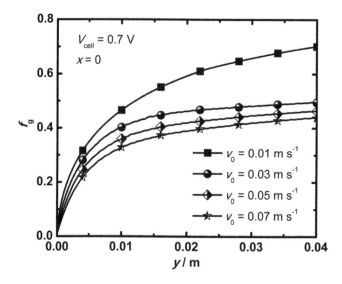

Figure 12.7.    Volumetric hydrogen fraction along the anode at different electrolyte velocities.

of the current density, a more accurate computation of the current density profile requires the incorporation of appropriate lateral dispersion mechanisms.

Figure 12.7 depicts the hydrogen distributions at different electrolyte velocities (with the inclusion of concentration polarization). Increasing the inlet velocity from $0.01\,\mathrm{m\,s^{-1}}$ to $0.07\,\mathrm{m\,s^{-1}}$ decreases the maximum hydrogen volume fraction at the anode from 0.701 to 0.438. This is mainly because the faster flowing electrolyte generates a thinner viscous layer and thus a larger velocity magnitude near the anode, which facilitates the removal of these parasitically produced bubbles. The decreasing trend of gas fraction becomes less apparent with the increase of inlet velocity.

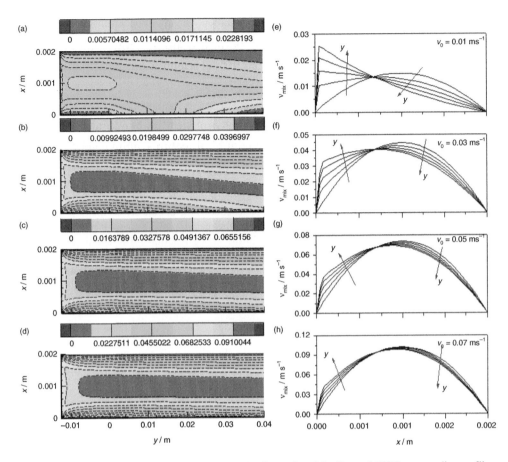

Figure 12.8.   Mixture velocity distribution in the aluminum-air cell for $V_{cell} = 0.7$ V. Corresponding profiles at the vertical positions $y = 0, 0.01, 0.02, 0.03, 0.04$ m are shown in figures in the right column.

### 12.4.3   *Velocity distribution*

The mixture velocity field inside the aluminum-air cell is presented in Figure 12.8. For each prescribed inlet velocity, it can be seen from both contours and profiles in Figure 12.8a–h, that the parabolic Poiseuille velocity profile exists at the entrance of the electrode region ($y = 0$), indicating a fully developed laminar flow. Because of the parasitic hydrogen release, the parabolic profile becomes asymmetric after passing the electrodes. This is due to the density difference between the gas and the electrolyte, which drives a buoyant flow near the anode (where the gas volume fraction is large). Progressing up the height of vertical positions in the cell, the asymmetry of the velocity profile becomes more significant. As compared in Figures 12.8e–h, the velocity profile distortion is alleviated with the increase of inlet velocity. This bubble-induced velocity change was not addressed by the previous studies on the aluminum-air cell, though it plays an important role in the species transfer described in the following subsections.

### 12.4.4   *Species distribution*

#### 12.4.4.1   *Single-phase flow*
##### 12.4.4.1.1   *Ionic species concentration distributions*
For a better understanding of the underlying mass transfer factors, a simpler single-phase situation is adopted in this section by assuming that the hydrogen generated is fully dissolved without

changing the physical properties in the electrolytic solution. The ionic species that participate in electrode reactions are $OH^-$ and $Al(OH)_4^-$. As the discharge of the aluminum-air cell proceeds, $OH^-$ is consumed at the anode according to Equation (12.1) and Equation (12.2), while it is formed at the cathode according to Equation (12.3), and $Al(OH)_4^-$ is formed at the anode according to Equation (12.1) and Equation (12.2). Although $Al(OH)_4^-$ does not take part in the cathodic reaction in Equation (12.3), it diffuses and migrates in the cathode region to maintain the electro-neutrality condition. Figure 12.9 illustrates the concentration variations of $OH^-$ and $Al(OH)_4^-$ at different velocities. Note here, concentrations have been made dimensionless by scaling to their inlet values. Figure 12.9 clearly shows that the concentrations of both species vary only in a thin diffusion layer adjacent to the electrodes. Their diffusion layers grow thicker towards the top of the cell with lower species transfer rates in that direction.

Because there is a strong dependence of the reaction rate on the anode $OH^-$ concentration, as indicated by the kinetics of the anode reaction (Equation (12.26)), an in-depth study on $OH^-$ transport is needed. Figure 12.9a shows that in the absence of the bubble effect, the $OH^-$ concentration along the anode drops by 47%. This is because the mass transfer process is not fast enough to replenish the continuous $OH^-$ consumption in the anode reactions. The anodic decreasing $OH^-$ concentration, corresponding to an increasing overpotential and decreasing reaction current density, is unfavorable for the cell performance. A species transport pattern in a pressure-driven flow is commonly characterized by the Graetz number ($Gz$), defined as the ratio of time scales for diffusion and convection (Gervais and Jensen, 2006; Yoon et al., 2006):

$$Gz^{-1} = \bar{t}_{conv}/\bar{t}_{diff} \tag{12.41}$$

Taking account of the species migration velocity, we slightly modify the above expression to be the time scale ratio between transverse and longitudinal transport:

$$Gz^{-1}_{mod} = \bar{t}_{long}/\bar{t}_{trans} = \frac{4LD_{i,0}}{vS^2} \pm \left| \frac{2z_i D_{i,0} FLj}{vS\kappa_0 RT} \right| \tag{12.42}$$

where the sign before the last term denotes the direction of the migration movement. For the $OH^-$ transport, the values of $Gz^{-1}_{mod}$ calculated for possible ranges of cell voltages, electrolyte velocities and electrode heights, are well below 1, indicating that the cell always operates in the "entrance region" where the longitudinal transport timescale is shorter than the transverse transport timescale, and a portion of $OH^-$ in the bulk cannot reach the electrode before leaving the cell. Species transfer in the entrance region is, therefore, usually enhanced by managing the diffusion layers (Yoon et al., 2006). The most direct way to diminish the thickness of the $OH^-$ depletion layer is to increase the velocity as demonstrated in Figures 12.9a–d. The change of diffusion layer thickness, reflected by the average Sherwood number ($\sim d_e/\delta_{avg}$) versus the Reynolds number, is plotted in Figure 12.10. With the increase of the velocity and thus the Reynolds number, the average Sherwood number increases, implying a thinning diffusion layer. Also shown in Figure 12.10 is the curve for a documented Sherwood-Reynolds number correlation (Roy et al., 2001). The discrepancy between the two curves in Figure 12.10 is less than 7.4% of the correlated data, which again confirms the reliability of the present model.

### 12.4.4.1.2  Migration contribution to transverse species transport

We evaluate the migration effect separately as it is crucial to the transverse transfer of ions in the cell, together with the diffusion process. Migration is driven by the electrical potential gradient, and is peculiar to the charged ion moving in an electrical field. Figure 12.11 compares the $OH^-$ distributions along the electrodes with and without migration. At the cell outlet ($y = 0.04$ m), the dimensionless concentration of $OH^-$ is 0.53 at the anode and 1.22 at the cathode in the presence of migration, compared to that of 0.30 at the anode and 1.52 at the cathode without migration. To quantify the proportion occupied by migration in the transverse $OH^-$ transfer, a ratio of migration flux to the flux from both migration and diffusion is defined and plotted along the $x$ axis for different $y$ positions in Figure 12.12. The ratio is less than 50% near the anode,

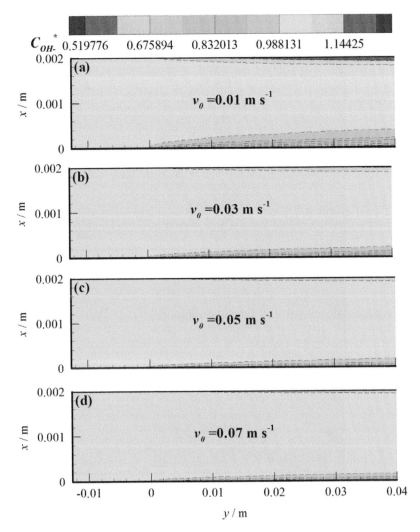

Figure 12.9. Dimensionless concentration distributions of species at different electrolyte velocities (a–d: $OH^-$; e–h: $Al(OH)_4^-$; $V_{cell} = 0.7$ V, single-phase flow).

indicating that the diffusion transfer is dominant in this region. The migration process gradually overcomes the diffusion process in a direction away from the electrode surfaces, attributed to the weakening concentration gradient along that direction. No diffusion is found in the bulk where a concentration gradient does not exist. Figure 12.12 also shows that the ratio at a lower $y$ position is slightly larger than that at a higher $y$ position.

### 12.4.4.2 *The effect of bubbles*

Contours of the $OH^-$ distribution, as illustrated in Figure 12.13, show the effect of the gas phase on $OH^-$ distribution. By comparing Figure 12.13a with Figure 12.9a, it is found that the diffusion layer at the anode becomes much thinner with the bubble release, indicating that the bubbles contribute to a promotion of $OH^-$ transport at the anode. At the cell outlet, the inclusion of the gas effect leads to an anode $OH^-$ concentration 26% higher than that of the previous single-phase situation. This is because of (i) an enhanced convection in the anode region raised by the buoyant

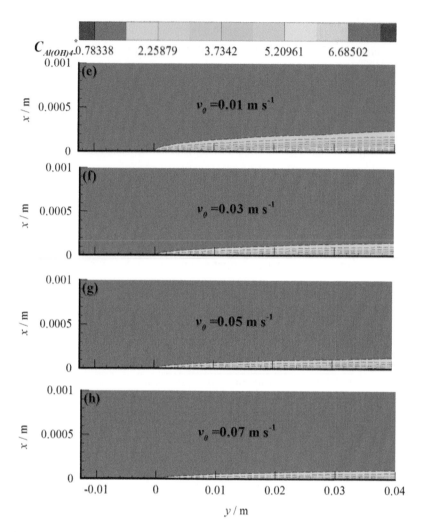

Figure 12.9.   Continued.

gas layer and (ii), a bubble-induced reduction of species diffusivity as expressed in Equation (12.14). The mechanisms behind the bubble-induced enhancement can be far more complex than this (Eigeldinger and Vogt, 200), which is beyond the scope of the present study. The above analysis supports the point that an aluminum-air cell with the parasitic hydrogen release is unlikely to suffer from the OH$^-$ transfer limitation as the gas bubbles are expected to prevent the OH$^-$ depletion at the anode even at an extremely low electrolyte velocity. In contrast, the diffusion layer at the cathode observed in Figure 12.13a thickens, hindering the OH$^-$ transport. Fortunately, the resultant ion transfer inhibition at the cathode seems not to affect the cell performance, according to the cathode kinetics used in this study. It is worth mentioning that the gas-retarded species transport might be of importance for situations in which the kinetics of the non-gas-evolving electrode are sensitive to the mass transfer. Increasing the electrolyte velocity can minimize the bubble impact on the concentration field as shown from Figures 12.13a to 12.13d.

## 12.4.5   Current density and potential distributions

For single-phase situations, variations of current density and electric potential along the electrodes in an aluminum-air cell primarily depend on the OH$^-$ transfer, as illustrated respectively in Figures

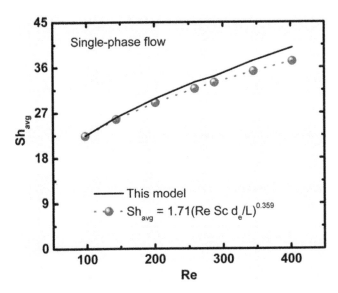

Figure 12.10. Average Sherwood number *versus* Reynolds number. Results of an empirical relation (Roy et al., 2001) are also shown in the figure using symbols with a dotted line.

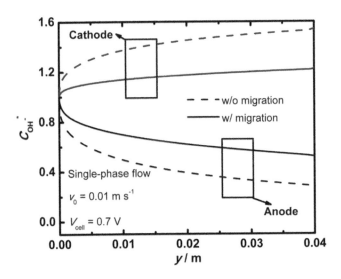

Figure 12.11. Dimensionless concentration of $OH^-$ along the electrodes with and without the migration.

12.14a and 12.14b. With the diffusion layer thickening and the anode $OH^-$ diminishing in the flow direction, the current density decreases along the electrodes as shown in Figure 12.14a. A higher electrolyte velocity corresponds to a more uniform current distribution due to its promoting effect on the mass transfer. The inset in Figure 12.14b relates the electrolyte potential to overpotentials according to Equations (12.33) and (12.34) to give an intuitive view. It can be seen from the inset that the contribution to the total voltage loss descends in the order of anode overpotential, cathode overpotential, and electrolyte ohmic loss for the given condition. Along the electrodes, the anodic polarization in Figure 12.14b increases as a result of the developing $OH^-$ depletion layer, while the cathodic polarization slightly decreases due to the decline in the current density. Similar to

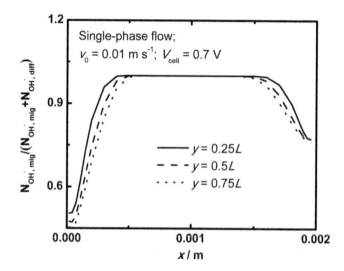

Figure 12.12.   Fraction of OH⁻ transport to electrodes by migration.

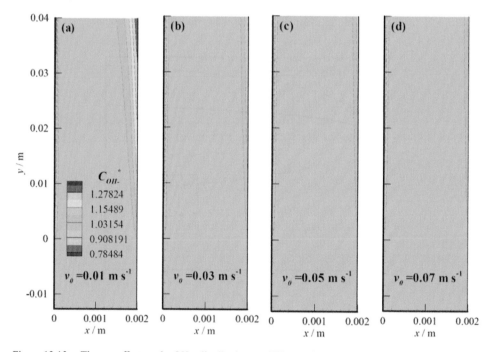

Figure 12.13.   The gas effect on the OH⁻ distributions at different electrolyte velocities ($V_{cell} = 0.7$ V, two-phase flow).

the current, the dash line in Figure 12.14b shows a more uniform potential distribution at a higher velocity.

The gas impact on the electric field can be manifold. Figure 12.15a demonstrates an increased current density with bubbles, contrary to our expectation. Nevertheless, an explanation might be found in Figure 12.15b where the bubble-mixing effect on the anode OH⁻ transport reduces the anodic polarization. Although the inter-electrode ohmic drop is also intensified by bubbles,

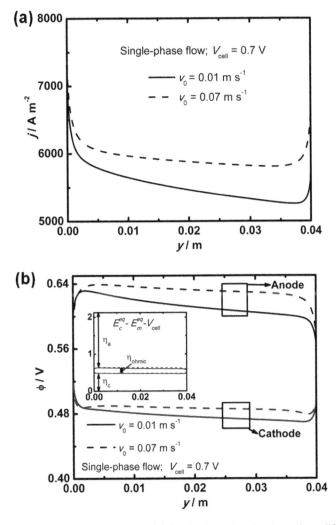

Figure 12.14.   (a) Current density, and (b) potential distributions along the electrodes at different electrolyte velocities.

its increment is small compared to the anodic polarization reduction for the given operating conditions. This trend may no longer be true if Equation (12.2) is excluded for the single-phase computations, and the operating parameters are changed.

By considering the gas impact, the velocity effect on the current density, averaged over the area of the electrode surface, is also examined under different cell voltages as illustrated in Figure 12.16. A higher average current density for a faster electrolyte flow rate is observed as the latter reduces both concentration and ohmic overpotentials.

## 12.5   CONCLUSIONS

In this study, a two-dimensional model is developed for an aluminum-air cell for the first time, based on a CFD approach. The model takes into account the parasitic gas evolution, fluid flow, electrochemical kinetics, and ion transfer by three mechanisms of diffusion, migration, and

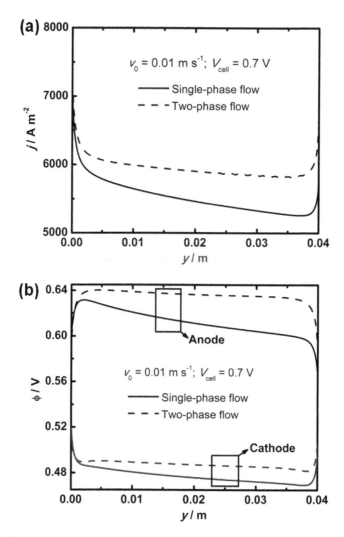

Figure 12.15.   The bubble effect on the (a) current density, and (b) potential distributions along the electrodes.

convection. A good agreement with experimental data was reached over the range of cell voltages studied. The model was further employed to investigate the distribution of hydrogen bubbles, velocity, ionic species, current density, and potential in an aluminum-air cell.

For the gas and velocity distributions, the non-uniformity of the transverse distribution of para-sitically evolved hydrogen and the gas-induced distortion of the velocity field in the aluminum-air cell, have been visualized and addressed. The confined gas distribution was found to lower the predicted current density.

For the species distribution, the roles of three transport mechanisms, particularly migration, were identified under a single-phase situation. Besides, an enhancement of OH$^-$ transfer at the anode was suggested to improve the cell performance because the OH$^-$ depletion at the anode, results in a decreasing reaction current density. The Graetz number estimated in the single-phase flow revealed that the OH$^-$ transfer in the aluminum-air cell occurs in the "entrance region" where the anode depletion is confined in a boundary layer, implying its enhancement is mainly achieved by controlling the depletion layer thickness. In addition to increasing velocity, our two-phase

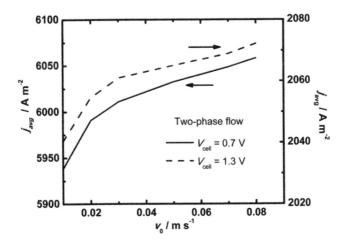

Figure 12.16.   The velocity effect on the cell performance.

simulations showed that the release of hydrogen bubbles effectively reduces the $OH^-$ depletion layer thickness at the anode. The bubble-mixing effect helps to prevent the occurrence of the $OH^-$ transfer limited cell performance. The model also suggested an inhibition of $OH^-$ transfer at the cathode induced by bubbles, though it does not affect the cell performance.

For the current and potential distributions, the anode $OH^-$ transfer is the main influencing factor without the consideration of the gas effect. It decreases the current density and increases anode overpotential along the electrodes. After the inclusion of the bubble effect, the increase of anode overpotential was found to be alleviated, but the ohmic drop in the electrolyte solution was also increased. Increasing the electrolyte velocity can reduce both concentration overpotential and electrolyte ohmic resistance, and thus increase the average output current density.

In general, the results and findings concluded from the present study give an insight into the fluid dynamics and reaction properties of an aluminum-air cell and demonstrate the powerful predictability of the present model. This model may act as an effective tool for future design of aluminum-air electrochemical systems.

## ACKNOWLEDGEMENTS

This project was supported by the ICEE and GRF #714313 of the Hong Kong Research Grants Council.

## REFERENCES

Abdel-Gaber, A.M., Khamis, E., Abo-ElDahab, H. & Adeel, S.: Inhibition of aluminum corrosion in alkaline solutions using natural compound. *Mater. Chem. Phys.* 109:2–3 (2008), pp. 297–305.

Beck, F. & Rüetschi, P.: Rechargeable batteries with aqueous electrolytes. *Electrochim. Acta* 45:15–16 (2000), pp. 2467–2482.

Boissonneau, P. & Byrne, P.: An experimental investigation of bubble-induced free convection in a small electrochemical cell. *J. Appl. Electrochem.* 30:7 (2000), pp. 767–775.

Bruggeman, D.A.G.: Berechnung verschieder physikalischer Konstanten von heterogenen Substanzen. *Ann. Phys.* 416 (1035), pp. 636–664.

Carbonio, D.J., Tryk, D. & Yeager, E. (eds): Electrode materials and processes for energy conversion and storage. In: S. Srinivasan, H. Wroblowa, & S. Wagner (series eds): *The Electrochemical Society Softbound Proceedings Series.* E. Pennington, NJ, 1987.

Chan, K.Y. & Savinell, R.F.: Modeling calculations of an aluminum-air cell. *J. Electrochem. Soc.* 138:7 (1991), pp. 1976–1984.

Cheng, F. & Chen, J.: Metal-air batteries: from oxygen reduction electrochemistry to cathode catalysts. *Chem. Soc. Rev.* 41:6 (2012), pp. 2172–2192.

Dahlkild, A.A.: Modelling the two-phase flow and current distribution along a vertical gas-evolving electrode. *J. Fluid Mech.* 428 (2001), pp. 249–272.

Eigeldinger, J. & Vogt, H.: The bubble coverage of gas-evolving electrodes in a flowing electrolyte. *Electrochim. Acta* 45:27 (2000), pp. 4449–4456.

Ferrando, W.A.: Development of a novel composite aluminum anode. *J. Power Sources* 130:1–2 (2004), pp. 309–314.

Gervais, T. & Jensen, K.F.: Mass transport and surface reactions in microfluidic systems. *Chem. Eng. Sci.* 61:4 (2006), pp. 1102–1121.

Gu, W.B., Wang, C.Y. & Liaw, B.Y.: Numerical modeling of coupled electrochemical and transport processes in lead-acid Batteries. *J. Electrochem. Soc.* 144:6 (1997), pp. 2053–2061.

Li, Q. & Bjerrum, N.J.: Aluminum as anode for energy storage and conversion: a review. *J. Power Sources* 110:1 (2002), pp. 1–10.

Manninen, M. & Taivassalo, V.: *On the mixture model for multiphase flow.* Valtion Teknillinen Tutkimuskeskus, Espoo, Finland, 1996.

Newman, J. & Thomas-Alyea, K.E.: *Electrochemical systems.* A John Wiley & Sons, Inc., NJ, 2004.

Ni, M.: On the source terms of species equations in fuel-cell modeling. *Int. J. Hydrogen Energy* 34:23 (2009), pp. 9543–9544.

Philippe, M., Jérôme, H., Sebastien, B. & Gérard, P.: Modelling and calculation of the current density distribution evolution at vertical gas-evolving electrodes. *Electrochim. Acta* 51:6 (2005), pp. 1140–1156.

Pickett, D.J.: *Electrochemical reactor design.* Elsevier, Amsterdam, The Netherlands, 1979.

Riegel, H., Mitrovic, J. & Stephan, K.: Role of mass transfer on hydrogen evolution in aqueous media. *J. Appl. Electrochem.* 28:1 (1998), pp. 10–17.

Roy, S., Gupte, Y. & Green, T.A.: Flow cell design for metal deposition at recessed circular electrodes and wafers. *Chem. Eng. Sci.* 56:17 (2001), pp. 5025–5035.

Rudd, E.J.: Final Report for the Department of Energy. Contract SNLA 02-8199, 1976.

Savinell, R. & Chase, G.: Analysis of primary and secondary current distributions in a wedge-type aluminum-air cell. *J. Appl. Electrochem.* 18:4 (1988), pp. 499–503.

Shah, A.A., Al-Fetlawi, H. & Walsh, F.C.: Dynamic modeling of hydrogen evolution effects in the all-vanadium redox flow battery. *Electrochim. Acta* 55:3 (2010), pp. 1125–1139.

Tang, Y., Lu, L., Roesky, H.W., Wang, L. & Huang, B.: The effect of zinc on the aluminum anode of the aluminum–air battery. *J. Power Sources* 138:1–2 (2004), pp. 313–318.

Tobias, C.W.: Effect of gas evolution on current distribution and ohmic resistance in electrolyzers. *J. Electrochem. Soc.* 106:9 (1959), pp. 833–838.

Wang, C.Y.: Fundamental models for fuel-cell engineering. *Chem. Rev.* 104:10 (2004), pp. 4727–4766.

Wang, X., Hou, Y., Zhu, Y., Wu, Y. & Holze, R.: An aqueous rechargeable lithium battery using coated Li metal as anode. *Scientific Reports* 3 (2013), p. 1401.

Yang, S., Yang, W., Sun, G. & Knickle, H.: Secondary current density distribution analysis of an aluminum–air cell. *J. Power Sources* 161:2 (2006), pp. 1412–1419.

Yang, S.H. & Knickle, H.: Modeling the performance of an aluminum–air cell. *J. Power Sources* 124:2 (2003), pp. 572–585.

Yoon, S.K., Fichtl, G.W. & Kenis, P.J.A.: Active control of the depletion boundary layers in microfluidic electrochemical reactors. *Lab on a Chip* 6:12 (2006), pp. 1516–1524.

Zhong, X., Zhang, H., Liu, Y., Bai, J., Liao, L., Huang, Y. & Duan, X.: High-capacity silicon–air battery in alkaline solution. *ChemSusChem* 5:1 (2012), pp. 177–180.

Zhuk, A.Z., Sheindlin, A.E., Kleymenov, B.V., Shkolnikov, E.I. & Lopatin, M.Y.: Use of low-cost aluminum in electric energy production. *J. Power Sources* 157:2 (2006), pp. 921–926.

Zu, C.X. & Li, H.: Thermodynamic analysis on energy densities of batteries. *Energy Environ. Sci.* 4:8 (2011), pp. 2614–2624.

NOMENCLATURE

$A$    constant in Equation (12.29) [A m$^{-2}$ V$^{-1}$]
$B$    constant in Equation (12.29) [A m$^{-2}$]

| | |
|---|---|
| $C_D$ | drag coefficient |
| $C_i$ | concentration of species $i$ [mol m$^{-3}$] |
| $C_i^{ref}$ | reference concentration of species $i$, equal to the inlet concentration of the species [mol m$^{-3}$] |
| $C_i^{SA}$ | surface concentration of species $i$ at the anode [mol m$^{-3}$] |
| $C_i^*$ | dimensionless concentration of species $i$, $(C_i/C_i^{ref})$ |
| $d_e$ | equivalent diameter, $(2S)$ [m] |
| $d_g$ | gas bubble diameter [m] |
| $D_i$ | diffusivity of species $i$ [m$^2$ s$^{-1}$] |
| $D_{i,0}$ | gas-free diffusivity of species $i$ [m$^2$ s$^{-1}$] |
| $E^{eq}$ | equilibrium potential [V] |
| $f$ | volume fraction |
| $F$ | Faraday's constant (96485.34 C equiv.$^{-1}$) |
| $g$ | gravitational acceleration (9.81 m s$^{-2}$) |
| $Gz$ | Graetz number |
| $J$ | volumetric current density [A m$^{-3}$] |
| $j$ | area-based current density [A m$^{-2}$] |
| $j_0$ | area-based exchange current density [A m$^{-2}$] |
| $L$ | electrode height [m] |
| $L_e$ | hydrodynamic entrance length [m] |
| $M_{H_2}$ | molar mass of hydrogen [kg mol$^{-1}$] |
| $n$ | number of electrons transferred in the rate-limiting step of the reaction |
| $N$ | species flux density [mol m$^{-2}$ s$^{-1}$] |
| $R$ | ideal gas constant (8.314 J mol$^{-1}$ K$^{-1}$) |
| $Re$ | Reynolds number, $(\rho_l v_0 d_e/\mu_l)$ |
| $Re_g$ | gas Reynolds number, $(\rho_l \lvert v_{lg}\rvert d_g/\mu_l)$ |
| $S$ | cell gap width [m] |
| $S_i, S_q$ | source terms for species $i$ and phase $q$ [mol m$^{-3}$ s$^{-1}$, kg m$^{-3}$ s$^{-1}$] |
| $Sh$ | Sherwood number, $(kd_e/D)$ (where $k$ is mass transfer coefficient) |
| $\bar{t}$ | time scale [s] |
| $t_i$ | transference number of species $i$ |
| $T$ | absolute temperature [K] |
| $v_o$ | inlet velocity [m s$^{-1}$] |
| $v_{drift,q}$ | drift velocity of phase $q$, $(v_q-v_{mix})$ [m s$^{-1}$] |
| $v_{mix}$ | mixture velocity, $((\sum_q f_q\rho_q v_q)/\rho_{mix})$ [m s$^{-1}$] |
| $v_{rq}$ | slip velocity, $(v_q-v_r)$ [m s$^{-1}$] |
| $V_{cell}$ | cell voltage [V] |
| $z$ | number of electrons transferred |

*Greek symbols*

| | |
|---|---|
| $\alpha, \beta$ | anodic and cathodic transfer coefficients for the electrode reaction |
| $\gamma, \lambda$ | coefficients involved in the electrode kinetic expressions that depend on the mechanism and stoichiometry of the reaction |
| $\delta$ | diffusion layer thickness [m] |
| $\eta$ | overpotential [V] |
| $\kappa$ | conductivity [S m$^{-1}$] |
| $\kappa_0$ | gas-free solution conductivity [S m$^{-1}$] |
| $\mu$ | viscosity [kg m$^{-1}$ s$^{-1}$] |
| $\rho$ | density [kg m$^{-3}$] |
| $\rho_{mix}$ | mixture density, $\left(\sum_q f_q\rho_q\right)$ [kg m$^{-3}$] |

| | |
|---|---|
| $\tau$ | stress tensor [N m$^{-2}$] |
| $\tau_{mix}$ | mixture stress tensor, $(\sum_{q} f_q \tau_q)$ [N m$^{-2}$] |
| $\phi$ | electric potential [V] |

*Subscripts*

| | |
|---|---|
| avg | average value over the area of the electrode surface (i.e. $\psi_{avg} = (\int_0^L \psi(y)dy)/L)$, where $\psi$ denotes an arbitrary variable) |
| $c$ | cathode |
| conv | convection |
| diff | diffusion |
| $g$ | gas |
| $i, h$ | species index |
| $l$ | liquid |
| long | longitudinal |
| $m$ | main anode reaction |
| mig | migration |
| mix | mixture |
| mod | modified |
| $p$ | parasitic reaction |
| $s$ | solid phase |
| trans | transverse |

*Superscripts*

| | |
|---|---|
| eff | effective |

# Subject index

# Sustainable Energy Developments

*Series Editor: Jochen Bundschuh*

ISSN: 2164-0645

Publisher: CRC Press/Balkema, Taylor & Francis Group

9. Advanced Oxidation Technologies – Sustainable Solutions for Environmental Treatments
   Editors: Marta I. Litter, Roberto J. Candal & J. Martín Meichtry
   2014
   ISBN: 978-1-138-00127-5 (Hbk)

10. Computational Models for $CO_2$ Geo-sequestration & Compressed Air Energy Storage
    Editors: Rafid Al-Khoury & Jochen Bundschuh
    2014
    ISBN: 978-1-138-01520-3 (Hbk)

11. Micro & Nano-Engineering of Fuel Cells
    Editors: Dennis Y.C. Leung & Jin Xuan
    2015
    ISBN: 978-0-415-64439-6 (Hbk)

Printed and bound by CPI Group (UK) Ltd, Croydon, CR0 4YY

22/10/2024

01777614-0003